Spon's Landscape Handbook

Spon's Landscape Handbook

Edited by
Derek Lovejoy Partnership

Fourth edition

E & FN SPON
An Imprint of Chapman & Hall

London • Weinheim • New York • Tokyo • Melbourne • Madras

**Published by E & FN Spon, an imprint of Chapman & Hall,
2–6 Boundary Row, London SE1 8HN, UK**

Chapman & Hall, 2–6 Boundary Row, London SE1 8HN, UK

Chapman & Hall GmbH, Pappelallee 3, 69469 Weinheim, Germany

Chapman & Hall USA, 115 Fifth Avenue, New York, NY 10003, USA

Chapman & Hall Japan, ITP-Japan, Kyowa Building, 3F, 2-2-1
Hirakawacho, Chiyoda-ku, Tokyo 102, Japan

Chapman & Hall Australia, 102 Dodds Street, South Melbourne, Victoria
3205, Australia

Chapman & Hall India, R. Seshadri, 32 Second Main Road, CIT East,
Madras 600 035, India

First edition 1997

© 1997 Derek Lovejoy Partnership

Typeset in 10/12pt Times by Saxon Graphics Ltd, Derby

Printed in England by Richard Clays Ltd, Bungay, Suffolk

ISBN   0 419 20490 3

Contents

Introduction to the fourth edition

There have been so many changes in the legislation, official guidance and techniques relating to landscape and external works since the third edition was written 10 years ago, that it was decided to rewrite the whole book and also take the opportunity to change the format of the work sections to comply with SMM7. This will make it easier for those already conversant with SMM7 to find specific items in the book, and has the added advantage that it can be read in conjunction with the current edition of *Spon's Landscape and External Works Price Book*, thus keeping abreast with the latest materials and prices as well as with important changes in the legislation and official guidance.

The opening sections of the book are outside SMM7 work sections. The first of these is a guide to planning and landscape law and official guidance, with brief abstracts of Acts, Statutory Instruments and Department of the Environment Guidance Notes. For any changes to the law after Autumn 1996 the reader is advised to check the current edition of *Spon's Landscape and External Works Price Book*.

The second section gives a review of computer-aided design techniques for landscape designers. The third section gives brief guidance on the type of information to be collected during the first site visit and where further documentary evidence may be sought; the final part of this section gives brief coverage of the responsibilities of the design team under various pieces of legislation, including the Construction Design Management Regulations.

The SMM7 work sections that comprise the main part of this edition have been divided into general guidance notes, British Standards, data, and outline specifications and drawings. The general notes will be of particular use to students, but will also act as a reminder to practitioners. Changes in the attitude to property and the owners' and users' duty of care to the community at large have inevitably made their mark on this edition, and readers will notice an increase in warnings on the subject of public safety and vandalism in the general notes. Often the two are incompatible, and it must be left to the design team to make final decisions on these matters.

The data include abstracts from the relevant current British Standards, together with abstracts from other technical sources. There have been many changes in the British Standards since the last edition, as many are now written in conjunction with member states of the European Union, and others have been altered to comply with European Union Law. The current standards are listed in each work section, but for any changes after Autumn 1996 readers should refer to that work section in the current edition of *Spon's Landscape and External Works Price Book*.

The specifications in each section do not set out to give comprehensive coverage of all items likely to be required in any contract document, but they do include the more frequently occurring items. They have been written in note form, and include references to other relevant work sections.

Note to readers

One intention of this publication is to provide an overview for those involved in landscape planning, design, construction and management and for students of these disciplines. It is not intended to be exhaustive or definitive, and it will be necessary for users of the information to exercise their own professional judgement when deciding whether to abide by or depart from it.

It cannot be guaranteed that any of the material in the book is appropriate to a particular use. Readers are advised to consult all current Building Regulations, British Standards or other applicable guidelines, Health and Safety codes and so forth, as well as up-to-date information on all materials and products.

Acknowledgements

The following manufacturers and suppliers provided photographs used in this handbook:

CED Ltd
728 London Road
West Thurrock
Grays
Essex RM20 3LU
UK

Chilstone
Victoria Park
Fordcombe Road
Langton Green
Kent TN3 0RE
UK

Claydon Architectural Metalwork
Units 11/12
Claydon Industrial Park
Gipping Road
Great Blakenham
Ipswich
Suffolk 1P6 0NL
UK

CU Phosco Street Furniture
Chartes House
Great Amwell
Ware
Hertfordshire SG12 9TA
UK

Haddonstone Ltd
The Forge House
East Haddon
Northampton NN6 8DB
UK

Ibstock Building Products Ltd
Leicestershire LE67 6HS
UK

Kingfisher Lighting
Unit 1 Portland Drive
Shirebrook
Mansfield
Nottinghamshire NG20 8TY
UK

Loren von Ehren Nurseries
Postfach 90 08 55
21048 Hamburg
Maldfledstrasse 4
21077 Hamburg
Germany

Marshalls Mono Ltd
Southowram
Halifax HX3 9SY
UK

Shap Concrete Products
Penrith
Cumbria CA10 3QQ
UK

SMP (Playgrounds) Ltd
Pound Road
Chertsey
Surrey KT16 8EJ
UK

Stenoak Fencing
Stenoak House
New Town
Uckfield
East Sussex TN22 5DL
UK

Woodhouse Co.
Spartan Close
Tachbrook Park
Warwick CV34 6RR
UK

Current legislation 1

The legislation in this book is current up to Autumn 1996.

It covers the main town and country planning legislation for England, Wales and Scotland. Readers will appreciate that the legislation abstracted is the editor's choice; it does not set out to be complete in any way, and the omission of any legislation must not be construed as meaning that it is of no importance to landscape designers. The legislation of the European Union and Northern Ireland has not been covered.

In order to be consistent with *Spon's Landscape and External Works Price Book*, the same subheadings have been kept. This will also enable the reader to note new legislation by referring to the current *Price Book*. The final section notes Department of the Environment circulars, Planning Policy Guidance Notes (PPGs), Mineral Policy Guidance Notes (MPGs) and Regional Policy Guidance Notes (RPGs), together with a few other useful publications.

1.1 A very brief introduction to the law

Acts or statutes are the main arm of the legal system. Each Act has a title, the year it received Royal Assent and a chapter number. The last two are a quick and infallible way of identifying a specific Act. For example, the Town and Country Planning Act 1990 chapter 8 can be identified by merely quoting 1990 c. 8; all marginal or footnotes within Acts refer to other Acts in this way.

Statutory Instruments – that is, regulations or orders – are produced by ministers or government departments, who are empowered to do so by a clause or clauses within the relevant Act. For example s.333 1990 c. 8 allows the Secretary of State to make regulations or orders. Statutory Instruments (SIs) also have a shorthand method of identification. For example, Town and Country Planning (Use Classes) (Amendment) Order 1995/297 can be referred to as 1995/297. As for Acts, SIs are always referred to in this abbreviated way in footnotes within other legislation.

The UK does not have a totally unified legal system, although much of the law applies to England, Wales, Scotland and Northern Ireland. Scotland, for example, has the Town & Country Planning Act (Scotland) Act 1972 c. 52 as the principal Act, although it has been amended by subsequent Acts.

There are also local Acts, or even by-laws, which may affect those involved in the planning or landscape work; these should be checked with the local authority.

The Department of the Environment's Regional Policy Guidance Notes, Planning Policy Guidance Notes and Mineral Policy Guidance Notes are growing in importance as the practical interpretation of legislation that many local planning authorities turn to when pursuing their local policies.

1.2 Acts and Statutory Instruments

1.2.1 TOWN AND COUNTRY PLANNING

England

Town and Country Planning Act 1990 c. 8
This is currently the principal Act. It is a consolidation Act, which with a few minor exceptions came into force on 24 August 1990. On that date the former principal Town and Country Planning Act 1971 was repealed; many other Acts were also repealed (see Planning (Consequential Provisions) Act 1990 below). Most of the sections formerly in the Town and Country Planning Act 1971 are embodied in the 1990 Act; it contains 15 parts and 17 schedules, while for clarity some parts are further divided into chapters.

The extracts cover sections likely to be of most use to landscape designers:

Part I Planning authorities
Defines the general policy functions delegated to local planning authorities, and lists authorities dealing with specific areas such as National Parks (NPs), Enterprise Zones (EZs), Urban Development Areas (UDAs) and Housing Action Areas (HATs).

Part II Development plans
Chapter I: Unitary Development Plans (UDPs) and metropolitan areas (including London). It requires the Local Planning Authorities (LPAs) to prepare and adopt and where necessary alter UDPs, and makes provision for joint UDPs. (Many local authorities have these in place, and have held public inquiries allowing the alteration stage to take place).
Chapter II: Structure and local plans: non-metropolitan areas

Part III Control over development
ss. 55–69   Includes a definition of development and new development, the requirement to obtain planning permission, and the methods by which it may be granted. It requires applicants to notify agricultural tenants of any relevant proposal, and LPAs to keep registers of planning applications available to the public.

ss. 70–75   Covers the determination of applications, and gives the LPAs power to impose conditions on the grant of permission and to alter or remove conditions previously imposed.

s. 76   Requires the LPAs to consider the needs of the disabled when applications involve public places.

ss. 77–81   Gives the Secretary of State (SOS) powers to deal with certain applications, and allows for the right of appeal to the SOS against a planning decision or the failure to make one within the allotted time.

s. 91   General conditions limiting the duration of planning permission.

s. 92   Covers 'outline planning permission' and 'reserved matters' (the latter is often where the landscape designer is approached to carry out the matters listed).

s. 106   Allows LPAs to enter into agreements relating to the use of the development of the land.

Part IV Compensation for effects of certain orders and notices

Part V Compensation for restrictions on new development in limited cases

Part VI Rights of owners etc. to require purchase of interest

Part VII Enforcement

Part VIII Special controls
Chapter I Trees:

s. 197   Local planning authority's duty to include provision for the preservation and planting of trees when granting planning permission.

s. 198   Local planning authority's power to make Tree Preservation Orders for individual trees, groups of trees or woodlands: deals with control of tree work, and the action to be taken on dead or dangerous trees.

s. 199   Procedures applicable to Tree Preservation Orders.

s. 200   Forestry Commission land.

s. 201   Power of local planning authority to make provisional TPOs in urgent cases.

s. 202   Power of the Secretary of State to make TPOs.

ss. 203, 204, 205   Compensation in respect of TPOs.

s. 207   Enforcement of duty to replace trees.

s. 210   Makes it an offence to fell or damage trees subject to TPOs.

s. 211   Trees in conservation areas deemed to be subject to TPOs.

s. 212   Secretary of State's power to disapply s.211.
s. 213   Enforcement of control over trees in conservation areas.
s. 214   Local planning authority's duty to keep registers of trees sub-
                ject to TPOs. These must be available for public inspection.

Chapter II Land adversely affecting amenity of the neighbourhood:
ss. 215–217   Gives the LPAs power to serve a notice requiring the owner or
                occupier to remedy the condition of the land.

Chapter III Advertisements:
s. 220   Allows the Secretary of State to make regulations to restrict or
                regulate the display of advertisements for the benefit of
                amenity or public safety. These regulations may provide for
                dimensions, appearance and the position and manners of fixing
                on site, as well as the consent of the local planning authority.
                They may also allow for an appeal procedure.
s. 221   Allows for specific regulations relating to different areas,
                including Conservation Areas, Experimental Areas and Areas
                of Special Control, as may be defined by the Secretary of State.
                It also makes provision for advertisements in place before this
                Act came into force.
s. 222   Any advertisements displayed that comply with the regulations
                made under s. 220 do not require planning permission.
s. 223   Where the removal of advertisements in place before 1 August
                1948 is required, a claim may be made to the local planning
                authority for reimbursement of any reasonable expense
                incurred.
s. 224   Enforcement of control over advertisements: any person dis-
                playing an advertisement in contravention of the regulations
                and who does not remove it when required to do so by the
                local planning office is guilty of an offence.

Part IX Acquisition and appropriation of land for planning purposes

Part X Highways
ss. 247–261   Power of SOS to: authorize the stopping-up or diversion of a
                highway or the making of a secondary highway; prohibit the
                access of vehicles over minor highways to improve the
                amenity of the area; extinguish or transfer public rights of
                way on land acquired for planning purposes; compulsorily
                purchase land for providing or improving highways or altern-
                ative rights of way.

Part XI Statutory undertakers

Part XII Validity

Part XIII Applications of the Act to Crown land

Part XIV Financial provisions

s. 303   Power of Secretary of State to make regulations for the pay-
            ment of fees to the LPA for matters dealing with planning
            applications.

Part XV Miscellaneous and general provisions

Deals with British Coal, ecclesiastical property and the Isles of Scilly.

ss. 320–323   Covers local planning inquiries, together with the procedures
               and time limits.
ss. 326–335   Settled land and land belonging to universities and colleges:
               procedure for serving notices and power of the SOS or an LPA
               to obtain information on their interest in the land.
s. 336   Interpretation of the terms used in this Act.

There are 17 schedules.

Planning (Consequential Provisions) Act 1990 c. 11

Lists the repeals, amendments, transitional and transitory matters and sav-
ings resulting from the new planning Acts. It is in force from 24 August
1990. The Acts that are wholly repealed include: the Town and Country
Planning Acts of 1947, 1962, 1968, 1971 and 1984; the Town and Country
Planning (Amendment) Acts of 1972 and 1977; the Local Authorities
(Historic Building) Act 1962; the Local Government and Planning
(Amendment) Act 1981; and the Planning Inquiries (Attendance of Public)
Act 1982.

Town and Country Planning (Minerals) Act 1981 c. 36

An important Act for landscape designers involved in mineral working,
restoration and rehabilitation, as it enhances the powers of the planning
authorities to control the landscape of mineral workings and spoil heaps
during and after extraction of minerals.

Housing and Planning Act 1986 c. 63

Parts of this Act that deal with planning matters include:

Part II   Simplified Planning Zones, amending the Town and Country
             Planning Acts 1971 and (Scotland) 1972;
Part V   Opencast coal workings;
Part VI,   which includes changes in dealing with listed buildings and
             conservation areas, local plans etc., and the control of adver-
             tisements in conservation areas.

Planning and Compensation Act 1991 c. 34

This is an important amendment Act relating to town and country planning
law, and extends the powers set out in the Town and Country Planning Act
1990.

Note: Changes to the sections in 1990 c. 8 are given in the order of that Act and not in the order in the new Act. Many of these items also apply to Scotland as amendments to their principal Act, the Town and Country Planning (Scotland) Act 1972.

s. 55	Meaning of development to include demolition of buildings, rebuilding, structural alterations or additions and other operations carried out by a builder. Also now included: fish farming including shellfish, and any mollusc including those in inland waters or tanks.
ss. 65–68	Notices of applications for grant of planning permission are redrafted.
s. 70	Power of local authority to decline to determine applications. Additionally, the LPA may refuse to consider planning applications that are repetitive or parallel to other applications that have been refused by the Secretary of State.
s. 71	Gives the Secretary of State power to require environmental assessments; substitutes time limits; and allows a development order to allow the LPA not to determine an application before the end of the prescribed period.
s. 79	Provides for the dismissal of an appeal where the appellant is responsible for the delay.
s. 106	Planning obligations: substitutes new section.
ss. 171–173	Enforcement amended by giving LPAs power to require information about activities on land; and allows LPAs to issue 'planning contravention notices', and to penalize for non-compliance. No enforcement allowed after four years from the start of operations on site, and allows LPAs to set time limits.
s. 179	Amends penalties for non-compliance with an enforcement notice.
s. 187	Adds new enforcement conditions, including time limits for enforcement of breaches of planning conditions and powers to obtain injunctions.
ss. 191–194	Substitutes sections on Established Use Certificates.
s. 192	Certificate of Lawfulness of proposed use or development allows the applicant to apply to the LPA by giving the information specified.
s. 196	Amends rights of entry for enforcement purposes.
s. 207	Replacement of trees: enforcement to take place not less than 28 days from serving the enforcement notice.
s. 210	Tree preservation orders: maximum penalty £20 000.
s. 214	Allows for authorized right of entry to check TPOs.
s. 299	Crown Planning Obligations amended; gives powers to enforce planning obligations, including Crown Planning Obligations.
s. 316	Land of interested LPAs etc. and development by them: substi-

tuted to allow certain provisions of Parts III, VII and VIII to apply.

Sch. 1 & 2   Alters the mining and waste permissions, and 'old mining permission' under the Planning Act 1947 c. 51.

In addition, the schedules to this Act make further amendments:

Sch. 3   To listed buildings, conservation areas and hazardous substances.

Sch. 4   Streamlining of development plan system.

Sch. 5   Modifies the procedure for making Simplified Planning Zone representations.

Town and Country Planning (Development Plan) Regulations 1991 SI 1991/2794

Set out the form and content of Structure, Local, Minerals Local, Waste Local and Unitary Development Plans. Revokes most previous development plan legislation, and provides for sorting out conflicts between such plans.

Town and Country Planning (Simplified Planning Zones) (Excluded Development) Order 1987 SI 1987/1849

The power to create Simplified Planning Zones (SPZs) lies with the London boroughs, district councils and urban development corporations; this order prevents them giving permission for a development that is a 'county matter'.

Town and Country Planning (Simplified Planning Zones) Regulations 1992 SI 1992/2414

These regulations set out the procedure for making and altering Simplified Planning Zone schemes (SPZs) in England and Wales. These include the requirement to make them available for public inspection, and to place notices in the *London Gazette* and the local press.

Town and Country Planning (Simplified Planning Zones) (Amendment) Regulations Order 1994 SI 1994/267

Amends the principal regulations 1992/2414 by requiring local planning authorities to consult the Urban Regeneration Agency when they propose to make or alter Simplified Planning Zones.

Town and Country Planning (Applications) Regulations 1988 SI 1988/1812

Deal with procedures for time-elapsed planning applications. Local planning authorities may now ask the applicant for further information to enable them to determine an application.

Town and Country Planning (Environmental Assessment and Permitted Development) Regulations 1995 SI 1995/417

Enable applicants to seek an opinion from the planning authority as to whether the proposed development would come under the environmental assessment regulations. Opinions under dispute may be referred to the Secretary of State. Where the planning authority has an interest in the development, an opinion may be sought from the Secretary of State.

Town and Country Planning (General Permitted Development) Order 1995 SI 1995/418

This order came into force on 3 June 1995. It revokes most of the Articles in the General Development Orders from 1988/1813 to 1995/298: a detailed list is given below. Allows for grant of planning permission for certain classes of development without an application being required under Part III of the 1990 Act. Schedule 2 (Permitted Development) is now subject to the Conservation (Natural Habitats etc.) Regulations 1994.

Orders revoked: 1990/457, 1990/2032, 1991/1536, 1991/2268, 1992/609, 1992/1280.

Orders partially revoked: 1989/603, 1989/1590, 1991/2805, 1992/658, 1992/1493, 1992/1563, 1994/678, 1994/2595, 1995/298

Town and Country Planning (General Development Procedures) Order 1995 SI 1995/419

This order came into force on 3 June 1995. It replaces the former principal order and its amendments. Deals with applications for planning permission, including outline permission and reserved matters and appeals; it allows for consultation with the relevant planning authority. The schedules contain the standard letters, notices, certificates etc. to be used.

Orders revoked: 1988/1813, 1989/603, 1989/1590, 1991/2805, 1992/658, 1992/53, 1992/1493, 1992/1563, 1992/2450, 1994/678, 1994/2595, 1995/298.

Town and Country Planning (General Development Procedure) (Amendment) Order 1996 SI 1996/1817

Amends the 1995 Order by requiring consultation with the Sports Council for England or Wales before grant or planning permission is given for development that may be detrimental to the provision of playing fields. The definition of "playing field" is given as: (1) the whole site which encompasses at least one playing pitch, (2) a delineated area which, together with any run-off area, is 0.4 hectares or more, and which is used in association football, American football, rugby, cricket, hockey, lacrosse, rounders, baseball, softball.

Town and Country Planning (Use Classes) Order 1987 SI 1987/764

This is the principal order, and divides the Use Classes into four main parts. Very briefly, these are:

Part A   A1 shops; A2 professional services; A3 sale of food and drink.
Part B   B1 offices; B2 general industrial; B3–7 special industrial processes; B8 storage and distribution.
Part C   C1 hotels, hostels and boarding houses; C2 residential institutions, hospitals, residential schools, colleges and training centres; C3 dwellings.

Part D   D1 non-residential institutions, medical centres, schools, libraries, public halls, churches; D2 assembly and leisure, including indoor and outdoor sports.

Current amendments of interest to the landscape designer are:

Town and Country Planning (Use Classes) (Amendment) Order 1992
SI 1992/610
Deals with Class B3.

Town and Country Planning (Use Classes) (Amendment) Order 1992
SI 1992/657
Deals with hazardous substances. These are also controlled by a concurrent order.

Town and Country Planning (Determination of Appeals by Appointed Persons) (Prescribed Classes) Regulations 1989 SI 1989/1087
Enable inspectors to determine appeals on listed building consent and listed building enforcement notices related to demolition of Grade II listed buildings. The Secretary of State continues to determine appeals concerning Grade I and Grade II* listed buildings.

Town and Country Planning (Determination by Inspectors)(Inquiry Procedure) Rules SI 1992/2039
Revoke 1988/945. They alter the time-scale for serving statements, and allow the appellant and the local planning authority to call for any document intended as evidence. Proof of evidence sent to the inspector containing more than 1500 words requires an accompanying written summary; this will normally be read out at the inquiry.

Town and Country Planning (Enforcement Notices and Appeals) Regulations 1991 SI 1991/2804
Set out the procedure to be followed for serving enforcement notices of planning conditions, and the information to be given by the planning authority.

Town and Country Planning (Control of Advertisements) Regulations 1992 SI 1992/666
The principal regulations, which deal with all types of advertisement, but the landscape designer is mainly concerned with advertisements and flag-poles in forecourts, and advertisements in Areas of Special Control. They revoke SIs 1989/670, 1990/881 and 1990/1562

Town and Country Planning (Control of Advertisements)(Amendment) Regulations SI 1994/2351
Amend the principal regulations; include allowing the use of on-site flags to advertise new housing, restrictions on height of flag-pole (4.6 m) and flag size (2 m²); not allowed in conservation areas, National Parks, AONBs etc.

Town and Country Planning General Regulations 1992 SI 1992/1492

Supersede the 1976 regulations. Bring local authorities under parts of the Town and Country Planning Act 1990 where they have an interest in the land. Correlate with other legislation the procedures for claiming compensation, extinguishment of rights of way, blight notices, and revoking planning permission.

Town and Country Planning (Special Enforcement Notices) Regulations 1992 SI 1992/1562

These regulations set out the procedure to be followed for serving enforcement notices of planning conditions, and they describe the information to be given by the planning authority, with special reference to Crown land.

Town and Country Planning Appeals (Inquiries Procedure) Rules 1992 SI 1992/2038

These rules set out the procedures to be followed for local planning inquiries and hearings when heard by an approved inspector. Covers an application referred to the inspector, or appeals made to him relating to planning permission, consent under TPOs, listed building consent or consent to demolish unlisted buildings in conservation areas. This last is commonly known as Conservation Area Consent.

Town and Country Planning Appeals (Determination by Inspectors) (Inquiries Procedure) Rules 1992 SI 1992/2039

Set out the procedure to be followed by inspectors appointed by the Secretary of State to determine appeals in relation to planning permission, listed building consent and consent for the demolition of unlisted buildings in conservation areas.

Town and Country Planning (Fees for Applications and Deemed Applications) (Amendment) (No. 2) Regulations 1992 SI 1992/3052

These regulations, which cover fees for planning applications, are updated regularly to set out the increases in the fees payable to the planning authority for planning applications, or for submissions that are deemed to be planning applications.

Scotland

Town and Country Planning (Scotland) Act 1972 c. 52

Parts of this Act have since been amended, but it is currently the main Act, and consolidates earlier Acts. It covers general planning control, including determination, deemed planning permission, duration, revocation and modification of planning permission, together with specific controls applying to conservation areas, buildings of special architectural or historic interest, trees, and Tree Preservation Orders.

Town and country planning Acts repealed:

- Town and Country Planning (Scotland) Act 1945 c. 33;
- Town and Country Planning (Scotland) (Amendment) Act 1951 c. 19;
- Town and Country Planning (Scotland) Act 1953 c. 16;
- Town and Country Planning (Scotland) Act 1963 c. 17.

Town and country planning Acts partially repealed:
- Town and Country Planning (Scotland) Act 1947 c. 53;
- Town and Country Planning (Scotland) Act 1954 c. 73;
- Town and Country Planning (Scotland) Act 1959 c. 70;
- Town and Country Planning (Scotland) Act 1969 c. 30;
- Town and Country Planning (Scotland) (Amendment) Act 1972 c. 42.

Other repeals include the Tree Act 1970 c. 43.

Town and Country Planning (Scotland) Act 1977 c. 10
Amends the principal Act 1972 in respect of local plans and stop notices.

Town and Country Planning (Amendment) Act 1985 c. 52
Now applies only to the Town and Country Planning (Scotland) Act 1972 s. 60; amended to require the replacement of trees destroyed or uprooted in a woodland, either in the same place or in a nearby area agreed by the local planning authority.

Designation of Structure Plan Areas (Scotland) Order 1995/3002
Lists the Structure Plan areas together with the relevant planning authority.

Town and Country Planning (Inquiries Procedures) Scotland Rules 1980 SI 1980/1676
Set out the procedure for local inquiries and hearings in Scotland.

Town and Country Planning Appeals (Determination by Appointed Person) (Inquiries Procedure) (Scotland) Rules 1980 SI 1980/1677
Set out the procedures to be followed for local planning inquiries and hearings when heard by an approved inspector.

Town and Country Planning (General Development) (Scotland) Order 1981 SI 1981/830
Covers planning procedures not laid down specifically in the Act, and is important in that the schedules list certain classes of permitted development; any development so listed does not require express planning permission. The orders (SI 1980/830 and SI 1984/237) also set out the requirements for advertising developments, any notification required to be given to neighbours, methods of appeal, etc.

Note: Short summaries of subsequent amendments are given below:

SI 1984/237
Redefines neighbouring land, and those cases that are required to be advertised. The schedule sets out the form for various forms and certificates.

SI 1985/2007
Minor changes, together with four new classes of permitted development.

SI 1990/508
Requires the appellant to notify the planning authority of their appeal, and takes into account written appeals.

SI 1991/147
Deals with permitted development of electric lines.

Town and Country Planning (Compensation for Restrictions on Mineral Workings) (Scotland) Regulations 1987 SI 1987/433
Modify the provisions for paying compensation where changes in planning permission or orders on mineral workings are imposed. Compensation now takes into account the expenditure voluntarily incurred to alleviate damage to amenity, or restoration. The NCB is exempt.

Town and Country Planning (Determination of Appeals by Appointed Persons) (Prescribed Classes) (Scotland) Regulations 1987 SI 1987/1531
These regulations set out appeals that may be determined by a person appointed by the Secretary of State (the inspector). Include appeals against planning decisions or in default of planning permission, and appeals against determinations as to whether a use or operation constitutes or involves development. They also deal with appeals against refusal of consent or conditional consent in respect of felling trees subject to a tree preservation notice.

Amendment Regulation SI 1989/557 deals with appeals relating to advertisements.

Town and Country Planning (Simplified Planning Zones) (Scotland) Regulations 1987 SI 1987/1532
Set out the procedures to be taken by district and general planning authorities who have established Simplified Planning Zones (SPZs).

Town and Country Planning (Simplified Planning Zones) (Scotland) Regulations 1995 SI 1995/2043
Amends the powers given to planning authorities to make simplified planning zone schemes under the Town and Country Planning (Scotland) Act 1972. The 1987 Regulations SI 1987/1532 are revoked except for the saving listed in s.23 of the 1995 regulations.

Town and Country Planning (Simplified Planning Zones) (Scotland)
Order 1995 SI 1995/2044
No SPZ scheme may have effect to grant planning permission for development that requires an environmental assessment.

Town and Country Planning (Appeals) (Written Submissions
Procedure) (Scotland) Regulations 1990 SI 1990/507
Set out the procedures and timetable for appeals to be heard by written submission. If the appellant, or the appellant and the local planning authority together, wish to waive the public inquiry procedure, and the Secretary of State agrees, the appeal may be decided on the basis of written submissions. Appeals against Tree Preservation Orders prohibiting felling are covered by the regulations. Appeals questionnaires are available for sites in Scotland from St Andrews House, Edinburgh.

Town and Country Planning (General Permitted Development)
(Scotland) Order 1992 SI 1992/223
Revokes the whole of 1983/1620, 1985/1014, 1988/977, 1989/148, 1991/147, and parts of 1981/830, 1984/237, 1985/2007 and 1990/508. The schedules give the types of permitted development, including change of use, caravan and forestry buildings, and certain mineral workings. 'Bad neighbour' developments are listed; of interest to landscape designers are wildlife parks, many sports buildings, and those that alter the character of an amenity area, bring crowds into a quiet area, or interfere with a homogeneous area. This is referred to as the 1992 Order in the Amendment Orders listed below:
    1992/1078; 1992/2084; 1993/1036; 1994/1442; 1994/2586 includes an amended clause for the 'restoration schemes' for coal-mining operations; 1994/3294 amends the status of the demolition of buildings.

Town and Country Planning (General Development Procedure)
(Scotland) Order 1992 SI 1992/224
The regulations amend the 1981 regulations. They cover procedures for planning applications, reserved matters, appeals, planning registers, listed buildings, public notices, and Established Use Certificates.

Town and Country Planning (General Development Procedure)
(Scotland) Amendment Order 1993 SI 1993/1039
Requires planning authorities to consult the Health and Safety Executive before granting planning permission for specific developments within a stated area of installations holding or using hazardous substances.

Town and Country Planning (General Development Procedure)
(Scotland) Amendments (No2) Order 1994 SI 1994/3293
Includes required procedures over peat extraction and tolls.

Town and Country Planning (General Development Procedure) (Scotland) Amendments (No3) Order 1994 SI 1994/3294
Includes required procedures over demolition and tolls.

Town and Country Planning (General Development Procedure) (Scotland) Amendment Order 1996 SI 1996/467
Amends the 1992 Order (SI 1992/224) as amended by SI's 1992/2083, 1993/1039. 1994/2585 and 1994/3293. Article 12 now requires planning authorities to send community councils a weekly list giving specified information on planning applications and be consulted on specific planning applications; it also requires the Scottish Environment Protection Agency to be consulted on matters dealing with developments concerning water and sewerage.

Town and Country Planning (Control of Advertisements) (Scotland) (Amendment) Regulations 1992 SI 1992/1763
These regulations amend the principal regulations 1984/467, and deal with all types of advertisement, but the landscape designer is mainly concerned with advertisements and flag-poles in forecourts, and advertisements in Areas of Special Control.

Town and Country Planning (Enforcement of Control) (No 2) (Scotland) Regulations 1992 SI 1992/2086
These regulations set out the procedure to be followed for serving enforcement notices of planning conditions. They cover waste disposal, listed buildings, conservation areas, and appeals against enforcement notices.

Town and Country Planning (Special Enforcement Notices) (Scotland) Regulations 1992 SI 1992/478
These regulations set out the procedure to be followed for serving enforcement notices of planning conditions, and they describe the information to be given by the planning authority, with special reference to Crown land.

Town and Country Planning (Fees for Applications and Deemed Applications) (Scotland) (Amendment) (No. 2) Regulations 1992 SI 1992/3137
These regulations, which cover fees for planning applications, are updated regularly to set out the increases in the fees payable to the planning authority for planning applications, or for submissions that are deemed to be planning applications.

1.2.2 TREES

Civic Amenities Act 1967 c. 69
Nearly all revoked, and now under the main Town and Country Planning Act 1990.

Town and Country Planning (Tree Preservation) Order 1969 SI 1969/17

Procedures for making and revoking Tree Preservation Orders. Includes schedules and forms for orders.

Wildlife and Country Act (Amendment) Act 1985 c. 31

Section 1 amends the Badger Act 1973 by presuming guilt on the part of those found digging, injuring badgers etc.

Section 2 amends the principal Act dealing with notification of Areas of Special Scientific Interest.

Section 3 amends the requirement for maps of National Parks.

Section 4 amends the Forestry Acts 1967–71 by requiring the Forestry Commission to achieve a balance between afforestation and the production of timber, and the conservation and enhancement of natural beauty, the conservation of flora, fauna and landform of special interest.

Town and Country Amenities Act 1974 c. 32

Now applies only to Scotland, and amends the Town and Country Planning (Scotland) Act 1972 by requiring changes in the law referring to Tree Preservation Orders and trees in conservation areas.

Forestry Act 1986 c. 30

Gives the Forestry Commissioners the power to require the restocking of land with trees after unauthorized felling. There is a right to appeal against restocking. Does not apply to trees with a Tree Preservation Order on them, or to trees felled before 8 September 1986.

Forestry Act 1991 c. 43

Not applicable; increases committee members.

Forestry (Felling of Trees) Regulations 1979/791

Deal with the tree owner selling trees or land; prescribe the notices required.

Forestry (Exemptions from Restriction of Felling) Regulations 1979/792

Cover procedural matters and prescribed forms.

Forestry (Modification of Felling Restrictions) Regulations 1985/1958

Reduce in any quarter the cubic content of trees felled without a licence from 30 m^3 to 5 m^3; also reduce the cubic content for selling.

Forestry (Felling of Trees) (Amendment) Regulations 1987/632

Amend 1979/791 on procedural matters.

Forestry (Exemptions from Restriction of Felling) (Amendment) Regulations 1988/970

Deal with exemptions for those participating in approved plans of operation for the following grant schemes: forestry, broad-leaved woodland and woodland.

Farm Woodland Premium Scheme 1992/905

Deals with administrative procedures, including available grants for wood-lands in Great Britain, and lists categories of land acceptable and those excluded.

Countryside Access Regulations 1994/2349

Allow for payments to farmers who agree to allow public access for a five-year period over a specific area of set-aside land for the purpose of quiet recreation. Maps required to show special features of the area; access routes to be 10 m wide to whole or parts of a field or fields.

Town and Country Planning Act 1990 c. 8

Part VIII Special Controls: Chapter 1 deals with trees under sections 197–214. These are summarized at the beginning of the chapter under the main Act.

Town and Country Planning (Tree Preservation Orders) Amendment Regulations 1981 SI 1981/14

The original regulations (SI 1969/17) have been amended by this SI; together they set out the procedure for making Tree Preservation Orders, and require all such preserved trees to be marked on a map, with a copy made available to the public for inspection. Local planning authorities are responsible for confirming Tree Preservation Orders in England and Wales, and all representations and objections should be made to them.

Town and Country Planning (Tree Preservation Order) (Amendment) Regulations 1988 SI 1988/963

These regulations enable the local authorities (for the purposes of compensation and except where an application is referred to them under the Forestry Act 1967) to treat trees in woodlands that have an outstanding or special amenity value in the same way as other trees.

Town and Country Planning (Tree Preservation Orders and Trees in Conservation Areas) (Scotland) Regulations 1975 SI 1975/1204

Prescribe form and content of TPOs.

Town and Country Planning (Tree Preservation Orders and Trees in Conservation Areas) (Scotland) Amendment Regulations 1984 SI 1984/329

The original regulations (SI 1975/1204) have since been amended by SI 1981/1385 and this SI. All describe the content of Tree Preservation Orders, and the procedure for making and preserving them. Local planning authorities are now responsible for confirming Tree Preservation Orders in Scotland, but a copy of any order must be lodged with the Conservator of Forests for Scotland.

Forestry (Felling of Trees) (Amendment) Regulations 1987 SI 1987/632

Set out revised forms for applications for a licence to fell growing trees, including Part 2, which deals with types of trees to be felled, and Part 3 dealing with existing areas (National Parks, AONBs, SSSIs etc.). Part 4 deals with Tree Preservation Orders. The application must cover information on types of tree to be felled, their age and condition, total number of each species, proposed date of felling, and the proposed treatment of the land after felling. Applications must be accompanied by an Ordnance Survey map showing location to a scale of 1:10 000 or 6 in to the mile.

Hedgerows Regulations 1997 SI 1997/1160

Increases the protection of important hedgerows. Section 3 gives the criteria for hedgerows that are to be protected. Planning permission is required for removal of protected hedgerows on the same lines as tree preservation legislation; the LPA may serve a Hedgerow Retention Notice to protect the hedgerow.

1.2.3 ENVIRONMENTAL CONTROL

Derelict Land Act 1982 c. 42

Amends parts of existing Acts relating to derelict land, including the National Parks and Access to the Countryside Act 1949. The main object is to bring derelict and unused land back into use, both in urban and rural areas. A register of vacant land is now kept, and there are substantial grants available to those who carry out reclamation work.

Planning (Hazardous Substances) Act 1990 c. 10

This Act lists the hazardous substances authorities, and sets out their powers. These authorities are generally district authorities or the London boroughs, though there are rather complex exceptions for non-metropolitan land within a National Park, and for the winning and working of minerals. In England (but not Wales), where the land is used for refuse or waste material the matter is referred to the appropriate authority. The definition of 'land' includes a wharf or harbour as described in the Harbours Act 1964. The Act sets out the complicated procedures dealing with the presence of hazardous substances on, over, or under land.

Environmental Protection Act 1990 c. 43

A comprehensive Act, mainly intended to clarify the responsibilities of various authorities with regard to the control of environmental harmful actions. Much of the Act will be brought into force by future regulations. It contains 9 Parts and 16 schedules. The most important Parts are:

- Part I Integrated pollution control and air pollution control.
- Part II Waste on land.
- Part III Statutory nuisances and clean air.

- Part IV Litter etc.: sets out the requirements of local authorities, the Crown and occupiers to keep land as clear as practicable of litter.
- Part V Amendment of Radioactive Substances Act 1960.
- Part VII Nature conservation in Great Britain: sets up three new nature conservancy bodies.

Environment Act 1995 c.25

A comprehensive Act which establishes Environment Protection Agencies in England, Wales and Scotland. It transfers various functions to it including that of waste authorities and all the functions, property and so forth from the National Rivers Authority.

Water and Sewerage (Conservation, Access and Recreation) (Code of Practice) Order 1989 SI 1989/1152

Control of Noise (Code of Practice for Construction Sites) Order 1984 SI 1984/1992

Approves British Standard Code of Practice for noise control on construction and open sites.

Control of Noise (Code of Practice for Construction Sites) (Scotland) Order 1985 SI 1985/145

Approves British Standard Code of Practice for noise control on construction and open sites.

Control of Noise (Code of Practice for Construction and Open Sites) Order 1987 SI 1987/1730

Under the Control of Pollution Act 1974 s. 71, the Secretary of State may give guidance on appropriate methods of minimizing noise by approving codes of practice for such operations as building, roadworks, demolition, dredging or other works of engineering construction.

Environmental Information Regulations 1992 SI 1992/3240

Implement EC Directive 90/313 giving public access to environmental information held by public authorities and others. Set out types of information that must be made available, and certain exclusions. Override some other statutory provisions for restricting access to information.

Town and Country Planning (Assessment of Environmental Effects) Regulations 1988 SI 1988/1199

Implement EC Directive 85/337. Environmental assessment (EA) is required for large-scale or environmentally sensitive projects that require planning permission under the Town and Country Planning Act 1971. Projects coming under other Acts will be the subject of separate legislation. Schedule 1 projects are large-scale engineering works for which an EA is compulsory.

Schedule 2 projects are extractive industries, and developments that may have a significant environmental effect; in this case the LPA determines the need for an EA. Schedule 3 sets out the information to be provided in an EA. In the case of any dispute between the developer and the LPA the Secretary of State will give a ruling. The LPA must state within three weeks whether an EA is required, and if so the LPA has 16 weeks instead of 8 to consider the planning application. Local authority development is subject to the same control, but implemented under Town and Country Planning General Regulations 1976 (SI 1976/1419); Crown development is not bound by these regulations, but the Crown will submit an EA to the LPA at the consultation stage.

Town and Country Planning (Assessment of Environmental Effects) (Amendment) Regulations 1990 SI 1990/367

Revise the Town and Country Planning (Assessment of Environmental Effects) Regulations 1988 (SI 1988/1199) to exclude projects listed in a new sub-para 2(a), which covers nuclear power stations or other nuclear reactors, except for research installations with a maximum power output of not more than 1 kW continuous thermal load. This amendment brings the regulations into line with the EC Directive on Environmental Assessment.

Town and Country Planning (Assessment of Environmental Effects) (Amendment) Regulations 1992 SI 1992/1494

Transfer requirements for publicity for environmental statement accompanying planning application to the Town and Country Planning General Development (Amendment) (No. 4) Order 1992 SI 1992/1493. The 1988 regulations will now apply where the local planning authority is also the applicant.

Town and Country Planning (Assessment of Environmental Effects) Regulations 1994/677

Deal with procedures for the preparation of environmental statements in connection with planning applications. The following projects are added to Schedule 2: a wind generator, a motorway service area, and coast protection works.

Environmental Assessment (Afforestation) Regulations 1988 SI 1988/1207

Deal with applications for grants or loans in respect of forestry planting. Where the Forestry Commission considers that such a forestry project is likely to have significant environmental or adverse effects on the ecology, it must take the environmental information into account before making a grant or loan.

Land Drainage Improvement Works (Assessment of Environmental Effects) Regulations 1988 SI 1988/1217

Implement EC Directive 85/337 for land drainage improvement works proposed by drainage authorities, including local authorities acting as drainage authorities. Affect all watercourses except sewers. The drainage authority may decide whether an environmental statement is required, but if the public asks for an environmental statement then the Secretary of State decides the need.

Land Drainage Improvement Works (Assessment of Environmental Effects) Regulations 1995 SI 1995/2195

Amends the 1988 Regulations SI 1988/1217 (the Principal Regulations). Modifies the procedure by a drainage body where it considers that an environmental assessment is or is not required in respect of improvement works, or where they are required to provide further information to the Secretary of State.

Environmental Assessment (Scotland) Regulations 1988 SI 1988/1221

These regulations cover applications made under the Town and Country Planning (Scotland) Act 1972 and the Town and Country Planning (Development by Planning Authorities) (Scotland) Regulations 1981. The schedules list projects for which an environmental assessment is required, and give a list of information required. Schedule 1 includes major installations such as oil refineries, power stations, steel, chemical and asbestos works, major airports, roads and ports; an environmental assessment is mandatory for these. Schedule 2 covers most industrial, extractive, infrastructure and noxious trade projects; an environmental assessment is required only if the environmental effects are likely to be significant.

Highways (Assessment of Environmental Effects) Regulations 1988 SI 1988/1241

Amend the Highways Act 1980 c. 66 by inserting s. 105A, which requires an environmental assessment to be made on certain new highways or major improvements to existing ones.

Electricity and Pipeline Works (Assessment of Environmental Effects) Regulations 1989 SI 1989/167

These regulations apply to the following:

- The construction or extension of a generating station on any land;
- The placing of an electric line other than a service line above ground;
- The construction of a pipeline for oil or gas, or the diversion of a pipeline that has been used for oil or gas.

The schedule gives the statutory contents of the environmental statement.

Electricity and Pipe Line Works (Assessment of Environmental Effects) Regulations 1989 SI 1990/442

These Regulations re-enact SI 1989/167 and apply to the following:

- The construction or extension of a nuclear or non-nuclear generating station on any land.
- The placing of some electric lines other than service lines above ground.

- The construction of a pipe line for oil or gas, or the diversion of a pipe line that has been used for oil or gas.

The schedule gives the statutory contents of the Environmental Statement. Revokes SI 1989/167 and is amended by SI 1996/422.

Harbour Works (Assessment of Environmental Effects) Regulations 1988 SI 1988/1336

Implements the EC Directive 85/337 EEC on Environmental Assessment to include Harbour Works. Projected Harbour Works will now have to be subjected to an Environmental Assessment, and the Secretary of State may determine whether the project falls under Schedule I or Schedule II of the Directive.

Harbour Works (Assessment of Environmental Effects) (No. 2) Regulations 1989 SI 1989/424

Covers harbour works below low water mark of medium tides, and which do not come under planning control or other harbour regulations. In England fishery harbours come under MAFF, in Wales they are under the Secretary of State for Wales. Marine works in Scotland come under the Secretary of State for Scotland; other harbour works come under the Secretary of State for Transport. Amended by 1996/1946.

Electricity and Pipeline Works (Assessment of Environmental Effects) Regulations 1990 SI 1990/442

These regulations apply to the following:

- The construction, extension and operation of a generating station on any land;
- The placing of an electric line other than a service line above ground;
- The construction of a pipeline for oil or gas, or the diversion of a pipeline that has been used for oil or gas.

The Secretary of State must take environmental information into consideration, but if the generating station is under 300 MW, a Statement is required only if the Secretary of State considers that the effects are likely to be significant. The schedule gives the statutory contents of the Statement.

Transport and Works (Assessment of Environmental Effects) Regulations 1995 SI 1995/1541

Amends the Transport and Works Act 1992 to implement an EC Directive on the assessment of the effects on the environment of certain public and private projects including the construction of railways, tramways and other guided transport systems.

1.2.4 WILDLIFE AND COUNTRYSIDE

Wildlife and Countryside Act 1981 c. 69

This Act affects existing legislation on nature conservation, the countryside, National Parks, the Countryside Commission and rights of way. It also lists 39 species of animals, insects and plants that are protected, and allows for the list to be altered. Bat roosts are given total protection.

Limestone pavements are protected, and removal of stone normally prohibited.

Sites of Special Scientific Interest (SSSIs) are designated by the Nature Conservancy Council, and must be notified to owners, occupiers and the local planning authority. SSSIs are registered as a local land charge. National Nature Reserves may be designated by the Nature Conservancy Council, who may apply by-laws to the reserve. Marine Nature Reserves are designated in a similar way. Grants or loans towards nature conservation may be made by the Nature Conservancy Council.

Management agreements may be made between the planning authority and the landowner for conserving or enhancing the natural environment; such agreements form a covenant.

Moor or heath areas in National Parks may be the subject of restrictive orders preventing conversion into agriculture or forestry uses, and the park planning authority must keep maps of areas to be conserved. Water authorities and internal drainage boards are required to consult the NCC before carrying out harmful operations in areas notified as being of special interest.

Conservation of natural beauty includes the conservation of flora, fauna, geological and physiological features.

Rights of way that have been ploughed over must be reinstated within two weeks of ploughing; the highway authority may reinstate the path and reclaim the cost for non-compliance.

Signposting and waymarking of footpaths, bridleways and byways is a duty of the highway authority, and may be supported by voluntary organizations.

Walls of regional character may be constructed by the highway authority in place of the normal fences protecting highways.

Hedges may be subjected to a trimming order between September and April.

The following orders vary the flora and fauna to be protected under the Act:

- Wildlife and Countryside Act 1981 (Variation of Schedules) Order 1988/288;
- Wildlife and Countryside Act 1981 (Variation of Schedules) Order 1989/906;
- Wildlife and Countryside Act 1981 (Variation of Schedules) Order 1991/367;
- Wildlife and Countryside Act 1981 (Variation of Schedules) Order 1992/320;

- Wildlife and Countryside Act 1981 (Variation of Schedules) Order 1992/2350;
- Wildlife and Countryside Act 1981 (Variation of Schedules) Order 1992/2674;
- Wildlife and Countryside Act 1981 (Variation of Schedules) Order 1992/3010.

Wildlife and Countryside Act 1981 (Amendment) Regulations 1995 SI 1995/2825

Amends the 1981 Act in relation to wild birds, inland waters, damage to livestock their food, crops, vegetables, fruit, growing timber or fisheries.

Wildlife and Countryside (Amendment) Act 1985 c. 31

This Act increases the power of the Countryside Commission, who can now direct county planning authorities on the conservation of areas of natural beauty. It also deals with notification of Areas of Special Scientific Interest. Badgers are given greater protection.

Wildlife and Countryside (Amendment) Act 1991 c. 39

Amends s. 5 and 11 of the 1981 Act. Section 5 deals with the killing and taking of wild birds and s. 11 with that of animals. It allows for the lawful killing or taking of certain species, in the interests of public health, agriculture, forestry or nature conservation, and without causing injury to protected species.

Wildlife and Countryside (Service of Notices) Act 1985 c. 59

This Act repeals s. 28(3) of the Wildlife and Countryside Act 1981 and s. 2(3) of the Wildlife and Countryside (Amendment) Act 1985. It alters the service of notices under the 1981 Act.

Rights of Way Act 1990 c. 24

This Act amends the law relating to rights of way and the disturbance or restoration of the surface of land over which any right of way passes. Where land is disturbed by the plough etc. the surface is to be made good for those wishing to use it, and the line of the way indicated. It is an offence not to comply with these requirements. Where a right of way is to be excavated or subject to engineering works the occupier may apply to the highways authority for an order to disturb the land for a period not exceeding 3 months; these orders may require temporary diversions and notices posted indicating the temporary route.

Protection of Badgers Act 1992 c. 51

Protects badger setts from interference and sets out exceptions to the Act for development and forestry, drainage, or agricultural work.

Wildlife and Countryside (Claims for Compensation under Section 30) Regulations 1982 SI 1982/1346

These regulations cover the procedure for claiming compensation from the Nature Conservancy Council.

Set-Aside Regulations 1988 SI 1988/1352

Implement EC Directive 85/797 on improving the structure of agriculture. Set out payments for farmers who set aside land for at least 5 years, which has been used for specified arable crops. Land must be fallowed, planted to woodland, or used for non-agricultural purposes. Set-aside land may not be used for building for industry, sale of goods, or residential purposes including hotels or hostels, offices, storage or distribution. The use of other land for arable crops is restricted, and the regulations include a compulsory management schedule.

Set-Aside (Amendment) Regulations 1989 SI 1989/1042

Add further crops to those listed in Schedule 1 of the 1988 regulations; widen the definition of arable land, together with other minor amendments.

Set-Aside (Amendment) Regulations 1990 SI 1990/1716

These regulations amend the Set-Aside Regulations 1988 (SI 1988/1352). They deal with the types of seed to be sown and the management during growth without the use of non-natural fertilizers and pesticides etc. Also deal with the general management of hedges, ditches and other rural features.

Set-Aside (Amendment) Regulations 1991 SI 1991/1993

Deal with the meaning of 'designated maps', and the definition of 'less favoured areas'.

Farm Woodland Premium Scheme 1992 SI 1992/905

This scheme allows for payments to alleviate loss of income when agricultural land is converted to woodland. There are conditions controlling the type of land that may be converted (mostly poor marginal farmland) and the species of tree that may be grown.

Town and Country Planning (Public Path Orders) Regulations 1993 SI 1993/10

Set out procedures for local authorities making public path orders.

Public Path Orders Regulations 1993 SI 1993/11

Set out procedures for creating, diverting, and extinguishing public paths. Give persons to be notified and types of map to be submitted.

Wildlife and Countryside (Definitive Maps and Statements) Regulations 1993 SI 1993/12

These regulations are made under the Wildlife and Countryside Act 1981, and define the scales and types of map showing public rights of way. Provide procedures for making and confirming rights of way orders.

Conservation (Natural Habitats Etc.) Regulations 1994 SI 1994/2716

Implement EC Directive 92/43/EEC. Schedules 2 and 3 list European protected animals and plants respectively. These are extra to those listed in the Wildlife and Countryside Acts and orders.

- Part I implements the EC Directive on Habitats EC 2078/92, and lists the competent authorities.
- Part II deals with the selection and registration of habitats in Britain of European importance and their control.
- Part III lists species of plants and animals protected under the Directive.
- Part IV lists the large number of Acts and regulations that are affected by the Directive.
- Part V deals with management agreements, compulsory purchase powers, and compensation.

There are various regulations dealing with special habitats; these involve an agreement to keep the land as a specified habitat for a set number of years.

The habitat regulations implement part of EC 2078/92 dealing with the relationship of agriculture and the environment. Aid is payable to farmers who undertake an agreed management programme for their land.

- Broadleaved woodland (Wales) SI 1994/3009
- Coastal belt (Wales) SI 1994/3101
- Former Set-aside Land SI 1994/1292
- Species-rich grassland (Wales) SI 1994/3102
- Water Fringe (Wales) SI 1994/3100
- Water Fringe 1994/1291, 1996/1480
- Scotland SI 1994/2710
- Former set-aside land SI 1994/1292, 1996/1478
- Salt Marsh SI 1994/1293, 1995/2871 and 2891, 1996/1479

Rail Crossing Extinguishment and Diversion Orders, Public Path Orders and the Definitive Maps and Statements (Amendments) Regulations 1995/451

Minor drafting amendments and new forms for closing public paths etc.

Regional Parks (Scotland) Regulations 1981 SI 1981/1613

Describe the procedure to be followed where an area in Scotland is intended to be designated as a National Park.

National Heritage (Scotland) Act 1991 c. 28

Part I sets up Scottish National Heritage and defines its functions and powers. Takes over the functions previously exercised by the Nature Conservancy Council for Scotland, including setting up nature reserves. It may promote developments that enhance the natural heritage of Scotland and acquire land compulsorily for this purpose. SNH is empowered to designate Natural Heritage Areas (these replace National Scenic Areas) where special protection measures may be required. It may make grants or loans to further its aims.

Part II enables the Secretary of State to make Control Orders restricting the abstraction of water for irrigation purposes from ground or inland water.

Part III enables the Secretary of State to make Drought Orders restricting the disposal and use of water in major drought conditions.

Habitats (Scotland) Regulations 1994 SI 1994/2710

Partially implement EC Directive 2078/92. Deal with arrangements for set-aside and the establishment and maintenance of habitats, and appropriate management agreements.

1.2.5 HERITAGE AND CONSERVATION

Agriculture Act 1986 c. 49

This Act deals with agriculture, horticulture and food production and the conservation and enhancement of the natural beauty and amenity of the countryside. This includes flora, fauna, geological and physiographical features, and features of archaeological interest. Section 18 allows Environmentally Sensitive Areas to be designated in consultation with the Countryside Commissions, Nature Conservancy Councils of England and Wales, and of Scotland.

It also addresses the need to balance an economic agricultural and horticultural policy against the conservation and enhancement of the natural beauty and amenity of the countryside. This includes the flora, fauna, geological and physiographical features, as well as features of archaeological interest. The promotion of the enjoyment of the countryside by the public as far as it is commensurate with economic and social interests in rural areas and the promotion and maintenance of a stable and efficient agricultural industry is also dealt with. Designation and management of Environmentally Sensitive Areas comes under this Act, and it lists those bodies to be consulted before designations can be made.

This Act applies to Crown Land, the Duchy of Lancaster, the Duchy of Cornwall, and government departments.

Water Act 1989 c. 15

As well as dealing with the administrative side of the water and sewerage authorities, this Act also deals with three aspects of water control: control of water supplies to the consumer, the control of rivers and other water bodies, and sewerage. Sets up the National Rivers Authority (NRA). The Director General of Water Services must ensure that private water authorities take into account conservation, protection of the environment, historic buildings and monuments etc., and the use of water and land for recreation.

Water Resources Act 1991 c. 57
Unlikely to be of interest to the landscape designer.

Planning (Listed Buildings and Conservation Areas) Act 1990 c. 9
An Act dealing with the special controls over listed buildings and buildings within conservation areas.

- Part I deals with listed buildings.
- Part II deals with conservation areas.
- Part III covers general matters including sections on local authority and Crown land, British Coal and ecclesiastical property.
- Part IV covers interpretation and other legal matters.

There are four schedules, and a table indicating the derivation of the provisions, many of which stem from the Town and Country Planning Act 1971.

Natural Heritage (Scotland) Act 1991 c. 28
Sets up Scottish Natural Heritage, and defines its powers and functions. Includes the power to designate Natural Heritage Areas, to make grants or loans, to prepare development proposals, to make access orders to the countryside, to control irrigation water, to make drought orders, to manage nature reserves, and to advise the Government on natural heritage matters.

Ancient Monuments and Archaeological Areas Act 1979 c. 46
Provides for the protection, acquisition and guardianship of ancient monuments. Allows for the designation of Areas of Archaeological Importance.

National Heritage Act 1983 c. 47
Amends the National Heritage Act 1979. Sets up Historic and Buildings and Ancient Monuments Commission for England; its functions do not extend to Scotland, Wales or Northern Ireland. Gives the Commission power to make grants and loans for the preservation of historic buildings, and to acquire funds for their acquisition.

Ancient Monuments (Class Consents) (Amendment) Order 1994 SI 1994/1381
Revokes SI 1981/302 and 1984/222. Increases the Classes from four to ten:

Class 1   Limits agricultural, horticultural and forestry work. Does not permit: subsoiling, drainage works, planting or uprooting trees, hedges and shrubs; any works likely to disturb the soil below 300 mm; demolition, removal, extension, alteration or disturbance to any building, structure or works or the remains thereof; the erection of any building or structure; laying paths, hardstandings, building foundations or the erection of fences and barriers.

Class 2   National Coal Board working restricted to 10 m or more below ground level.

Class 3   British Waterways limited to repair and maintenance.

Class 5   Work essential to health and safety.

Class 6   Works by the Commission are permitted.

Class 7   Works of Archaeological Evaluation permitted by approved persons following consent.

Class 8   Works for maintenance and preservation of scheduled monuments permitted by approved persons following consent.

Class 9   Grant-aided work permitted by approved persons following consent.

Class 10   Work undertaken by Royal Commission; permits survey markers to be placed no deeper than 300 mm.

Ancient Monuments (Class Consents) (Scotland) Order 1996 SI 1996/1507

Grants scheduled monument consent for the works set out in the schedule, these come under six headings as Class I to VI as follows:

Class I   Agricultural, Horticultural and Forestry works

Class II   Coal Authority or their Licensees

Class III   British Waterways Board

Class IV   Repair or Maintenance of Machinery

Class V   Urgent Health and Safety measures

Class VI   Archaeological Evaluation

All have restrictions on the work that they may carry out, for example: Class I excludes amongst other work, (c) sub-soiling, drainage works, planting or uprooting trees, hedges or shrubs, the stripping of top soil, tipping operations, the felling or removal of trees or the commercial cutting and removal of turf; and (f) for other than domestic gardening work, laying paths, hard-standings or foundations for buildings or the erection of fences or other barriers.

Operations in Areas of Archaeological Importance (Forms of Notice) Regulations 1984 SI 1984/1285

These regulations require the developer or his agent to notify the local authority of their intention to start work 6 weeks before work commences on site. Sample forms are given.

Areas of Archaeological Importance (Notification of Operations) (Exemption) Order 1984 SI 1984/1286

Operations that are exempt from notification are listed as:

- agriculture, horticulture or forestry not below 600 mm;
- landscaping not below 600 mm;

- repairs to roads, footpaths, water, drainage and any other services not below 600 mm.

Planning (Listed Buildings and Conservation Areas) Regulations 1990 SI 1990/1519

These regulations conform with the Planning (Listed Buildings and Conservation Areas) Act and associated Acts dated 1990. The original regulations have been amended to comply with the latest administrative procedures; the schedules contain forms of notice.

Environmentally Sensitive Areas – General Notes SI 1986/2249 and others

All these areas are made under the Agriculture Act 1986 c. 49. The aim is to:

- conserve and enhance the natural beauty of the areas referred to;
- conserve flora and fauna, geological and physiographical features;
- protect buildings and other objects of historic interest in the area.

Areas are defined on maps, which can be inspected at local MAFF offices or at MAFF headquarters.

Farmers who have entered into a management agreement with MAFF must obtain written advice from the Minister concerning the siting and materials for any proposed building, engineering work or roads other than Class VI of Town and Country Planning GDOs 1977–85.

There are currently 38 Environmentally Sensitive Area designations in England, Scotland and Wales; these are listed below in alphabetical order. The first SI under each area refers to the principal order; subsequent numbers are amendments. If you are working within one of these areas, it is important to be acquainted with both principal and amendment orders together with any subsequent orders as they are published.

- Argyll Islands: SI 1993/3136, 1994/3067, 1996/1996;
- Avon Valley: SI 1993/84, 1994/927, 1995/197;
- Blackdown Hills: SI 1994/707;
- Breadalbane: SI 1987/653, 1992/1920 and 2063, 1995/3096, 1996/738;
- Breckland: SI 1993/453, 1994/923, 1995/198;
- Cairngorms Straths: SI 1993/2345, 1996/1963;
- Cambrian Mountains: SI 1986/2257, 1987/2026, 1992/1359 and 2342, 1994/240, 1995/243;
- Central Borders: SI 1993/2767, 1996/1964;
- Central Southern Uplands: SI 1993/996, 1996/1969;
- Clun: SI 1993/456, 1994/921, 1995/190;
- Clwydian Range: SI 1994/238;
- Cotswold Hills: SI 1994/708, 1995/200;
- Dartmoor: SI 1994/710;
- Essex Coast: SI 1994/711;
- Exmoor: SI 1993/83, 1994/928, 1995/195 and 960;

- Lake District: SI 1993/85, 1994/925, 1995/193;
- Lleyn Peninsula: SI 1987/2028, 1994/241;
- Loch Lomond: SI 1987/654, 1992/1919 and 2062, 1995/3097;
- Machair of the Uists, Benbecula, Barra and Vatersay: SI 1988/495, 1993/3194, 1996/1962;
- North Kent Marshes: SI 1993/82, 1994/918, 1995/199;
- North Peak: SI 1993/457, 1994/922, 1995/189;
- Pennine Dales: SI 1986/2253, 1992/55 and 301, 1993/460, 1994/930, 1996/923;
- Preseli: SI 1994/239;
- Radnor: SI 1993/1211;
- Shetland Islands: SI 1993/3150, 1996/1965;
- Shropshire Hills: SI 1994/709;
- Somerset Levels and Moors: SI 1986/2252, 1988/176, 1992/53, 1994/932, 1996/920;
- South Downs: SI 1986/2249, 1987/2032, 1992/52, 1994/931, 1996/924;
- South Wessex Downs: SI 1993/86, 1994/924;
- South West Peak: SI 1993/87, 1994/926, 1995/192;
- Stewartry: SI 1993/2768, 1996/1967;
- Suffolk River Valleys: SI 1987/2033, 1993/458, 1994/920, 1995/194;
- Test Valley: SI 1993/459, 1994/919, 1995/191;
- The Broads: SI 1986/2254, 1988/175, 1992/54, 1994/929, 1996/921;
- Upper Thames Tributaries: SI 1994/712;
- West Penrith: SI 1986/2281, 1992/51, 1994/933, 1996/922;
- West Southern Uplands: SI 1993/997, 1996/1968;
- Ynys Mon: SI 1993/1210.

1.2.6 MISCELLANEOUS

Commons Registration Act 1965 c. 64
Provides for the official registration of common land, including public spaces such as village greens. Arrangements also made for local authorities to take over land not claimed under the Act.

Local Government (Access to Information) Act 1985 c. 43
Allows for public access to the meetings and documents of most local authority councils and committees, except where confidential or exempt information is to be discussed. Agendas and minutes (with certain exceptions) must also be made available. Exempt information includes financial information on tenders and contracts, legal matters, and notices to be served.

Local Government (Inspection of Documents) (Summary of Rights) Order 1986 SI 1986/854
Specifies enactments that confer upon the public the right to attend meetings, to inspect, copy, and be supplied with documents. Subjects included: town planning, highways, education, and control of pollution.

Local Government (Inspection of Documents) (Summary of Rights) (Scotland) Order 1986 SI 1986/1433

Local authorities are now required to give the public the right to attend certain meetings, inspect, copy, and be supplied with elected documents. Details are given in the schedule.

Control of Pesticides Regulations 1986/1510

Made under the Environmental Protection Act 1985 c. 48, and with certain specified exceptions require the approval of the Secretaries of State before pesticides can be used. Approved substances are updated on regular listings. These are covered in detail at the beginning of the chapter on planting.

Land Registration (Official Searches) Rules 1986 SI 1986/1536

These rules change the prescribed forms of application to allow them to be used by licensed and other recognized conveyancing bodies on the same basis as solicitors. They also include the procedure for an official search by a purchaser. Schedules include the prescribed forms and the information to be supplied when these forms are not used: e.g. electronic mail enquiries.

Withdrawal of Requirement upon Local Authorities to Compile and Publish Information about Unused and Underused Land DoE Circular 1996/3

Withdraws DoE Circular 1989/18 which required specified Local Authorities in England to follow the Code of Practice for the Publication of such information which was annexed to 1989/18. It also withdraws a similar Code of Practice issued to Government Departments and other owners of Crown land.

Compulsory Purchase of Land Regulations 1994 SI 1994/2145

Takes account of relevant changes in legislation. Deals with compulsorily purchased open land. Includes prescribed forms.

Petroleum Act 1987 c. 12

Though this Act deals mainly with offshore installations, s. 25 deals with pipelines, and amends the Pipelines Act 1962. The Pipelines Act Schedule 1 (Applications for Construction Authorizations) introduces a new paragraph 6a, which requires notice to be given of modifications to routes to every local planning authority on the route, and any persons specified by the Minister. Time is allowed for objections, appeals and public inquiries. The new paragraph 6b enables the Minister to call a public inquiry in respect of an application for a pipeline construction or any modifications thereto.

Access to Neighbouring Land Act 1992 c. 23

Allows a person on the dominant land to obtain a Court order for access to neighbouring (subservient) land to carry out essential building, or landscape repair where it is considered necessary for the preservation of the land and where it would be difficult to carry out the work from the dominant land. Landscape repair includes drainage, ditch repair or clearance, the removal of dangerous, damaged or diseased trees, shrubs or hedges. Compensation may be payable, and the Court may make conditions of access.

Town and Country Planning (Control of Advertisements) (Amendment) (No 2) Regulations 1990 SI 1990/1562

These regulations amend the 1989 regulations (SI 1989/670) in order to conform with the Town and Country Planning Act 1990 and its associated Acts.

Town and Country Planning (Control of Advertisements) (Scotland) Regulations 1984 SI 1984/467

These regulations are similar to those that concern England and Wales. They are made under the Town and Country Planning (Scotland) Act 1972.

Highways (Road Humps) Regulations 1996 SI 1996/1483

These new regulations revoke the 1990 regulations SI 1990/703 and the amendment SI 1990/1500. It is now permitted to construct road humps on any highway that has a 30 MPH speed limit. They may be of any height between 25 and 100 mm, and the minimum length is 900 mm, unlike the 1990 Regulations there are no diagrams or sections to give the permitted shape of a hump. There are no restrictions on the position of humps except at zebra, tramway or rail level crossings, or where there is a structure 6.5 m or less above the carriageway. Lighting should comply to BS 5489 or to the two previous requirements for lamps at max. 38 m or specific lighting to the humps. Traffic signs are to be placed by the highway authority.

Highways (Traffic Calming) Regulations 1993 SI 1993/1849

Allows traffic calming work to be carried out by highways authorities within the limits set out and provides for consultation with the police and other official users. Features may include: lighting, paving, grass; bollards, planters, walls, rails and fences; trees, shrubs and plants. Restrictions include the construction of overrun areas and restrictions on rumble devices. Warning signs must be provided unless the design is self explanatory.

Roads (Traffic Calming) (Scotland) Regulations 1994/2488

Allow road authorities to construct build-outs, chicanes, islands, rumble areas etc., and to provide lighting, paving, grass, fences, walls, trees and shrubs in the ground or in planters: all to be commensurate with the safety and environmental enhancement of the traffic calming works.

1.3 Department of the Environment publications

1.3.1 CIRCULARS

Town and Country Planning (Minerals) Act 1981: DoE Circular 1982/1
Explains the provisions of the Act. County planning authorities are designated as the mineral planning authorities, and they have powers to require restoration, aftercare, and maintenance of the land after extraction. Controls cover soft landscaping only, and apply for 5 years after extraction is completed. Restoration schemes must be agreed with the mineral planning authority, and are required only where the land is restored to agriculture, forestry, or amenity uses. Amenity use includes wildlife conservation.

Use of conditions in planning permissions: DoE Circular 1985/1
Sets out the types of planning conditions that it is reasonable to impose on a developer, and describes how these may be enforced. Includes landscaping conditions dealt with as reserved matters.

Planning control over oil and gas operations: DoE Circular 1985/2
Gives guidance to mineral planning authorities on control over on-shore oil and gas operations. The main recommendations are published in the Memorandum on the Control of Mineral Working. Mineral planning authorities have policies for the restoration and aftercare of sites.

Reservoirs Act 1975: DoE Circular 1985/5
The Act deals with large raised reservoirs containing over 25 000 m³ of water above natural ground level. General planning policies on conservation and landscaping also apply to raised reservoirs.

Aesthetic control: DoE Circular 1985/31
This circular suggests that local planning authorities should reduce the number of applications delayed, refused or sent to appeal on the grounds of detail design matters. They should recognize that aesthetics are an extremely subjective matter, and they should not impose their tastes on developments. However, in sensitive areas such as National Parks, AONBs, conservation areas, and similar places, design control is important, and authorities should reject obviously poor designs.

Local Government (Access to Information) Act 1985: DoE Circular 1986/6
This circular discusses the underlying principles set out in the Act, which allows access for the public and press to most local authority council meetings, and access to agendas, reports, minutes, and also to some background papers. Sets out the information that may or may not be made public.

Town and Country Planning (Minerals) Act 1981: DoE Circular 1986/11

This circular describes those parts of the Act that came into force in 1986; they include powers to control the environmental effects of mineral workings.

Town and Country Planning (Agricultural and Forestry Development in National Parks etc.) Special Development Order 1986: DoE Circular 1986/17

This circular discusses the discretionary control that the local planning authorities in the areas listed in the schedule have over the siting, design and external appearance of certain farm and forestry developments. The controls are similar to those already in force in the Lake District, the Peak District, and Snowdonia. These classes cover agricultural buildings and uses, and forestry buildings and works. Certain developments are now subject to prior notification; failure to notify may result in an enforcement notice being served.

Planning appeals by written representation: DoE Circular 1986/18

Describes the method by which decisions on planning appeals can be speeded up. Appendices give a format for written statements and a flow diagram of target dates in a timetable for written representation of planning appeals.

Housing and Planning Act 1986: Planning Provisions: DoE Circular 1986/19

Discusses: Part II simplified planning zones, Part IV hazardous substances, Part V opencast coal, Part VI miscellaneous provisions (listed building and conservation areas, local plans and unitary development plans etc.), s. 45 (Control of Advertisements Experimental Areas), s. 46 (land adversely affecting amenity of neighbourhood).

Town and Country Planning (Appeals Written Representation Regulations 1987: DoE Circular 1987/11

Explains the regulations, which are intended to speed up planning procedures by setting time limits.

Redundant hospital sites in green belts: planning guidelines: DoE Circular 1987/12

Mature landscape should be preserved, and public access improved. Existing buildings should be used where possible.

Development of contaminated land: DoE Circular 1987/21

Local planning authorities and developers have been encouraged to make full use of urban areas rather than greenfield sites. This circular gives advice on the identification, assessment and development of contaminated sites. These include: old sewage works and farms; land previously used for industrial purposes; completed landfill sites liable to settlement, emission of gases, and possible combustion. The sewage sites and, more particularly, the industrial sites may contain a wide range of hazardous substances. All these hazards must be identified and assessed in relation to those who will work on the site as well as the future occupiers and users of the new buildings and the land surrounding them. Particular care must be taken with regard to the services supplied to them.

Code of practice on preparing for major planning inquiries in England and Wales. Code of practice for hearings into planning appeals: DoE Circular 1988/10

This circular applies to local inquiries held under the Town and Country Planning Act 1971 ss. 36 and 37, which are decided by the Secretary of State. The first part discusses the rules in detail; Annexes 1 and 2 provide the Code of Practice; Annexes 3 and 4 give timetables for procedures; Annexes 5 and 6 give extracts from the Chief Planning Inspector's reports. The code should help the inspector and the parties involved to identify in advance all those who intend to participate, and the extent of their participation.

Environmental Assessment: DoE Circular 1988/15

Explains the Town and Country Planning (Assessment of Environmental Effects) Regulations 1988, which implement EC Directive 85/337. This circular discusses the procedure for submitting environmental statements (ES) with planning applications and their publicity. Generally the LPA decides whether or not an environmental assessment (EA) is required for a project. The Ministry of Defence will prepare relevant EAs; the Ministry of Transport will cover motorway service areas, and guidance on EAs in Enterprise Zones. Simplified Planning Zones and New Towns will follow. All EAs must be notified by the LPA to the Secretary of State, and the LPA must enter them in the planning register.

General Development Order Consolidation: the Town and Country Planning General Development Order 1988 and Town and Country Planning Regulations 1988: DoE Circular 1988/22

This circular sets out very clearly the changes in SI 1988/1813. Section I describes the purpose of General Development Orders. Section II sets out the main changes in the 1988 consolidated GDO. Section III comments on the contents of the GDO and applications regulations. Appendix A describes the practice that planning authorities should follow. Appendix B gives the statutory and discretionary procedures for publicity applications. Appendix C gives the consultations required before planning permission can be granted. Appendix D covers Article 4 directions restricting permitted development.

Environmental assessment of projects in Simplified Planning Zones and Enterprise Zones: DoE Circular 1988/24

Simplified Planning Zones (SPZs) schedule 1 projects – that is, those in Schedule 1 to the Town and Country Planning (Assessment of

Environmental Effects) Regulations 1988 (1988/1199) – must be the subject of environmental assessment. There are alternative procedures for sch. 2 projects (those in sch. 2 of 1988/1199).

Caravan sites and Control of Development Act 1960 Model Standards: DoE Circular 1989/14

This circular draws attention to the revised Model Standards; these are now divided into mobile homes on permanent residential sites and static holiday caravan sites. The main changes include improved space standards to prevent the spread of fire between vans; these are now based on the booklet *Fire Spread Between Park Homes and Caravans*, dated February 1989 and available from HMSO.

European Council Directives on public supply and works contracts tender documents: avoiding national discrimination within the European Community: DoE Circular 1990/7

Local authority tender documents must avoid naming particular products and using proprietary names for articles; they must also avoid descriptions and specifications that point to or favour a particular supplier. It contravenes Community law for a specification to be phrased so as to oblige a contractor to use only articles meeting a UK standard without it being made clear that articles meeting national standards of another Member State or an ISO standard recognized in another Member State and offering equivalent guarantees on safety, suitability and fitness for purpose will also be acceptable.

Electricity generating stations and overhead lines: DoE Circular 1990/14

Large-scale generating stations and major overhead lines cannot be installed without a consent granted by the Secretary of State for Energy. Nuclear generating stations require an environmental assessment. Environmental statements may be required for major overhead lines.

Rights of Way Act 1990: DoE Circular 1990/17

Rights of way are mainly obliterated by ploughing out and encroachment by crops. Cross-field paths may be ploughed, but not field-edge paths or unsurfaced highways. The occupier must replace paths after disturbance, and the local authority may issue enforcement notices and in default take direct action to restore the path. Maximum and minimum widths are specified for restoration and maintenance, and rights of way must be visible on the ground. Farmers have been issued with guidance notes by the Countryside Commission and MAFF.

Planning and Compensation Act 1991: planning obligations: DoE Circular 1991/16

Explains the new s. 106 on planning obligations, which replace the old s. 52 agreements. 'Planning obligations' also replace the old 'planning gain' conditions imposed by planning authorities, which had no legal status. Planning obligations are established in a more rigid legal framework than s. 52 agreements or planning gains, and must be formally recorded. They may be related to roads, land or buildings, but they must be directly connected with community needs arising from the proposed development, such as shortage of car parking, cheap housing, open space, or leisure facilities.

Planning and Compensation Act 1991: Implementation of main enforcement provisions: DoE Circular 1991/21
Useful background information on ss. 1–11 and other provisions in the Act. Annexes provide detailed guidance on enforcement procedures, including those for listed buildings, Tree Preservation Orders, mineral planning and hazardous substances control.

Town and Country Planning General Regulations 1992: Town and Country Planning (Development Plans and Consultation) Directions 1992: DoE Circular 1992/19
Explains how the regulations bring local authorities under the same planning controls as private applicants.

Development and flood risk: DoE Circular 1992/30
The National Rivers Authority and other drainage bodies have powers to require flood defence works to be carried out. The circular is aimed primarily at local authority development plans, but the landscape designer is advised to check flood risks before submitting a planning application.

Public rights of way: DoE Circular 1993/2
Gives guidance to planning authorities and others on the recording, maintenance and protection of the rights of way network, and explains in detail the procedures for altering or extinguishing rights of way. It gives advice on managing the rights of way network as a public amenity for recreation.

Environmental assessment: amendment of regulations: DoE Circular 1994/7
Adds further projects to Schedules 1 and 2, and correlates environmental assessment regulations with recent planning legislation.

Permitted development and environmental assessment: DOE Circular 1995/3
Explains the relationship between permitted development and the need to carry out an environmental assessment.

1.3.2 DEVELOPMENT CONTROL POLICY NOTES (DCPS)

Access for the disabled: DCP 16

Recommends that disabled people should have suitable access to buildings and car parking spaces. Disabled routes should be signposted. Special provision may be needed for disabled people to park in pedestrian areas.

1.3.3 PLANNING POLICY GUIDANCE NOTES (PPGS)

Green belts: PPG 2

This is a new edition dated 1995. It gives five reasons why green belts are created:

- to prevent neighbouring towns merging into each other;
- to prevent the unrestricted sprawl of large built-up areas;
- to prevent encroachment on the countryside;
- to assist in conserving the character and setting of historic towns;
- to encourage the recycling of derelict and other urban land for development.

The object of green belts is to:

- provide town people with access to the open countryside;
- provide outdoor sport and recreation near urban areas;
- retain and enhance landscape near towns;
- improve damaged and derelict land around towns;
- promote nature conservation interest while retaining land for farming, forestry and the like.

There are currently just over 1.5 million ha of land within green belts, in 14 different locations, and the landscape designer should check the actual boundaries with the planning authority.

Simplified Planning Zones: PPG 5

Explains the purpose of and planning procedures in SPZs.

Telecommunications: PPG 8

Recommends cooperation from planning authorities for telecommunication developments; these are deemed to be important for economic growth. Provides useful illustrations of antennae.

Nature conservation: PPG 9

A useful document that describes government policies on conservation and its relationship to land use planning. Sets out the statutory provisions for nature conservation; describes the various types of protected site.

Strategic guidance for the West Midlands: PPG 10

This provides a framework for Unitary Development Plans, and sets out the land use planning objectives.

Strategic planning for Merseyside: PPG 11

This provides a framework for Unitary Development Plans, and sets out the land use planning objectives.

Development plans and regional planning guidance: PPG 12

A useful guide that sets out the whole planning structure, and includes a code of practice on development plans. Sets out regional guidance that covers issues of regional importance, and indicates a development framework for 20 years or so. The purpose of the RPGs is to ensure consistency of policies throughout a region. Sets out content and range of development plans, and the relationship between different types of development plan.

A Guide to Better Practice: A practical guide for implementing PPG 13 dated October 1995

Sets out the need to reduce travel through land use and transport planning. Section 2 discusses planning principles and means of putting these into practice. Section 4 discusses planning at the different levels from Regional to urban and rural planning areas and recommends local authorities to consider these in conjunction with PPG13. This section contains a useful framework for land use and complementary transport measures. Section 5 discusses the location of development, tourism and recreation. Section 6 Transport measures. All sections are supported by studies of cities and other locations in the UK and Europe.

Development on unstable land: PPG 14 1990

This document describes the effect of instability on development and land use. It sets out the responsibilities of authorities to include problems of instability in the planning process. The Government wishes to make the best use of all available land, but problems have arisen where development has been allowed on unstable land; it is intended that planning control should be used to select appropriate uses for unstable land. While the developer and land owner are responsible for safe development, the planning authority should give guidance. Appendix A describes the causes of instability, and Appendix B lists useful sources of information. There are three main causes of instability: subsidence of underground cavities, natural or man-made; unstable natural or man-made slopes; ground movement due to shrinkage, or expansion, compression and changes in the water-table.

Development on Unstable Land: Landslides and Planning: Annex 1 to PPG 14

Discusses the problems associated with landslides and sets out strategies for dealing with them. Includes a map showing the distribution of the 8835 landslides recorded in the UK. This publication is dated March 1996.

Archaeology and planning: PPG 16 1990

This guidance note requires local planning authorities to include policies on preserving important archaeological remains in their local and unitary development plans, and to define these on maps. The note allows local planning authorities to make archaeological investigation a condition of planning consent, and gives them the power to forbid the start of a development until an agreed scheme has been undertaken. Archaeological excavation financed by the developer must not be treated as a planning gain. There is separate legislation for scheduled monuments. Developers and landscape designers should follow the British Archaeologists and Developers Code of Practice, obtainable from the British Property Federation, 35 Catherine Place, London SW1E 6DY.

Sport and recreation: PPG 17 September 1991

With the growing importance of all types of sport and recreation facilities in towns and cities, and the use of the countryside for those purposes, suitable open spaces should be protected from development. National Playing Fields Association minimum standards for outdoor play are given, together with examples of LPA standards for publicly accessible open space, by areas and distance from home. Refers to *Strategic and regional planning guidance*: PPG 12 (scheduled to be revised in 1992).

Enforcing planning control: PPG 18 December 1991

Describes the new improved enforcement powers under the Planning and Compensation Act 1991. Cancels advice in: PPG 1 (January 1988), paras 30, 31; PPG4 (January 1988), para 19; and DoE Circular 22/80, paras 15, 16 and Annex B.

Outdoor advertisement controls: PPG 19

Outdoor advertisements are very strictly controlled in conservation areas. In open countryside the quality of the surroundings will be considered in the planning control over advertisements.

Coastal planning: PPG 20 1992

Deals with the special needs of coastal planning, including provision for tourism, flood protection, water and sewage, water recreation, and coastal defences. Discusses the Heritage Coast designation and other designations affecting development and conservation.

Tourism: PPG 21

Deals with the economic place of tourism and its effect on the environment. Special consideration must be given to tourist development in environmentally controlled areas.

General policy and principles: PPG 1992/1

Replaces PPG 1/1988. Sets out general principles of operating the planning control system, including time limits for planning decisions and appeals. Mainly directed at planning authorities, but useful to landscape designers as a guide to applicants. A useful aid to understanding the whys and wherefores of planning control.

1.3.4 REGIONAL PLANNING GUIDANCE NOTES (RPGS)

The Strategic Guidance objectives include: contributing to revitalizing older urban areas; developing safe efficient transport systems; giving proper respect to the environment; maintaining the vitality and character of established town centres; sustaining and improving the amenity of residential districts; giving high priority to environmental maintenance; protecting green belts and metropolitan open land; preserving fine views, the surrounding countryside and the national heritage.

- Strategic Guidance for Tyne and Wear   RPG 1
- Strategic Guidance for West Yorkshire   RPG 2
- Strategic Guidance for London   RPG 3
- Strategic Guidance for Greater Manchester   RPG 4
- Strategic Guidance for South Yorkshire   RPG 5
- Strategic Guidance for East Anglia   RPG 6
- Strategic Guidance for Northern Region   RPG 7
- Strategic Guidance for East Midland Region   RPG 8
- Strategic Guidance for South East   RPG 9
- Thames Gateway Planning Framework   RPG 9a
- Strategic Guidance for South West   RPG 10
- West Midlands Region   RGP 11
- Yorkshire and Humberside   RGP 12
- North West   RPG 13

1.3.5 MINERALS PLANNING GUIDANCE NOTES (MPGS)

General Considerations and the Development Plan: MPG 1

Opencast coal mining: MPG 3

The review of mineral working sites: MPG 4

Minerals planning and the General Development Order: MPG 5

Reclamation of mineral workings: MPG 7

Provision of raw materials for the cement industry: MPG 10
Agricultural or green belt land, if used, must be restored to a high standard; this and aftercare is an important factor of any application.

1.3.6 OTHER DEPARTMENT OF THE ENVIRONMENT PUBLICATIONS

Environmental assessment: a guide to the procedures. DoE, 1992
Official guidelines describing the complex procedures required for the

preparation and submission of an environmental assessment. Includes extracts from schedules to the Town and Country Planning (Assessment of Environmental Effects) Regulations 1988, the 1985 EC Directive 85/337, and the Department of the Environment Circular 1988/15, which sets out the criteria for Schedule II projects that may require an environmental assessment. The guide also includes a flow chart for environmental assessment and planning permission application procedures.

Risk criteria for land-use planning in the vicinity of major industrial hazards. Health and Safety Executive, 1989

Sets out the types of process likely to cause major hazards, the kinds of development at risk, and the criteria for deciding the level of risk. Local planning authorities already receive notification and advice from the Health and Safety Executive on major hazards: this includes suggested 'consultation distances' round each installation, within which advice on development is needed. Suggested figures are given showing acceptable distances for developments from hazardous installations.

Policy appraisal and the environment. DoE, 1991

This is an excellent document, written in very good English, which sets out clearly and simply the relationship between central government policies and the environment. Although it is intended for central government policy-making officers, it is a first-rate introduction to the impact of development on the environment, and the considerations that should govern policies affecting the environment. The document is of great value to landscape designers and others who wish that consideration for the environment should be reflected in planning policies and development plans.

Amenity reclamation of mineral workings. DoE, 1992

A useful illustrated guide summarizing studies and projects from various sources on reclaiming mineral sites of different types. Covers recreation, wildlife conservation, and planning guidance. Gives advice on surveys and design.

Planning guidance: index of current planning guidance. DoE, 1990

The first part lists those DoE circulars that have been cancelled; the second part gives a chronological list of extant MOHLG and DoE circulars; the third part lists Development Control Policy Notes, PPGs, MPGs and RPGs. There is a subject index at the end of the document.

Computers in landscape architecture 2

Mark Martin, Associate, DLP and consultants
Miles Mathews and Han Shi

2.1 Introduction

Information technology (IT) systems are increasingly playing an important role in the way many landscape practices operate, with nearly all UK practices now using computers for certain aspects of work, and demand in other areas set to rapidly grow.

The introduction of an IT system can be a high-risk activity for any business, and therefore it is important that a good understanding of the technology and its appropriateness to specific business applications is gained before implementation.

This section has been written to give the landscape architect a greater understanding of the technology currently available, by first explaining what a computer system is, and then by examining the various stages that an organization's IT policy may need to address, from finance, health and safety, and specification, through to specific landscape applications that have been identified as of growing interest to the landscape professional (Fig. 2.1). Finally, this section focuses on system management, administration and communication, because these areas will become increasingly important as the use of computers within the office develops.

Although great care has been taken in the compilation and preparation of this section to ensure accuracy and quality of information, the rapidly changing nature of technology will result in some references becoming dated very quickly, and therefore independent professional advice should always be sought. The use of particular software and hardware manufacturers' names has also been limited, to avoid directing the reader to a particular product that may, over time, be no longer the market leader or the most suitable. Instead, comprehensive descriptions of the equipment and applications have been used to provide the reader with a thorough reference guide to the main principles employed when selecting and using IT within the landscape office.

2.2 Computer system

A computer system can be broadly defined as the combination of the **hardware** (including any peripheral parts like a printer) and the **software** (Table 2.1).

Table 2.1 Computer system components

Specification	Example
Hardware	
Input and output devices	Mouse, digitizer, monitor, printer
System box	Processor, memory, graphics card
Software	
Operating systems	DOS, Windows, Windows NT, Apple Mac, UNIX
Applications	Word processing, CAD, GIS

2.3 Financing IT

Most computer systems are obsolete within a relatively short time, because of advances in technology, and therefore the options available for financing the purchase of a system and the overall IT costs for implementation should be considered carefully.

2.3.1 TYPES OF FINANCE

Leasing, rental and outright purchase are the typical methods for funding the cost of new equipment (Table 2.2).

Table 2.2 Types of finance

Finance type	Advantages	Disadvantages
Leasing (Independent professional advice should always be sought before signing any lease)	1. Fixed payments over a specified term 2. Residual value can stay with purchaser depending on type of lease	1. More expensive than outright purchase 2. Ongoing cost during repayment period 3. Negotiations with lessor over final ownership or receipt of full residual value can be difficult and prolonged 4. The need to upgrade equipment can result in entering a new long-term lease to avoid cancellation penalties with the existing lease

Table 2.2 continued

Finance type	Advantages	Disadvantages
Rental	1. No long-term commitment 2. Good for addressing short-term needs 3. No maintenance costs	1. No return on investment 2. Expensive in the medium to long term
Purchase	1. Tax advantages 2. Full ownership and control 3. Least expensive in the long term if suitable cash funds are available	1. Depreciation may be faster than can be written off against tax 2. Tied to equipment over a long period 3. High initial capital cost 4. Maintenance costs not included after warranty period

2.3.2 COSTS ASSOCIATED WITH IMPLEMENTING IT

When calculating the costs of introducing a computer system, a budget should be drawn up that takes the following factors into consideration:

- hardware and software;
- installation, configuration and customization;
- training;
- maintenance and administration;
- supplies, security and insurance premiums.

These are discussed below.

Hardware and software

The cost of hardware and software is perhaps one of the easiest costs to quantify, although the purchaser should be clear as to whether the quotes received from suppliers include VAT, delivery, installation and any after-sales service.

Installation, configuration and customization

Once the equipment has been purchased it should be positioned within the office in accordance with the Health and Safety (Display Screen Equipment) Regulations 1992 and the Workplace (Health, Safety and Welfare) Regulations 1992 (see later). This may result in the need to purchase new furniture such as work surfaces, chairs and blinds. The landscape architect

Platform choice

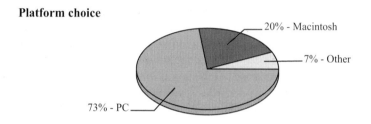

Use of and interest in software applications

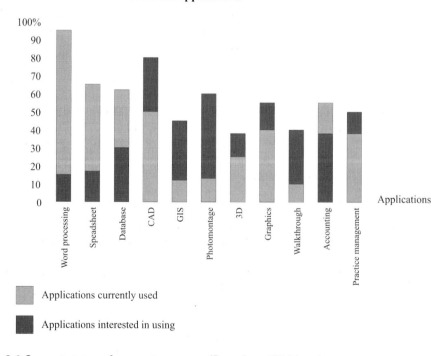

Fig. 2.1 Current status of computer usage. (Based on 1994 Landscape Institute survey on computers in landscape architecture)

may also need help in configuring the equipment correctly, and therefore adequate technical support should be incorporated within the IT budget. Some support companies also offer additional benefits to the user through customization of industry-standard software to improve ease of use and productivity.

Training

If the benefits of investing in IT are to be fully realized, the provision of a sufficient training budget for users is a necessity. The training of users generally takes longer than anticipated, and the hidden cost in loss of production is often underestimated.

The type of training may be either 'in house' by a particular member of staff, or through a suitably qualified external training centre. City and Guilds training courses at local colleges are also available, and can offer excellent value for money. If an appropriate person for training does not exist within the office, there may be the additional cost of recruiting a suitably skilled person.

Computer training is a suitable activity for part of an individual's annual Landscape Institute CPD training, and therefore may be budgeted as such if funds are limited.

Maintenance and administration

The majority of new equipment comes with a limited warranty. The implications of acting on this warranty if a system failure occurs should be assessed and discussed with the supplier before purchase. Some warranties require the equipment to be returned to the manufacturer, and therefore the cost to the practice in terms of inconvenience or lost business may be significant. Choosing a reliable supplier may help to minimize this risk. Computer suppliers accredited to **BS 5750/ISO 9001** have been acknowledged as having a management system in place that has the philosophy 'right first time', and therefore should prove more reliable than others, although it should be remembered that this standard is not a quality control guarantee.

In addition to warranties, support contracts with suppliers may also be entered into. These usually provide cover for all the practice's IT equipment. The type of service will depend on the contract entered into, but should include annual health checks and the fixing of any hardware problems within a specified time-scale, with the provision of suitable loan equipment if this is not possible.

Even with adequate external support, some cost provision should be made within the practice budget for administering the day-to-day running and support of the IT system.

Supplies, security and insurance premiums

Once a system is in place, additional costs may be incurred, such as the supply and storage of consumables like paper, computer disks and ink cartridges. Provision for increases in insurance premiums and extra security measures should also be considered.

2.4 Health and safety

The rapid growth of IT equipment within the office in recent years and new working practices have raised a number of health and safety issues for employers and users to address. Employers are obliged to identify and assess the risks of using IT and to take measures to prevent or reduce them, so far as is reasonably practical. The list below highlights some of the risks and safety standards, but is not intended to be exhaustive in content.

2.4.1 FIRE

Overloading of circuits or cables, short-circuits, loose or damaged connections, or IT equipment (particularly the processor) simply overheating during use, may create an electrical fire.

In the event of a fire, hand-held water-based extinguishers should not be used because of the potential risk of electrocution. A non-conductive gaseous medium, such as carbon dioxide or halon (BCF) is preferable; powder-based extinguishers will also work, but are likely to render the equipment completely useless afterwards.

2.4.2 ELECTRICITY

The Electricity at Work Regulations 1989 (EAWR) require employers to ensure that the electrical system and any appliance connected to that system are maintained in a safe condition. New office equipment should meet IEC/EN 60 950/BS 7002 *Specification for safety of IT equipment including electrical business equipment*. Multi-socketed extension cables should be used carefully, because they can create a trip hazard, as well as having the potential to overload a system. In some practices a 'clean' circuit to be used only by IT equipment is installed to reduce the risk of overloading and subsequent system failure.

2.4.3 EMISSIONS

Initial concerns about the use of computers were over the effects of radiation emissions from the use of visual display units (VDUs), although at present no scientific proof has been shown that the low levels emitted are a risk to health. Screen filters are not necessary for reducing emissions, but may be helpful in some circumstances for reflection or glare.

2.4.4 ERGONOMICS

The Health and Safety (Display Screen Equipment) Regulations 1992 provide guidance on how employers can reduce certain health and safety risks to computer users through careful workstation design. The main points of guidance are as follows:

- The work surface should allow flexible arrangements; it should be spacious and glare free.
- The work chair should be adjustable.
- A suitable footrest should be available if required.
- Distracting noise should be minimized.
- Legroom and clearances should allow for postural changes.
- Lighting should be adequate.
- Glare or distracting reflections should be eliminated.
- Reflective window covering should be used where necessary.
- Software should be appropriate to the task and provide feedback on system status, and there should be no undisclosed monitoring.
- The screen should have a stable image; it should be adjustable, readable, glare and reflection free
- The keyboard should be usable, adjustable, detachable and legible.

2.4.5 EYE AND EYESIGHT TESTS

These are covered under Regulation 5 of the 1992 Regulations, which give users – and those who are about to become users – entitlement to appropriate eye and eyesight tests on a regular basis to ensure that their vision is matched to their display screen work.

2.4.6 REST BREAKS

These should be taken at regular intervals. Short, frequent breaks are more satisfactory than occasional longer breaks. Users should also take 'micro-breaks', which involve looking away from the screen for about 10 seconds, to vary their eyes' focal range on a regular basis during the course of a day. Activities that involve postural change as well as visual change should be built into the daily working programme.

2.5 Specification

As in most other fields, computer systems in the landscape industry are dominated by the IBM-compatible PC. Apple Macintoshes are also used, particularly by those associated with architectural and graphic design practices. UNIX is the third most popular platform, and tends to be used by landscape architects working in conjunction with engineers.

This section describes some of the hardware and software components found within a standard computer system. It is intended only as a information guide, and does not attempt to recommend a particular level of specification, as this will be determined both by the application desired and by recent advances in technology. As a general rule, the more demanding an application is, the higher the specification will need to be, as will the budget. Graphic applications, for example, are more demanding on computer resources than text applications.

2.5.1 HARDWARE SPECIFICATION

See Fig. 2.2.

The main hardware components that a landscape architect will have to specify when purchasing a computer (if a computer specialist is not employed) are as follows:

- central processing unit;
- memory (RAM, hard disk);
- system architecture;
- standard input devices (keyboard, mouse, floppy disk, CD-ROM);
- network card;
- digitizer;
- monitor;
- printer.

Central processing unit (CPU)

The CPU is a single chip – a microprocessor – which processes instructions required by the computer software to run. The speed with which the software runs will depend partly on the speed of the processor, which in turn will depend on the following three factors:

- clock rate (measured in megahertz);
- processing efficiency;
- bus widths (measured in bits).

Clock rate

The majority of processors are now 'clocked', which refers to the fact that the operations are synchronized to the beat of an electronic clock. The clock emits millions of pulses each second, and its speed is recorded in megahertz (MHz). Typical speeds currently commercially available include 75 MHz at the lower end and 266 MHz at the higher end.

Processing efficiency

Instructions are rarely performed within one clock beat, and often require several operations to complete. Therefore the efficiency by which the instructions are processed can affect the performance speed. Most processors now use production-line techniques or 'pipelines' to process more than one instruction at a time.

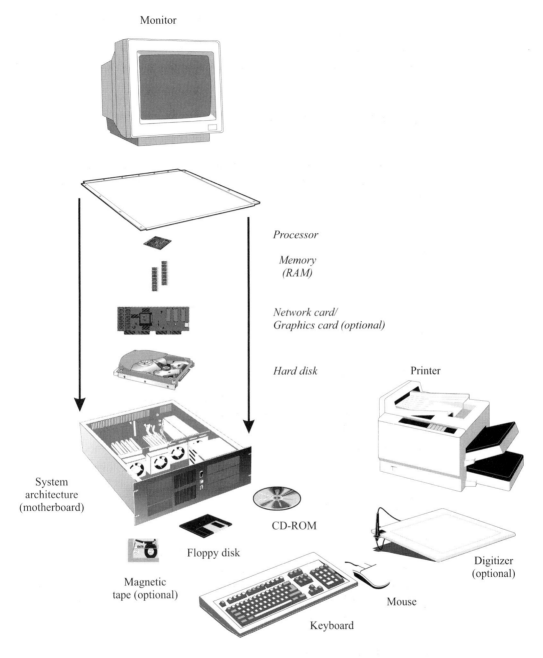

Fig. 2.2 Hardware specification.

Bus width

The amount of data that can be processed by each instruction is determined by the number of bits that a processor can handle at a time. 32-bit and 64-bit processors are currently two of the most popular standards.

There are several families of CPU available today; the most popular are indicated in Table 2.3. Of these the Intel x86 family, originally designed in the early 1980s, is the most widely used, because it is employed by IBM and IBM-compatible PCs. The recent 586 chip by Intel was renamed the Pentium after a court case ruled that numbers could not be registered as a trademark, and the latest 686 has now been named the Pentium Pro.

Table 2.3 Types of processor

Processor	Used by
Intel x86	PC-compatible architecture
Motorola 680x0	Apple Macintosh architecture
RISC	Sun SPARCStations
	Hewlett Packard's PA-RISC range
	IBM RS/6000 range
	IBM and PowerMac PC range
	DEC Alpha PC range
	Silicon Graphics Indigo range

The biggest challenge to the dominance of the x86 family over the next few years is likely to come from the newer **reduced instruction set chips** (RISC), which are both faster, and cheaper to make. Such processors have already been widely adopted in engineering workstations and other specialist sectors.

Memory

Computers contain a variety of types of memory, but the two most important types are **RAM (random access memory)** and **hard disk memory**. Other memory types used for data transfer and storage are also important, and are described later.

Because of the variety of memory types a hierarchy often exists within the computer, with the fastest and most expensive closest to the processor and the slowest further away. Data held in memory is moved from one to the other as necessary.

RAM is the fastest and most widely used form of microchip-based memory. However, anything held within this memory will be lost as soon as the power is removed. Many computer systems now have slots for RAM chips, packaged in the form of exchangeable SIMMs (single inline memory modules), which allow for easy memory upgrades. When purchasing a machine it should be specified that at least two of these slots remain empty to allow for future upgrades.

Hard disks provide the permanent memory for most computers, storing both programs and data for the user. Hard disk memory has improved greatly in terms of cost, reliability, capacity and physical size over the last few years, but so have demands by software programs and users.

Two common types of hard disk are:

- EIDE (enhanced integrated drive electronics);
- SCSI (small computer systems interface).

SCSI drives are often used as network servers, because it is possible to 'daisy-chain' together up to seven devices, thereby allowing the user an easy upgrade path to increased capacity as and when necessary.

System architecture

The performance of a computer is largely determined by how well the individual hardware components are matched and connected (and by how well the software exploits it). The composition of the system as a whole is much more significant than the theoretical performance of the processor. IBM uses mainly multi-channel architecture (MCA), whereas everyone else uses industry-standard alternatives:

- ISA (industry standard architecture);
- EISA (extended ISA);
- VESA (Video Electronics Standard Association);
- PCI (peripheral component connect).

PCI is currently the most popular of the above architectures, although with the current rate of technological change it will soon lag behind new hardware developments. Many computer manufacturers realize this, and therefore make a compromise when putting together a computer between performance and what they regard is a reasonable price for the total package.

Standard input devices

The **keyboard** is based on the conventional typewriter, and allows the input of letters and numbers. Additional 'control' and 'function' keys can be used to trigger specific operations within a software program.

The **mouse** is a more modern input device, which works by pointing and selecting commands from the screen. A mouse normally has two buttons, which can be used to facilitate commands.

The **floppy disk** is a standard input and output device, and works in the same way as the hard disk, recording data magnetically onto a rotating disk, which is protected by a plastic case. A 3.5 in high-density disk, which holds 1.44 Mb of data, is the industry standard.

CD-ROM (compact disc read-only memory) holds up to 650 Mb of data, and is similar in size and physical appearance to an audio CD. Information can be recorded onto the CD only with a special CD writer, after which, as the name suggests, the computer can only read information off the CD and never save to it.

Network card

The network card is a specialist card that is installed within the system box of a computer, and facilitates connection to other machines via special cabling. It achieves this by handling all the protocols associated with the network, and relieves the CPU of loading caused by the handling of network traffic. The type of card and cabling will depend on the type of network being used, and therefore specialist advice should be sought at the time of purchase.

Advantages of networking machines within an office include the ability to share data and expensive resources such as printers and high-speed communications. The various types of network cabling arrangement will be discussed later.

Digitizer

Digitizers, or digital tablets, are used mainly for converting original hard-copy drawings into a suitable computer format. This is achieved by placing the work flat on the tablet and then accurately tracing the points of the drawing with a puck (an input device that is similar to a computer mouse). Tablets may vary from A5 to A0 in size. Special pens may also be used instead of a puck when working on graphic work.

Monitor

The monitor is a television-like device that enables the computer to display information to the user. Three key factors in selecting a monitor are:

- **resolution** – this determines how much detail can be seen on the screen;
- **dot pitch** – this determines how sharp the image is;
- **size** – diagonal measurement expressed in inches.

Resolutions are usually stated in **pixels**, each pixel representing one dot on the screen. The following types of resolution are likely to be available:

- Video graphics array (VGA) is 640 pixels wide by 480 pixels high.
- Super VGA (SVGA) generally denotes 800 pixels wide by 600 pixels high.
- Extended graphics array (XGA) can usually achieve 1024 pixels wide by 728 pixels high. XGA can also cover the highest resolution commonly found, which is 1280×1024 pixels.

Dot pitch is usually set at 0.28 mm for most displays, but some offer 0.26 mm, which will be sharper.

A standard size for a desktop screen is 14 in, which will allow reasonable strain-free working at 800×600 pixel resolution. For higher resolutions a larger screen is recommended, usually at least 17 in, although 20–21 in screens are preferable for graphic and CAD work. As screen sizes increase so does the housing, and therefore placement on a suitable work surface is important.

Higher-resolution screens may require a graphics accelerator card to be added to the computer to provide the resolution required and allow for an acceptable speed of drawing. The graphics accelerator card contains display or video memory (DRAM or VRAM), which increases the amount of visual graphics that can be displayed, and reduces the refresh time.

Printers and plotters

Until recently most standard office productivity work, such as word processing, was output through a printer, whereas drawing work was output through a plotter. Recent advances in technology, however, now make the distinction between the two less clear.

Plotters such as the pen plotter originally worked slightly differently from printers in that they drew an object at a time (e.g. a title border), whereas printers drew everything at once, working in bands, lines or pages. The plotter would also use special pens mounted in a carousel, the pens being automatically changed during the printing process to represent the various colours or line thickness specified in the drawing. Although this method was excellent at producing drawings, users found that typical problems with it included the length of time it took to produce a drawing, pens drying out, pens ripping through the paper on detailed work, poor text reproduction, and noise from the machine. To address these problems, some manufacturers turned to printer technology, although the term 'plotter' is still widely used, particularly when describing devices that are capable of producing large-format drawings.

Three of the most common types of printer technology in use are:

- inkjet;
- laser;
- dye sublimation.

Inkjet printers

These are currently the industry standard for outputting large-format drawings. They work by squirting ink from a cartridge through holes in the print head, which moves over the paper. The printers are sometimes called bubble jet printers, depending on the print head technology used.

Colour versions use either a special three-colour cartridge or three separate cartridges in addition to the black. This type of colour reproduction is known as four-colour CYMK (cyan, yellow, magenta and key black).

Resolution available is also determined by the print head technology being used. The two current industry-standard manufacturers, Hewlett Packard and Canon, offer 300 dpi (dots per inch) and 360 dpi for colour output and 600 dpi and 720 dpi for black and white output respectively, although in practice resolution is also determined by the type of paper medium used.

Laser printers

These are currently the industry standard for outputting black and white A4 work. Colour A3 printer versions are also available, but are very expensive to purchase, and therefore many landscape architects prefer to use a bureau for this service. Laser printers use technology based on photocopiers; depending on the make, they can print at very high-quality continuous tone resolutions, up to 600 dpi, on standard media. Standard desktop models print at 4 ppm (pages per minute), although there are network printers available that can print at 20 ppm and on both sides of the paper. The main constraints in using a laser printer for all landscape work are the purchase cost and the paper size, with A3 format being the maximum standard size currently available.

Dye sublimation printers

These work by heating and mixing special coloured wax in the print head before depositing this onto the print medium. Thermal wax transfer printers work in a similar way but deposit coloured dye from a heated wax 'ribbon'. Both are capable of producing photographic-quality images, although these types of printer are very expensive, and the size of the output is limited to A3.

2.5.2 SOFTWARE SPECIFICATION

See Fig. 2.3.

A computer is useless unless it can be given instructions. These instructions are contained within computer software programs. There are two types of software program: application programs and operating system programs.

Application software allows the computer to perform specific tasks: for example, word processing, computer-aided design, or photomontage. Underneath the application programs is another layer of software called the **operating system**, which acts as a intermediary between the application and the hardware itself: for example, DOS, Windows NT, AppleMac.

Application programs cannot run without the appropriate operating system, and most operating systems are tied to a particular processor. Hence the massive catalogue of existing DOS (and Windows) applications can be run only on a PC with an Intel x86 processor. This is now beginning to change with initiatives such as the PowerMac, which can operate with both PC and Mac applications.

2.6 Software applications

WORD PROCESSING

See Fig. 2.4.

Word processing offers the ability to compile, store, edit, retrieve and manipulate text (and occasionally graphics) electronically.

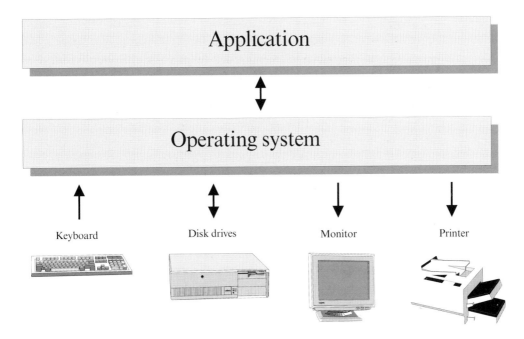

Fig. 2.3 Software specification.

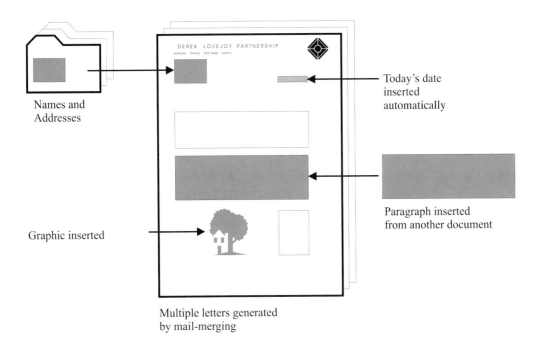

Fig. 2.4 Word processing.

The most basic types of word processor are **text editors**. These normally have limited text manipulation facilities, and require each line to be typed separately. Their use is mainly for writing and editing computer programs.

Standard **word processing** packages work at paragraph level, wrapping text automatically from line to line within a document. Additional functions normally include the ability to change text fonts and sizes, justify paragraphs, indent them, change the line spacing, check the spelling, and embed graphics.

Future trends are to produce programs that make existing features more user-friendly. Automatic checking of spelling, by highlighting unknown words, is one of the new features. Integration with other programs, such as spreadsheets and databases, is also likely to become easier.

Typical landscape applications include the preparation of standard letters, specifications, contracts and reports.

2.6.2 SPREADSHEETS

See Fig. 2.5.

Spreadsheets are derived from the manual spreadsheets used in simple accounting systems. They can be used in almost any application where calculations are required, and typically appear as an electronic table divided into rows and columns. Modern spreadsheets also allow pages as well as rows, which allow a third dimension to be added to the data. Each intersection of a column and row is known as a **cell**, and a cell can contain text, a number, or a formula to work on data held in other cells.

Landscape applications include, for example, bills of quantities, where the landscape element quantities are recorded in one column of cells (e.g. number of benches), the cost of each element in another column, and the formula to multiply the two in a third cell.

Users should be aware of the GIGO (garbage in, garbage out) rule when

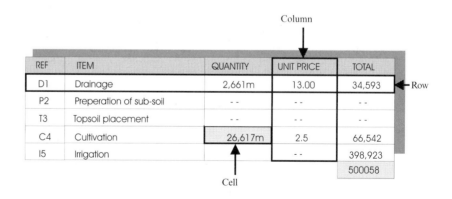

Fig. 2.5 Example of a spreadsheet.

using spreadsheet programs. Mistakes can easily be made when applying a formula in a cell (usually by referring to the wrong cell), which is not immediately apparent because it is normal for the results of the formula to be displayed, and not the formula.

When used correctly, spreadsheets are very versatile, and are excellent at testing 'what if' situations where a number of input values are variable. They are easy to understand, although setting up a spreadsheet model can be time consuming, as is amending an existing model.

Spreadsheets are also capable of holding simple databases, but as soon as complex data relationships need to be recorded a dedicated database program will be necessary. Advanced spreadsheet packages contain modules to access external databases, both locally and on network machines.

Typical landscape applications include preparation of bills of quantities, plant schedules, project management costings and office budget monitoring.

2.6.3 DATABASES

See Fig. 2.6.

a) EXAMPLE OF FLAT FILE

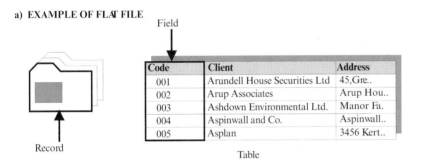

b) EXAMPLE OF RELATIONAL DATABASE MODEL

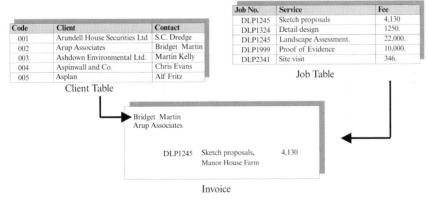

Fig. 2.6 Examples of database types: (a) flat file; (b) relational.

A database is a collection of related information or data about a subject, which is organized in a way that allows controlled retrieval of information. Any collection of information that serves this purpose may be regarded as a database. It does not need to be stored on a computer; for example, a telephone directory is a database, as it is an organized collection of names, addresses and telephone numbers.

A **database management system** (DBMS) is a computer tool used for the systematic management of a large collection of information. There are four basic types:

- flat file;
- relational;
- extended network or entity relationship;
- object oriented.

A **flat file** database displays information in a similar way to a spreadsheet, in rows and columns, and is simply an electronic card filing box. Each card in the box constitutes a **record**, and the various categories of data, such as name, address and telephone number, constitute the **fields**. The whole collection of records is referred to as a **table**.

If you were to use a single table to store invoice details, for example, each record would need not only the details of each order but also the name and address of each customer. Ten different orders placed by the same customer would mean entering the same name and address ten times, which is both inefficient and unreliable.

An improved solution is to store the customer names and addresses in one table and the other details in another. The two tables can then be linked through a common field or **primary key**, such as a customer ID number, to generate the desired invoice. This method of handling data is known as a **relational** database model.

Sometimes the links between a variety of tables are too complex for a relational database to handle, and this is where **extended network** or **entity relationship** databases are used. Examples include personnel, accounts, and medical record databases.

Object-oriented databases use features found in relational and extended network databases, but treat items of data as objects, which allows them to store pictures, sounds, or complete documents as individual items. This makes this database type ideal for multimedia.

The right database needs to be chosen to suit a particular application. The standard Windows databases are easy to set up and use, and are ideal for relatively small applications. If the data is likely to be accessed by a number of users concurrently, or is likely to grow very large in size, then more heavyweight databases should be considered. The additional costs of the database will be recouped by the savings in users' time and frustration.

Typical uses within the landscape office include client lists, job lists, marketing initiatives, plant materials, library indexes, archiving and landscape product information.

2.6.4 DESKTOP PUBLISHING

Desktop publishing (DTP) employs the concept of using frames to create page layouts. The user can insert either text or pictures into each frame. These frames give the user more creative control over the positioning of the page elements, such as paragraphs of text, than can be achieved with a word processor.

This distinction, however, is becoming increasingly blurred as the latest versions of word processing packages begin to include many of the features previously available only in dedicated DTP software. Word processing packages may therefore be sufficient for average DTP needs, although, for advanced functions such as the production of very large documents, the high-precision placement of text and graphics and the ability to apply sophisticated print options such as colour separations, a leading DTP package is preferable.

Landscape applications include the preparation of reports, marketing material, public display boards and final artwork.

2.6.5 COMPUTER-AIDED DESIGN

See Figs 2.7 and 2.8.

Computer-aided design (CAD) applications are powerful drawing tools, which use **objects** to construct drawings. An object is a drawing element such as a line, circle, or text string. To instruct the CAD package which object to draw, a series of command menus exist, which can be selected from the screen by either the mouse or keyboard. Once a command is selected, the user supplies properties for the chosen object by responding to prompts on the screen. These properties usually include the exact size, location and colour of the desired object. Once drawn, the object appears on the screen, and then the process is repeated for a new object. Tools are provided to allow the user to modify the objects within the drawing in a variety of ways, including rotating, scaling, moving and erasing. When the drawing is complete, a hard copy can be obtained by sending a plot out to a printer.

Table 2.4 lists some of the advantages and disadvantages of using CAD.

Table 2.4 Advantages and disadvantages of using CAD

Advantages	Disadvantages
Drawing accuracy	Learning curve
Ease of use	Specialized skills required
Flexibility	Non-availability of specific landscape software
Ease of electronic communication and sharing of files	Can be difficult to monitor/maintain drawing discipline on multidisciplinary projects

| Excellent at dealing with repetitive work | Inconvenient to use as a pure design tool |
| Good at making revisions | System performance power places limitations on the types of graphics that can be achieved. DTP applications may sometimes be more appropriate. |

CAD risk management

This area is very important, because it relates to consultancy service agreements. More and more clients are demanding CAD as a standard deliverable, and sometimes have unrealistic expectations about what they will be able to do with CAD files once they are delivered. A few of the issues that need consideration when using CAD on a project are as follows:

- professional liability of CAD deliverables;
- the right to use/edit CAD files – protection of subsequent changes after formal delivery;
- ownership of files;
- impact of using CAD on construction schedule;
- CAD file translations;
- long-term archives – data death;
- CAD file acceptance period;
- compensation.

(Data death occurs when stored information, particularly in magnetic storage, becomes corrupted over time because of natural degradation of the physical properties of the storage medium.)

2.6.6 THIRD-PARTY ADD-ON LANDSCAPE CAD PACKAGES

A number of landscape software packages have been developed to increase the efficiency and productivity of the landscape architect when working on CAD. These packages often work in association with industry-standard CAD packages, and include tools that address the specific needs of the landscape architect. Typical features that may be provided within a package are:

- project management tools;
- symbols libraries;
- softworks routines;
- hardworks routines and databases;
- cost-estimating tools;
- surface-modelling capabilities;
- three-dimensional design and visualization tools;
- visibility analysis routines;
- specifications.

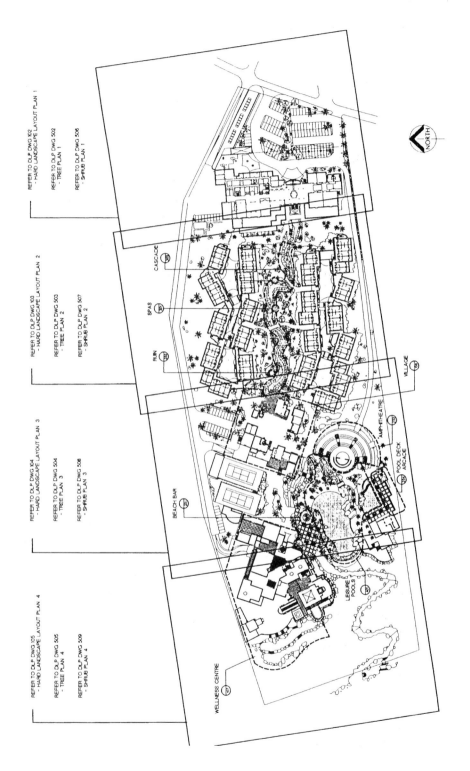

Fig. 2.7 Example of CAD-generated key plan.

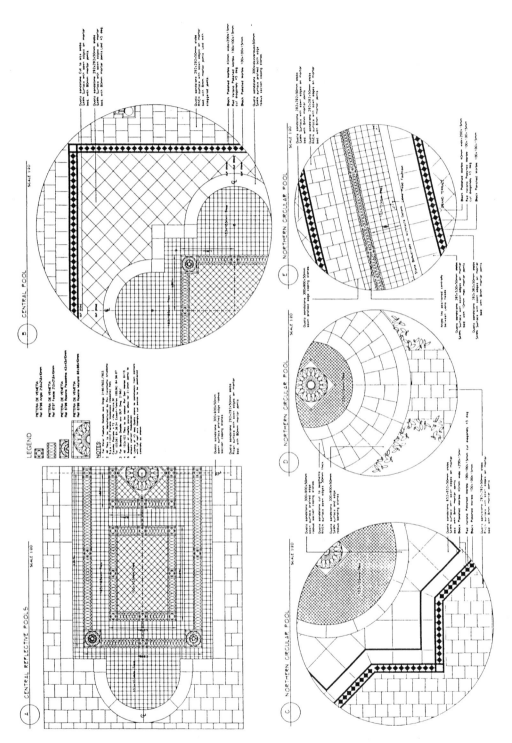

Fig. 2.8 Example of CAD-generated details.

Exactly what features are included within a package will depend on the manufacturer, and therefore the landscape architect should always ask for a demonstration of the software before purchase. The ability of the supplier to provide full support, training and upgrades should also be considered.

Two of the most important functions offered to the landscape architect by third-party add-on software are softworks routines and surface modelling.

Softworks routines

See Fig. 2.9.

The ability to embed data within drawings and then process that information quickly and accurately makes CAD an ideal medium for the production of softworks contract drawings.

To produce a planting plan, most commercially available programs use a combination of three key computer applications:

- graphics;
- database;
- spreadsheet.

Graphics offered by the program must be able to convey information clearly and accurately to the contractor, as well as demonstrate the design intent of the landscape architect. Labelling of plants within a drawing is often either by individual species name (if space is available), a key (for more complex plans), or by plant mixes. Not all planting programs offer all these options.

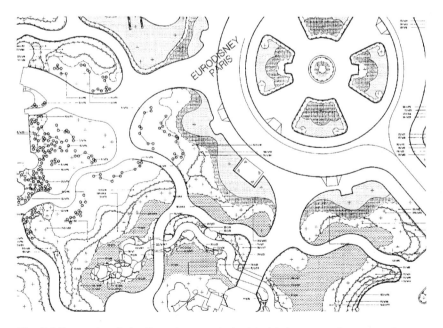

Fig. 2.9 Example of planting plan created with third-party software package.

To aid the architect when selecting plant material, most CAD programs use a link to a **database**. In addition to detailing the names and types of plants, databases are increasingly detailing more specific information on the species as well as illustrative material. Any database should be easy to customize by the landscape architect.

The planting schedule is essentially a **spreadsheet**, which is directly linked to information attached to a drawing, such as species and planting density and total quantities. The advantage of this is that as the drawing is modified and updated the schedule automatically updates, thereby making plant numbers within the drawing and the specification consistent.

Surface modelling

See Figs 2.10 and 2.11.

There are a number of software packages available that provide the landscape architect with the ability to create accurate three-dimensional models of the earth's surface. Once created, these models can be used for a variety of tasks, including cut-and-fill calculations, slope analysis, perspectives, three-dimensional design and the production of zones of visual influence.

A typical surface modelling project will involve the following stages:

1. data collection and input;
2. creation of the surface model;
3. analysis and design.

Data collection and input
Three-dimensional contours and spot heights of the existing landscape and any proposed changes are normally the only information that is required for data input.

Fig. 2.10 Example of three-dimensional model created with surface-modelling software.

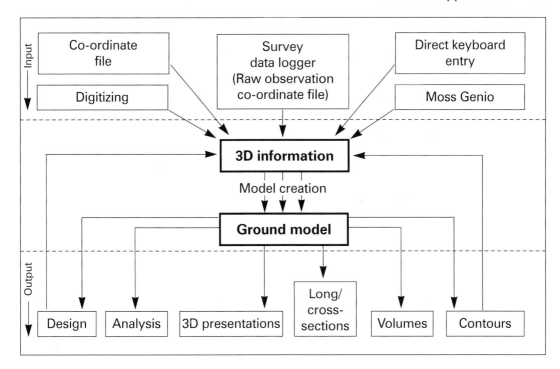

Fig. 2.11 Diagrammatic analysis of the processes involved when using ground-modelling software.

The existing information may be available in a suitable digital format (e.g. DXF) from either a map supplier (e.g. Ordnance Survey) or a land surveyor if a site survey has taken place. Alternatively, the information can be entered manually by digitizing, scanning or typing in the relevant data. Manual methods of input are usual for entering proposals; however, if paper maps are being used to enter existing surface information, it is important that an appropriate copyright licence has been gained before digitizing or scanning the information in. Ordnance Survey can provide guidance on this matter.

Once the three-dimensional data has been collected, it is converted into an XYZ data file, where X represents the easting, Y represents the northing and Z represents the elevation of the points being entered.

At this stage it is important to check that there are no rogue data lines or points (for example, spot heights that are not at the correct height), as these will cause problems during the modelling and analysis stages.

Creation of the surface model

There are three typical methods used by standard programs to calculate the surface model:

- **Rectangular** or **cells**, where data is 'mapped' to a grid. Modern products tend not to use this method, because of its limited resolution.

- **Triangular** – by far the best-established method, in which the ground surface is calculated as a series of triangles.
- **String**, in which linear features are used (may be thought of as the 'bent metal coathanger' method).

Leading systems available today tend to use both triangular and string methods to create a surface model. Once created, the surface model may be described as a DTM (digital terrain model), a DGM (digital ground model), a topographical model, or a TIN (triangulated interpolation network), depending on the method of calculation and software used.

Analysis and design

Most surface-modelling programs also include powerful analysis and design tools. These provide a variety of functions:

- contours;
- three-dimensional surface meshes and faces;
- cross-sections and profiles;
- volumetrics (cut and fill);
- surface area;
- slope and gradient analysis;
- visibility analysis;
- embankment design;
- 'draping' of two-dimensional features onto the three-dimensional model;
- report levels at any location.

Of these, the four most popular functions are the production of cross-sections, three-dimensional surface meshes for perspectives and walk-throughs, calculation of cut-and-fill volumes, and visibility analysis.

Cross-sections and profiles

Generating a section from a surface model allows the user easily to visualize and analyse the topography of a site.

The procedure varies according to the software being used, but generally involves defining the end points of the cross-section within the surface model and then specifying the horizontal and vertical increments as well as the vertical scale factor required. The section is then automatically calculated and exported to the CAD drawing, where it is manually positioned.

Many profiling routines deal only with the surface model, and will therefore ignore components or features that are not an integral part of the surface model being utilized. For example, hedge and house profiles must be added separately to the section, which can be a cumbersome and time-consuming activity.

Three-dimensional surface meshes and faces

The ability to create three-dimensional surface meshes and faces is extremely useful in the production of perspectives and walkthroughs, as they enable the CAD package to use 'hidden' line and solid rendering techniques when presenting the results.

The procedure is generally quite straightforward, and involves simply selecting the surface model file and then specifying the density and type of three-dimensional face or mesh to be applied.

The main limitation on the use of this application is the density of the mesh, which if set too high causes problems during manipulation of the information as well as significant delays in the regeneration time of the display.

Volumetrics

The ability to calculate the cut-and-fill volume of a proposal accurately, and to apply bulking and compaction factors, makes this feature of great benefit to the landscape professional.

The procedure normally involves the creation of two surface models, one of the existing situation and one of the proposed. One model is then subtracted from the other to give the resultant cut or fill volume.

The results of the volume calculation are dependent on the accuracy of the original data.

Visual analysis

See Fig. 2.12.

Zones of visual influence (ZVI) used in planning applications and environmental statements are increasingly being generated by surface-modelling packages. These packages, with their ability to process large numbers of iterative calculations, provide a more rigorous test of potential visibility of a proposed development than either a field survey alone or the drawing of a small number of cross-sections.

The method of production of a ZVI will vary according to the particular software used, but in principle all require the XYZ data of the viewpoint or object under analysis (e.g. the top of a proposed pylon or overburden mound) to be entered into the ZVI program along with the observer's eye level (often assumed to be 1.8 m) and surface model. The program then analyses the intervisibility between the specified object, the observer and the ground model according to a resolution specified by the user: the higher the resolution the greater the accuracy, but the longer it will take to generate the results of the analysis. The results are then illustrated on the computer screen as zones of visibility, which can be exported into a CAD package and overlaid onto an appropriate map so that the implications of the development can be fully understood.

When using ZVI packages, the areas of visibility identified are indicative only; they are not a categorical statement of visibility. This is because the computer-generated ZVI is based generally only on the topography of the land surrounding a development, and does not therefore take into account screening elements, such as woodland blocks or settlements, type of development, distance of view from development, curvature of earth, population

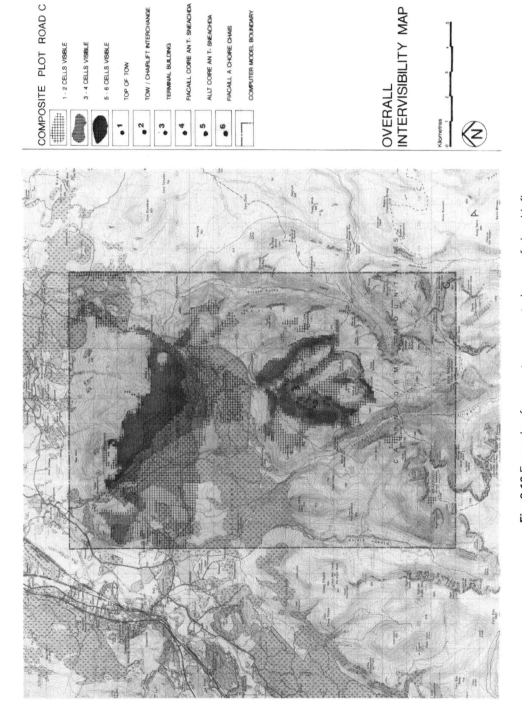

COMPOSITE PLOT ROAD C

1 - 2 CELLS VISIBLE
3 - 4 CELLS VISIBLE
5 - 6 CELLS VISIBLE

●1 TOP OF TOW
●2 TOW / CHAIRLIFT INTERCHANGE
●3 TERMINAL BUILDING
●4 FIACALL COIRE AN T- SNEACHDA
●5 ALLT COIRE AN T- SNEACHDA
●6 FIACALL A CHOIRE CHAIS

COMPUTER MODEL BOUNDARY

OVERALL
INTERVISIBILITY MAP

Kilometres
0   1   2   3   4   5

N

Fig. 2.12 Example of computer-generated zone of visual influence.

density, type of view (e.g. kitchen window, footpath, car), mitigation measures, weather conditions, season or time of day. The results therefore tend to represent a worse-case scenario.

Certain manufacturers are beginning to tackle these issues, as are other software applications such as GIS.

2.6.7 GEOGRAPHICAL INFORMATION SYSTEMS

See Fig. 2.13.

Geographical information systems (GIS) have developed by linking database management system applications with digital cartography and CAD applications to create a new tool that can store, map, analyse and present spatially referenced information in a more powerful and efficient way than was ever possible by manual methods of map manipulation or database research.

The term 'spatially referenced' is used to describe digital data that is linked to a real-world coordinate system, thereby giving it a known point on the earth's surface.

GIS techniques are being applied to many fields. Retailers are using them to analyse catchment areas for proposed stores. Utilities such as electricity and gas suppliers use them to keep track of their property, plant and lines. The health authorities use them to analyse patterns of illness. In landscape they are being used in areas such as land-use planning, environmental assessment and environmental monitoring.

A GIS project will normally involve the following five stages:

1. data preparation and input;
2. data storage;
3. data management;
4. modelling and analysis of the data;
5. data presentation.

Data preparation and input

Data preparation and input tends to require the highest time investment and to be the most expensive component in the establishment of a GIS. The quality of the data input is very important, as this will ultimately affect the quality of the results obtained.

Sources of socio-economic and environmental data include field surveys, government surveys, commercial databases, topographical maps, digital elevation data (XYZ files), thematic maps and remotely sensed imagery. These are normally combined with additional information that has to be entered into the system manually.

The Ordnance Survey (OS) has perhaps done the most to facilitate the growth of GIS in the UK by its commitment to supply all its maps in a digital format, although it has also been criticized for the high price it charges for this service. Other major data holders include the Census Office, the Meteorological office and the DSS.

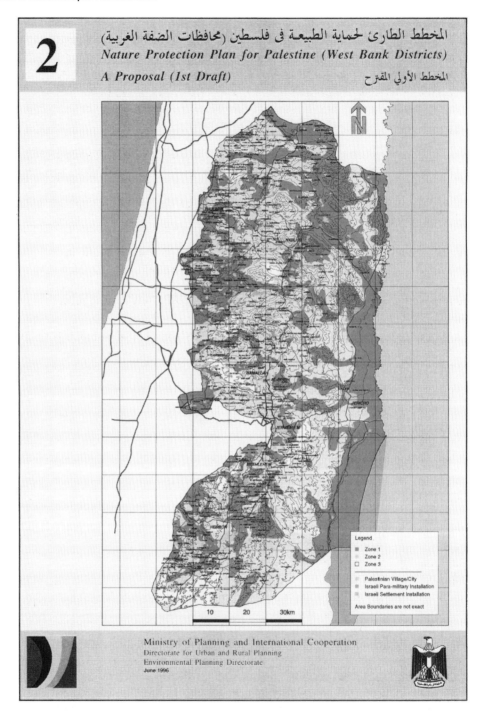

Fig. 2.13 Example of computer GIS mapping.

Data storage

Data is normally stored in raster or vector formats. These may include NTF, DXF, DWG, TIFF, JPEG, BMP and WMF formats depending on the software filters available with the GIS package being used. Database and spreadsheet links with industry standard packages are also normally included.

Data management

Physical constraints in terms of computer storage capacity and processing power make it extremely important to manage the large volumes of spatial data efficiently, especially if the GIS is to be cost effective.

Data can sometimes be more detailed than is required, and therefore a decision must often be made on the choice of source scale or resolution to be used. Ideally, the scale should not be so detailed as to generate too much data to handle, nor should it be so general that potential impacts are ignored. For example, if you are doing an analysis where the results will be presented at 1:10 000 scale there is little need in using data that is accurate to the millimetre. Accuracy of 1 m to 25 m may be sufficient, depending on what is being analysed.

Layering tools may also be used to assist the effective management of the system. Outlined in Table 2.5 is an extract from a layering convention developed by the Ordnance Survey and the Autodesk User Group.

Table 2.5 GIS layering convention

G801 0940	**Landscaping**
G801 0941	Existing trees
G801 0942	Proposed trees
G801 0943	Trees to be removed
G801 0944	Shrubs
G801 0945	Ground cover
G801 0946	Grass and lawns
G801 0947	Boulders
G801 0948	Rock, bark, and other landscaping beds
G801 0949	Schedules
G801 0950	**Roads, paving, car parking**
G801 0951	Road centre-lines
G801 0952	Kerbs
G801 0953	Parking bays

Data analysis and modelling

Often, when a GIS is used for planning purposes, it is merely for the production and presentation of maps. In these circumstances it is being used only as a sophisticated map-making tool, and the user would probably be better served by using a less complex and cheaper system. The true potential of GIS lies in their ability to integrate, manipulate and analyse various data sets, ask 'what if' types of question, and thereby improve the decision-making process. For example, in the case of visual analysis it is possible to import the zone of visual influence identified by a surface modelling package and then link this information with population density data (i.e. postcode data) so that the impact of proposals on settlements and individual dwellings can be examined in greater detail.

Data presentation

The ability to communicate the results of GIS analysis effectively is essential to the efficient use of the technology. Output can be in a variety of forms: from lists of data sets to fully rendered maps and graphs. The time needed to produce the final artwork is often underestimated, and this factor should be considered when organizing the work programme.

2.6.8 COMPUTER VISUALIZATION

Typical computer methods used for visualizing a project include:

- perspectives;
- photomontage;
- walkthroughs and animation;
- virtual reality.

Perspectives

In order to produce a perspective, a three-dimensional model of the scheme under study must first be created. The standard procedure then involves entering into the computer program coordinate information on where the observer is standing and also on the direction in which the observer is looking. The program then automatically calculates the distance and angle of view, and presents the results in a wire-line format. Hidden line removal is then often applied in order to make the interpretation of the perspective easier.

Photomontage

See Figs 2.14 and 2.15.

Computer photomontage is the combining and manipulation of existing photographic images with computer-generated elements to portray the visual effect of a proposed scheme.

The production of a photomontage typically involves the following stages:

Selection of viewpoints

These are generally based on the criteria of being viewpoints to which the public would have access, and which give a fair representation of the impact

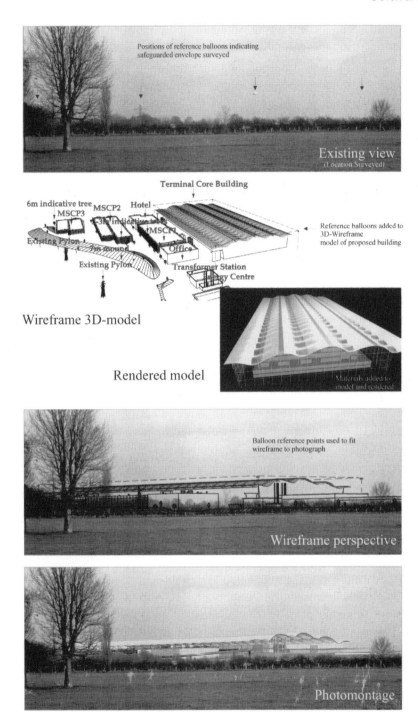

Fig. 2.14 Example showing the various stages in the production of a photomontage.

Fig. 2.15 Existing view and photomontages at year 1 and year 15, showing proposed road, lighting columns, balancing pond and screen planting.

of the scheme. It must also be possible to survey the position of the camera and any nearby reference features, such as electricity pylons, lighting columns and fence posts.

Site photography and field survey

Generally, photographs are taken with a 35 mm camera with a 50 mm focal length lens, as this gives an accurate representation of an observer's natural field of view, although larger-format cameras are now increasingly being used for improved clarity of image. Panoramic photography is useful for photo-montage context, provided that allowance is made for the difference between curved and flat picture plane construction. The field survey records the easting, northing and AOD elevation of the camera location and selected reference points. Markers such as pegs or spray paint may be needed to record the camera location if comparative photographs during different seasons are required.

Three-dimensional computer model creation

A three-dimensional computer model is then created of the proposals, into which the surveyed reference points are added. A wire-line perspective view can then be generated by entering the known camera point and known target point (the centre of the photograph) into the CAD program.

Scanning

A flatbed or drum scanner is used to convert the photographic image into a digital format that can then be imported into the computer.

'Fitting' the wire-frame perspective to the existing photograph

Once it is in a suitable format, the wire-frame showing the proposals can be scaled to 'fit' the photograph, using the reference points as a guide.

Rendering

Having established the scale and form of the proposals, they can be rendered using colour-imaging software. A photographic library of features similar to those proposed is often used for this purpose to increase the realism of the final photomontage.

Printing

The resultant images are then printed, via a high-quality printer, to form the final 'hard copy' images. Printers used for this purpose include dye sublimation and laser.

Walkthroughs/animations

See Fig. 2.16.

In addition to two-dimensional paper-based presentations, three-dimensional walkthroughs and animations are increasingly being used to help a client and interested parties to visualize a project.

Fig. 2.16 Example of two frames from an animation sequence.

Animations are usually created by selecting a predetermined route through a three-dimensional model and then fully rendering the views along that path. To produce high-quality smooth animations, up to 30 rendered frames per second may be necessary. To achieve this quantity of frames, network rendering across a number of machines is often the only feasible option. Once the rendering is complete, the results are stored as an animation file, which can then be played straight from the computer or recorded onto CD-ROM or video. Table 2.6 lists the requirements for the production of animations.

Table 2.6 Requirements for the production of animations

3D model (drawing with x, y and z coordinates)
Rendered surfaces and materials
Animation software
Video-capture board or CD writer
Video-editing software
Large hard disk (at least 4 Gb)
Computer video recorder

Virtual reality

Virtual reality (VR) is the simulated visual and audio experience of a real environment, where the image and sound instantly change as you change your point of view. It is, in effect, a medium that provides participative three-dimensional visualization and simulation. Based on our perception of spatial reality, virtual reality uses the principles of perspective to 'fool' the eyes into believing the image to be real.

VR is still at an embryonic stage in development, but has the potential to become a very important design tool of the future.

Current VR systems are available as desktop versions using large monitors and submersive versions using head-mounted displays. Both types can run on the same computers that run CAD programs. Limitations in processing power and storage, however, often mean that the graphics remain quite basic.

2.7 System management and administration

Systems management and administration involves the control and organization of a company's human and technical IT resources, including staffing, networking, security and the implementation of standards, conventions and procedures.

Some of the concepts discussed in this section are suitable only for larger landscape practices, although it is hoped that certain items will be of benefit to all.

2.7.1 STAFFING

Staff with a good working knowledge or interest in IT are an important business asset. Not everyone needs to be a computer expert, but a good balance between strong and weak IT skills needs to be considered within an office.

It is important to designate a person to oversee the direction and management of the computer resource, as well as someone to support the day-to-day running of the system, a task commonly referred to as **housekeeping**. These are not mutually exclusive responsibilities, and may be carried out by the same individual (Table 2.7).

Table 2.7 Staff IT tasks

IT tasks	Individual actions
Housekeeping	1. Order of consumables 2. System maintenance 3. Backup 4. Archive
Development of IT Strategy	1. Reading journals, keeping up to date

Table 2.7 continued

IT tasks	Individual actions
IT management	2. Specification of hardware and software
	3. Planning/organizing current resources
	4. Planning/strategy for future resources
	1. Control of IT use
	2. Resourcing projects
	3. Liaising with other consultants on specific IT matters
	4. Developing standard systems
	5. Day-to-day IT support
IT training	1. Training of new personnel
	2. Improving skills of existing personnel
IT research and development	1. Learning, using and incorporating new software
	2. Developing bespoke software
	3. Responding to immediate job specific requirements
Raising practice profile	1. Informal talks
	2. Lectures
	3. Magazine articles

In larger organizations, the responsibility for the management and development of the IT resource can be shared amongst a group of individuals. This has the added advantage of providing a career structure for individuals working in a specialist field that is not directly recognized as part of the landscape industry (Table 2.8).

2.7.2 SECURITY

See Fig. 2.17.

Theft, viruses, power failure, vandalism and fraud are all possible IT security threats that a practice must consider. Personal information held within the office database may also place obligations on the employer under the Data Protection Act 1984.

To limit the impact of such events, a security policy should be introduced that addresses the threats by means of:

1. prevention (e.g. increased physical security);
2. timely detection (e.g. password protection);
3. defence (e.g. anti-virus software, power security);
4. recovery (e.g. backup).

Recovery is often the last form of security, and therefore it is essential that an effective backup and archive procedure is in place. This should include

Table 2.8 Example of staff IT structure in a large organization

	IT Director	System Manager	CAD manager/coordinator	Advanced user	Intermediate user	Beginner	Trainee
Initiate programme	✓						
Provide management support	✓						
Design programme	✓						
Set goals	✓						
Monitor programme	✓						
Research and development	✓	✓					
Implement office programme	✓	✓					
Monitor goals	✓	✓					
Purchase and maintain hardware and software	✓	✓					
Implement and maintain software configurations	✓	✓					
Coordinate user support	✓	✓					
Evaluate new applicants	✓	✓					
Coordinate manpower allocation	✓	✓					
Implement standards, conventions and procedures	✓	✓					
Support standards, conventions and procedures	✓	✓					
Minimum three years' IT experience	✓	✓	✓				
Demonstrate ability to organize and manage aspects of project	✓	✓	✓				
Understand project co-ordination issues	✓	✓	✓				
Understand overall office procedures and goals	✓	✓	✓				
Complete understanding of auxiliary computer systems	✓	✓	✓				
Complete understanding of plotting procedures and troubleshooting	✓	✓	✓				
Organize outside plotting services in coordination with project schedule	✓	✓	✓				
Supervise outside plotting services in coordination with project schedule	✓	✓	✓				
Complete understanding of project structure	✓	✓	✓				
Minimum two years' CAD experience		✓	✓	✓			
Basic understanding of CAD coordination responsibilities		✓	✓	✓			
Basic understanding of project coordination		✓	✓	✓			
Basic understanding of overall office procedures and goals		✓	✓	✓			
Basic understanding of auxiliary computer systems		✓	✓	✓			
Good understanding of plotting procedures		✓	✓	✓			

Table 2.8   continued

	IT Director	System Manager	CAD manager/coordinator	Advanced user	Intermediate user	Beginner	Trainee
Complete understanding of standards		✓	✓	✓			
Good understanding of project structure		✓	✓	✓			
Ability to support other users		✓	✓	✓			
High productivity level		✓	✓	✓			
Understand plotting procedures		✓	✓	✓	✓		
Good understanding of standards					✓		
Basic understanding of project structures					✓		
Good productivity level					✓		
Exposure to CAD						✓	
Good understanding of programme commands						✓	
Understands computer basics						✓	
Low productivity level						✓	
Understand basic commands							✓
Productivity liability							✓

storing one full set of backed-up information off site. Tape is the traditional backup medium, although hard disks and removable hard disks (magnetic optical) are increasingly being used.

Over a period of time all magnetic recordings deteriorate in quality, and therefore when archiving data it may be prudent to transfer the data to a more stable medium, such as CD-ROM.

2.7.3 NETWORK

A network is a group of computers linked together to improve efficiency by:

- sharing resources (e.g. a printer);
- sharing information (e.g. a report or drawing that several people are working on);
- improving communication (e.g. via electronic mail).

A network contained within an office is commonly known as a **LAN** (local area network). There are two basic types of network: **peer-to-peer** and **client/server**.

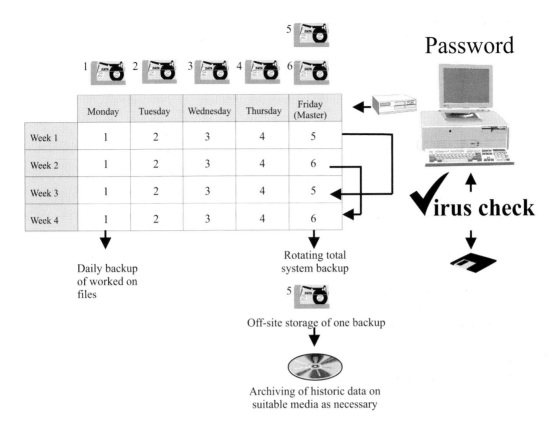

Fig. 2.17 Typical backup and security techniques.

Peer-to-peer networks

See Fig. 2.18.

This is the simplest network, and is best suited to small businesses. As the name implies, it is a group of computers that are hooked together; all have the same (peer) status in the network. Table 2.9 lists the advantages and disadvantages of a peer-to-peer network.

Table 2.9 Advantages and disadvantages of a peer-to-peer network

Advantages	Disadvantages
Cheapest form of network, requiring only a network card, cabling or infrared ports (for wireless signals) and network software	Limited capacity, and tends to slow down when 15 or more computers are connected together

Table 2.9 continued

Advantages	Disadvantages
Relatively easy to set up and maintain	Using shared facilities can slow down a computer that has peripherals connected to it
Can share files, printers and peripherals such as CD-ROM players	No central control and security; each user has to administer their own computer in terms of granting access to shared files and peripherals
Can send each other messages	No overall system backup – user dependent
	No server or boss that regulates network traffic or acts as a controller (hub) for managing the network.

Client/server networks

See Fig. 2.18.

A client/server network has a dedicated computer acting as a **server** or boss, which controls the activity of **clients** or computers connected to the network. Table 2.10 lists the advantages and disadvantages of a client/server network.

Table 2.10 Advantages and disadvantages of a client/server network

Advantages	Disadvantages
Acts as central control or hub for sharing peripherals such as printers and CD-ROM players	More expensive; requires a dedicated computer to act as a server
Provides central backup facilities	More complicated to set up and maintain
Provides global network security including access to files, passwords, virus protection	Requires standards, conventions and procedures
Allows centralized general management of network	
Greater number of computers can be connected together	
Allows sharing of applications from a central location	
Central storage of data	

a) Bus Network

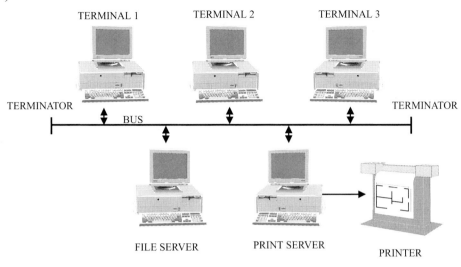

b) Star Network

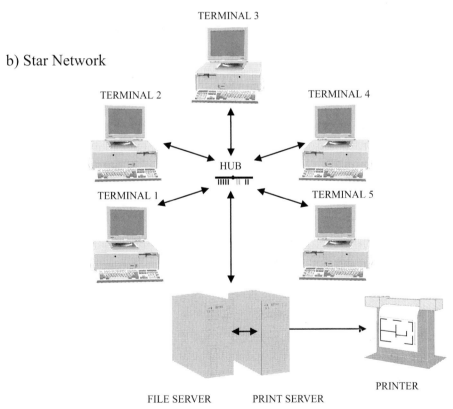

Fig. 2.18 Network topologies: (a) bus; (b) star.

Network operating systems

Common network platforms are:

- UNIX;
- Novell;
- Windows NT;
- Macintosh;
- OS/2.

The two most popular network operating systems are Novell and UNIX. Windows NT is rapidly gaining ground, especially in environments where CAD is used heavily.

Network topology

Two of the most commonly used cabling patterns or network topologies are **star** and **bus**.

Star topology is the most common configuration, and is the one most recommended, because it is the most fault tolerant, the easiest to modify, and excellent for centralized management and monitoring of network performance. However, it is more expensive, as it uses more cable than the bus topology. The cabling most used is unshielded twisted pair (UTP) using RJ45 connectors, and requires a hub to connect from the computer network card to the server. When one computer 'goes down' (fails), the rest of the system will still be up and running.

Bus topology is the simplest and the easiest to install; it requires less cable, and is easier to extend. However, networks problems are harder to isolate, and the network slows down during heavy use. The cabling used is coaxial using BNC connectors. The biggest disadvantage is that, when a connection is lost, this will bring down the entire network segment.

2.7.4 STANDARDS, CONVENTIONS AND PROCEDURES

In an office where there are more than a few staff, it is important to set up some standards, conventions and procedures. This will make working on computers more efficient, because staff will know where to find files, what the layout should look like, and so on. These rules are usually compiled into a manual, which should be made available to all staff.

General office procedures should include:

- a directory structure to establish where files will be saved;
- a file-naming convention to establish naming of files;
- text style and type to give continuity in presentation;
- templates for standardizing the layout of all practice documents, including letters, faxes, reports and drawings;
- security procedures to include backing up, virus detection and cleaning and archiving of data.

CAD-specific procedures should include:

- standard insertion point (0,0) and layer for blocks;
- attachment and location of cross-references;
- colours (by layer)
- standard pen settings for plotting;
- grid coordinates to orientate plans (national grid);
- layering convention (BS 1192 or custom);
- drawing scale (millimetres or metres);
- file-transfer checklist, including format, fonts and compression program;
- title block.

British Standards that may be of use in the IT field include the following.

BS 5750 Pt 13: 1991 *Guide to the application of BS 5750 Pt 1 to the development, supply and maintenance of software*

This document translates the requirements of BS 5750 Pt 1 into terminology recognized in the IT industry.

BS/ISO 10561

This international and British Standard covers the evaluation of printer operation by measuring printer throughput.

BS 1192

This British Standard covers drawing conventions.

BS 4783 (7 parts)

This British Standard gives recommendations for the storage, transportation and maintenance of media (disk cartridges).

BS 6616 and ISO 1831

These international and British Standards deal with the requirements of printing and paper quality. Specification for paper and print for optical character recognition (OCR).

2.7.5 COMMUNICATIONS

See Fig. 2.19.

Probably the most time consuming and difficult aspect of any project is finding out the right information, at the right time, from the right source: in other words, communicating with the client, consultants or contractors effectively.

File transfer

One of the benefits of using a computer system is that the data is in digital format, which can be stored, shared and manipulated very quickly. When

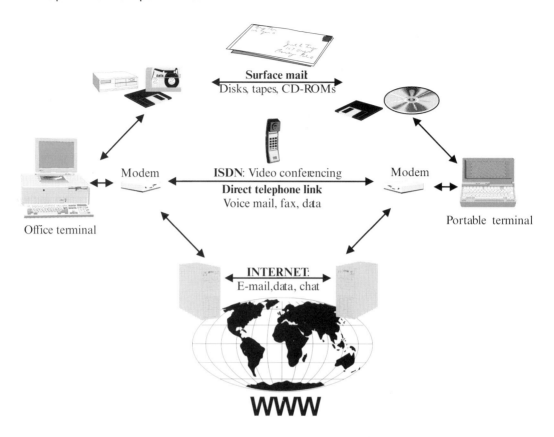

Fig. 2.19 Data communication techniques.

sharing large volumes of data between consultants a compatible form of data transfer should be agreed at an early stage (Table 2.11).

Table 2.11 Forms of transmitting data

Type	Transmission rate/ Maximum file size
Modem	Up to 28 800 bps (20 K per second) on standard telephone lines
ISDN	128 K per second
WAN (wide area network)	Private leased lines
3.5 in floppy disk	Files under 1.44 Mb (an individual file can be compressed across multiple disks if required)
Tape	DAT can hold up to 24 Gb in compressed mode

Table 2.11 continued

Type	Transmission rate/ Maximum file size
CD-ROM	Read/write 650 Mb usually for archiving; however, can be used for distributing data to other parties. Most other consultants/ teams will have a CD-ROM player
Optical disk	Magnetic optical disks can hold up to 2.6 Gb, and have access speeds comparable to those of hard disks
Syquest	5.25 inch 44 Mb/88 Mb/200 Mb disks 3.5 inch 105 Mb/270 Mb disks
Floptical disk	3.5 inch 21 Mb
External SCSI hard drive	Up to 9 Gb. Drive can be 'daisy-chained' to other SCSI devices

The Internet

The Internet is a public access network enabling anybody with a computer to connect to information on other computers on other networks all over the world.

The Internet began in the early 1960s, when four US military colleges were connected together via computers for the first time. By 1972, 37 university sites were allowed to access and share information with the military colleges. In the 1980s, individual users were allowed access to university sites. Today there are millions of users accessing the Internet all over world, and continued growth is still strong.

In order to access the Internet the following is required:

- a computer;
- a modem with a high transmission rate (ideally 28 800 baud or faster)
- a means of telephone access, either via a private network where costs are incurred on a pay-as-you-use basis, or via a 'service provider', which operates on a flat monthly fee;
- software such as Netscape Navigator to navigate the Internet.

The World Wide Web (WWW) is a web of interconnected computer sites. Each site is a computer holding information on or about one or more subjects. A user visits the site and is encouraged to go to another site on related subjects by just a click on the mouse.

The three most popular tools used on the Internet are electronic mail (e-mail), news and Web browsers.

E-mail allows messages to be sent from one computer to another via the Internet. Document or drawing files can be attached to these messages, giving the receiver immediate access to the digital information as well as the ability to make alterations or comments as required. When sending data electronically, it is a good idea to compress the files in order to preserve the integrity of data and reduce transmission time and therefore cost. Depending on the service provider being used, some attachments must also be encoded first by special software before being sent and then decoded upon receipt.

News or Usenet news provides a forum for conducting global conferences on any topic desired.

Browsers provides a way of accessing information that requires no technical knowledge at all. The information, whatever its form, is presented as a series of magazine-like pages.

Intranets

These are private networks based around the same software and protocols as the Internet, but with strict security to keep out outsiders. With an intranet, a company can publish internal web sites, for staff and clients, which can cover everything from staff newsletters to company brochures and common data, all accessed using a common web browser.

Desktop video conferencing and whiteboarding

With high-speed phone lines such as ISDN (integrated services digital network) becoming more commonplace, desktop video conferencing can provide an alternative to face-to-face meetings. For acceptable video refresh rates, 128k bandwidth is the minimum. To achieve this the following items are required: a digital camera; a special video card; a modem or terminal adapter if using ISDN; and appropriate software on each computer.

Whiteboarding is a means whereby a document shared between two users can be redlined at either end and the results made visible at the other end. (Redlining is similar to using a highlighter pen on paper. It allows you to bring attention to specific issues electronically without affecting the content of the document.) Whiteboarding is usually associated with desktop video conferencing where for example two people, one located in London and the other in Los Angeles, are discussing a project and in particular a certain drawing visible to both parties. The person in London writes some notes and the person in Los Angeles sees what has been written instantly. This is an invaluable tool in the situation where two offices in different locations are working on the same project.

2.8 Further information

British Computer Society
Sanford Street
Swindon SN1 1HJ
Tel: 01793 417417

British Standards Institution (BSI)
Enquiry Section
Linford Wood
Milton Keynes MK14 6LE
Tel: 01908 221166

CAD User
24, High Street
Beckenham
Kent BR3 1AY

Central Computer and Telecommunications Agency
Riverwalk House
157–161 Millbank
London SW1P 4RT
Tel: 0171 217 3338

Computing Services Association
Hanover House
73–74 High Holborn
London WC1 6LE
Tel: 0171 405 2171

Association of Electronics Telecommunications and Business Equipment
Industries
Russell Square
London WC1B 5EE
Tel: 0171 331 2030

Ergonomics Society
Devonshire House
Devonshire Square
Loughborough LE11 3DW
Tel: 01509 234904

Health and Safety Executive
Public Information Centre
Broad Lane
Sheffield S3 7HQ
Tel: 0114 289 2345

HMSO Publications Centre
PO Box 276
London SW8

Society of Information Technology Managers
c/o Computer Services
Lincolnshire County Council
Orchard House
Orchard Street
Lincoln LN1 1YG
Tel: 01522 552222

National Computing Centre
Oxford House
Oxford Road
Manchester M1 7ED
Tel: 0161 228 6333

Chartered Institute of Purchasing and Supply
Easton House
Easton on the Hill
Stamford PE9 3NZ
Tel: 01780 56777
Fax: 01780 51610

Ordnance Survey
Digital Services Division
Romsey Road
Southampton
SO9 4DH
Tel: 01703 792431
Fax: 01703 792962

Institute of Software Engineering
30 Island Street
Belfast BT4 1DH
Tel: 01232 738507

Pre-contract information 3

3.1 Site visits

These few notes are only a guide to the landscape designer when making site visits: they do not include a full land and vegetation survey with accurate levels, trial holes, full soil sampling, and so forth. Use maps or site plans of the area to record particular features when visiting the site, and once the client asks for further information draw up suitably scaled maps and detailed site plans in order to plot features more accurately. Information collected should include:

- aspect and prospect;
- general character, topography;
- soil types, and any rock outcrops;
- vegetation and rough trial hole for depth of top soil for future planting;
- specific landscape features of importance;
- woodlands, copses, trees and shrubs;
- usage of nearby properties;
- natural watercourses;
- overhead power lines and cables, underground pipelines, drains, tanks;
- any large man-made features not shown on the maps and plans;
- ease of access to the site for construction vehicles and equipment.

3.2 Desk survey

Check and, where applicable, plot on site plans:

- information from Ordnance Survey maps;
- ownership of site boundaries;
- local planning authority Tree Preservation Orders (TPOs);
- the presence of flora and fauna protected by legislation;
- easements and rights of way likely to interfere with proposals;
- piped services and overhead power transmission lines;
- sewers, and natural gas and oil pipelines.

3.3 Contaminated land

The landscape designer is often required to produce a scheme on contaminated land, and although the site will no doubt be cleaned up or sealed by experts, it is useful to be able to have a rough check of possible contaminants by noting signs of their existence in the state of the vegetation, or the unusual colour of the soil. It is assumed that the architect or developer will be responsible for all buildings erected on site, as these will require special precautions to be taken.

On known contaminated sites specialist site investigation is an essential part of the design process. If in any doubt about a site, the landscape designer should advise the client to commission site investigation and remedial work.

3.4 Information for the contractor

Provide sufficient written and drawn information to allow the contractor to give a fair and competitive tender. Particular points to include: underground and overhead cables, water, gas, sewer and surface water drains, and any other features liable to cause damage to people, animals or plants. The drawings must indicate the entrance for the contractor's vehicles and plant.

3.5 Protection

Provide temporary fencing to secure the site, protect trees, hedges and shrubs or any other special feature to be preserved. This is particularly important in conservation areas. In areas of little or no vandalism, cleft chestnut fencing 1.5 m high should be sufficient; for greater security choose one of the higher-security fences described in section Q40.

3.6 Negligence

Accidents are all too frequent an occurrence on any working site, and while the contractor is officially on site, he is responsible for general site safety, and for the safety of his workforce and all official visitors to the site. This makes temporary fencing of deep pits (for example, those required for swimming pools) a wise precaution at the end of each work day as well as on completion of the contract. See also the notes on health and safety below.

3.7 Access for emergency vehicles

Access for emergency vehicles as well as for the contractor's vehicles and plant must be considered.

3.8 Trees and other vegetation

Trees, hedges, shrubs, lawns and so forth to be preserved must not have spoil heaps, rubbish, vehicles, machinery, plant or fuel stored near to them. In the case of trees this prohibition extends to the spread of the tree. Trees – particularly those with TPOs on them – protected hedges and other features to be retained should be clearly marked on both the site and the plans.

3.9 Pesticides

This is a very important subject. The definition, use and prohibitions on the storage, application and disposal of pesticides are included in the notes in section Q30, and this should be referred to when writing the preliminaries to the contract. Pesticides should be applied only by those qualified and trained in their use.

3.10 Bonfires

If these are allowed by the local authority, they must be on specified sites, and never within 10 m of any vegetation to be preserved. Particular care must be taken with TPO trees. The Highways Act prohibits the lighting of bonfires within 12 m of the centre of a highway.

3.11 Noise

There are many regulations made under the Health and Safety at Work etc. Act 1974 as well as British Standards on the need to reduce the noise on sites and that made by specific equipment. This is the contractor's problem, but the landscape designer can stipulate any reasonable extra ban that he or she wishes in the preliminaries of the contract. This can range from a limit on the hours at which work with 'noisy' machinery or equipment is carried out, to a ban on the use of loud radios or recording equipment.

3.12 Health and safety measures; Construction, Design and Management Regulations

The Health and Safety at Work etc. Act and the Occupier's Liability Acts place a duty of care on the owner and/or occupier, so that failure to protect the occupants, the workforce or the public at large from foreseeable hazards is an offence, and negligent landscape designers may find themselves in considerable trouble if they have not considered the problems of hazards carefully. The Acts also apply to sites under construction.

The Construction, Design and Management Regulations 1994 (1994/3140) came into force on 31 March 1995. These are complex regulations, and to a great extent deal with large contracts; the landscape designer should contact the Landscape Institute for any current changes that have been made to the JCLI Form of Agreement for Landscape Works, or any other form of contract to accommodate these regulations.

The notes below are only a rough guide as to when the regulations apply or do not apply; they should not be used for taking a final decision.

The regulations do not affect:

- employing less than five people;
- projects taking less than 30 days or 500 person-days;
- projects for the householder;
- projects under local authority enforcement.

The regulations do affect:

- all demolition except that for the householder;
- site clearance for larger projects;
- all large contracts except those under LA enforcement.

They are intended to:

- ensure (within foreseeable reason) that the health and safety of all who carry out, will carry out work or maintenance or use the project is protected;
- ensure (within foreseeable reason) that the design includes sufficient information about the project, including the materials, so as not to adversely affect the health and safety of all who carry out, will carry out work or maintenance or use the project.

It is therefore more important than ever that the landscape designer discusses the purposes for which the project is to be used within the foreseeable future, specifies materials and equipment fit for their purpose and discusses their use with the manufacturer, and provides the client with a commissioning document describing the use for which the project was designed and the maintenance requirements of all surfaces.

3.12.1 HEALTH AND SAFETY EXECUTIVE

The Health and Safety Executive publishes numerous useful publications. Those listed below are most likely to be of interest to the landscape designer.

- CDM Regulations: Code of Practice;
- *Designing for Health and Safety in Construction* (green book);
- *Guidance for Clients, Planning Supervisors, Designers and Contractors: a Guide to Managing Health and Safety in Construction*;
- *Guide to Managing Health and Safety in Construction* (red book).

3.13 Drawings

3.13.1 GENERAL GUIDANCE

Whether the drawings illustrate the initial sketch design or the contract drawings that accompany the specification, the landscape designer must never forget that they are a means of communicating information, and the type of drawing will vary enormously according to the stage of the project.

At the initial stage, drawings should include rough section and perspective sketches, as the client will want to see the project presented in a colourful and interesting way, as well as having some indication of the cost involved. At this stage a computer-generated model will always enhance any presentation, and help to describe the scheme. Computer-generated models of part or the whole scheme are also desirable when lay committees or individuals are being asked for funds. Where the project is the subject of 'reserved matters', or where the scheme is in an area such as a National Park, AONB or one of the many types of conservation area and likely to be the subject of much official and public debate models, sketches and photographs are invaluable.

When preparing the initial design for early costings, the landscape designer will work up a schedule of materials, finishes and planting. Once the rough estimate has been presented to the client, the landscape designer must be prepared to make adjustments to the brief and re-present the scheme to the client for approval. It is far better to get full approval at this stage and so avoid having to present the client with an unwelcome invoice for extra fees for abortive work at the working drawing stage.

The landscape designer must ensure that both the working drawings and specification are clear, informative and unambiguous, so that the quantity surveyor can produce an accurate bill of quantities, and so that all the contractors invited to tender can then put in fair figures for the work, which are easily comparable with those of their competitors.

3.13.2 BRITISH STANDARDS

1192 Construction drawing practice:
    Pt 1: 1984 *Recommendations for general principles*
    Pt 3: 1987 *Recommendations for symbols and other graphic conventions*
    Pt 4: 1984 *Recommendations for landscape drawings*
    Pt 5: 1990 *Guide to structuring computer graphic information.*

3.13.3 BS 1192 CONSTRUCTION DRAWING PRACTICE PT 4: 1984 RECOMMENDATIONS FOR LANDSCAPE DRAWINGS*

1. Scope

This part of this standard gives recommendations for the preparation of landscape drawings and schedules.

---

*All Figure and Table numbers in section 3.13.3 are those given in BS 1192.

Recommended symbols and abbreviations are included and used in a series of typical drawings. The appendices include summaries of information commonly needed and of methods and calculations commonly used when landscape drawings are being prepared.

2. Definitions

For the purposes of this part of BS 1192, the definitions given in BS 1192 Pt 1 and BS 3975 apply.

3. Stages of work

General

The main stages of a project and the drawings that may be needed are described in the following sections. Figures 1–10 give typical examples of drawings. Table 1 gives preferred scales recommended for the different types of drawing.

A brief will normally have been agreed before the commencement of the data collection stage.

Copies of original drawings should be taken for future reference purposes at various important stages: e.g. the obtaining of statutory approvals, and the bill of quantity stage.

For ease of communication and handling, the size of all drawings should be standardized (see also BS 1192 Pt 1). The degree of detail to be presented, together with the size of the site, will influence the scale of a drawing.

The same scale should be chosen for all mapping, where possible, to assist cross-referencing. When site mapping is covered by more than one drawing, a precise method of reference between adjacent sheets is essential. A key mosaic is often used.

Information commonly needed at the various stages of drawing preparation work is given in appendix A.

Data collection

This stage, incorporating landscape survey information, comprises the recording in an objective and systematic format of all available factual data related to a site, including sections, diagrams or schedules as needed. Aerial surveys can offer extensive opportunities for data collection in larger landscape projects. Records should state the date of origin and the source of data.

Appraisal of site data

The objective of this stage is to identify important site characteristics before a brief is evaluated. A full appraisal of site data will demand the presentation of value judgements, normally by way of plans supplemented by notes.

Interpretation of a brief

Spatial requirements of a brief should be converted into schematic diagrams that can assist in its interpretation.

Preliminary design stage

Schematic diagrams used for the interpretation of a brief and for the appraisal of size data come together as a design in sketch plan form whereby the feasibility of proposals can be tested. This can be one of a series of preliminary drawings, which may culminate in a preferred version. Such drawings should be consecutively numbered and dated, recording the logical progression to the preferred version.

The preferred version of preliminary drawings should be tested against the criteria governing the project concerned and against the proposals of other specialists.

Final design stage

The preferred version of preliminary drawings is developed during the final design stage. The method of presentation (to a client) will influence the selection of the type of drawings to be prepared.

Production drawing stage

Production drawings should be legible and concise, with sufficient information about a design to enable it to be implemented, and should be sufficiently detailed to permit the preparation of bills of quantities. They should not duplicate information contained in specification or schedules. Layout drawings should be capable of clear interpretation, and in complex layouts it may be necessary for separate plans to be produced for various work aspects, such as site clearance, underground services, hard and soft landscape works. The separate plans should be prepared upon copies of a single basic or master outline drawing of the layout. The master drawing should be retained as a basis for drawings that may be needed for additional aspects of the work, or for any major revision.

Record drawings

During the course of construction, a set of drawings should be kept up to date with all variations and changes in design, sizes or setting out that have occurred. Upon completion of construction, a set of 'as-built' drawings may have to be prepared.

Maintenance drawings

Drawings to show the locations and frequency of maintenance operations are essential, to ensure the implementation of the design intentions over a long period. This information can take the form either of a drawing together with specification notes, or of an annotated plan.

Graphical symbols for drawings are shown in Table 2.

Appendix A: Information commonly needed

Note: This appendix is to be read in conjunction with Table 1.

A summary of information commonly needed in the various stages of drawing preparation is given in Table 7 (see page 117)

Table 1 Preferred scales commonly used for different types of drawing

Job stage	Subject	1:50000	1:25000	1:10000	1:2500	1:1250	1:500	1:200	1:100	1:50	1:20	1:10	1:5
1 Data collection	Geology and soils†	●	●	●	●	●	●						
	Contours and gradients: (vertical intervals) 10.0 m	●	●	●	●								
	5.0 m			●	●	●							
	2.0 m				●	●	●						
	1.0 m					●	●	●					
	0.5 m					●	●	●					
	0.25 m						●	●	●				
	Hydrology and drainage	●	●	●	●	●	●						
	Vegetation, habitat factors, climate and microclimate					●	●	●	●				
	Structures, roads and services					●	●	●	●				
	Access and communications, archaeological and local history data, land use, pollution factors, legal and planning aspects	●	●	●	●	●	●						
	Social and demographic factors	●	●	●									
2 Appraisal	Appraisal of site data, appraisal of client's brief, feasibility study		●	●	●	●	●						
3 Design	Development sketch, preliminary sketch plan, final sketch plan		●	●	●	●	●	●					
	Layout				●	●	●	●					
	Details					●	●	●	●				
4 Production	Location plan	●	●	●	●	●							
	Site plan (general location)			●	●	●	●						
	Setting out plan, preparatory works, groundworks				●	●	●	●					
	Hard construction, services co-ordination, planting plan						●	●	●	●			
	Construction details									●	●	●	●
5 Post construction	Record and maintenance drawings				●	●	●	●					

* It should be noted that increasing use is being made internationally of scales 1:1000 and 1:2000.

† Certain data is only available in imperial scales, e.g. 6" : 1 mile

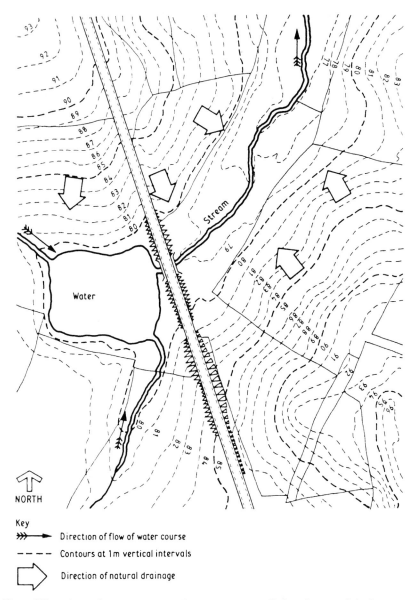

Key

>>>───▶   Direction of flow of water course

─ ─ ─ ─   Contours at 1m vertical intervals

⟰   Direction of natural drainage

Note: When based on a contoured survey map, relief and natural drainage plans should indicate heights of every fifth contour picked out in a heavier line. Values of contours should be printed 'up the slope'. Plans should also indicate all watercourses, lakes etc. and direction of flow, together with direction of natural drainage.

Fig. 1 Relief and natural drainage.

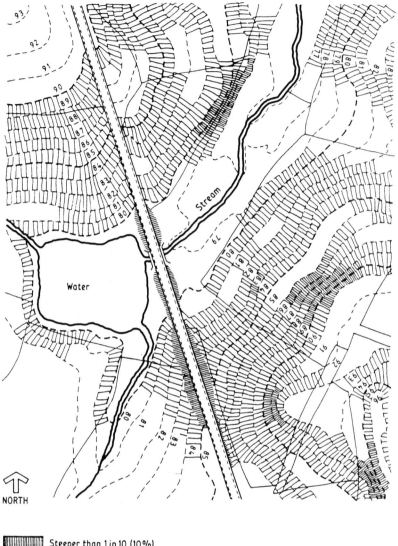

Steeper than 1 in 10 (10%)
1 in 10 to 1 in 30 (10% to 3.3%)
Flatter than 1 in 30 (3.3%)

Note: Slope analysis plans, based on contour surveys, should indicate by shading or hatching slopes of varying degrees of steepness. The choice of the ranges of slope indications will vary depending on the site and its proposed use.

Fig. 2 Slope analysis.

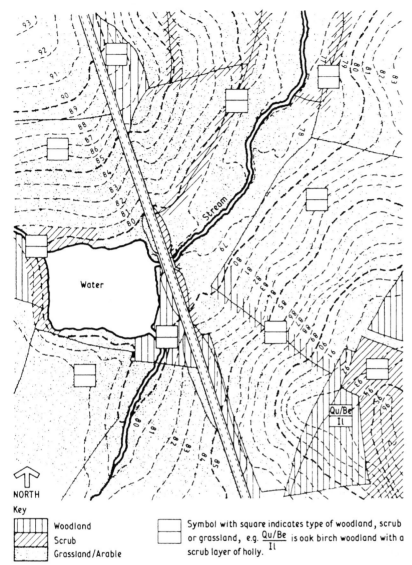

Key

	Woodland		Symbol with square indicates type of woodland, scrub
Scrub		or grassland, e.g. $\frac{Qu/Be}{Il}$ is oak birch woodland with a
Grassland/Arable		scrub layer of holly.

Note: Vegetation surveys are prepared following site investigation, and the use of aerial survey photographs when necessary. They should show at least three major plant divisions, e.g. woodland, scrub/shrub and grassland/arable, indicated by hatching. Further subdivision may be necessary. Within this broad category each area should be defined and catalogued according to the subtype, i.e. variety of dominant plants, presence of sublayers of vegetation and presence of other species.

Fig. 3 Vegetation survey.

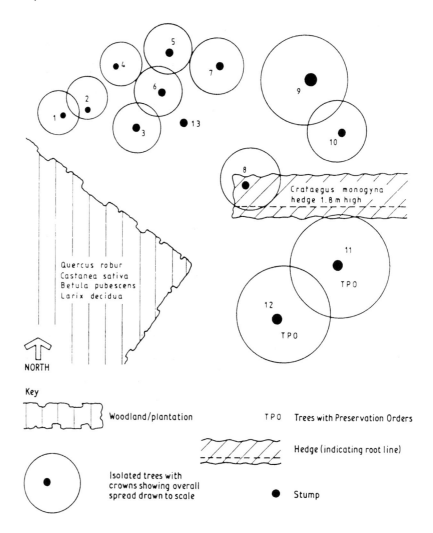

Key

Woodland/plantation

TPO   Trees with Preservation Orders

Hedge (indicating root line)

Isolated trees with
crowns showing overall
spread drawn to scale

Stump

Note: In limited surveys the scheduled information can be added to the drawings in the form of notes.

Fig. 4 Tree survey.

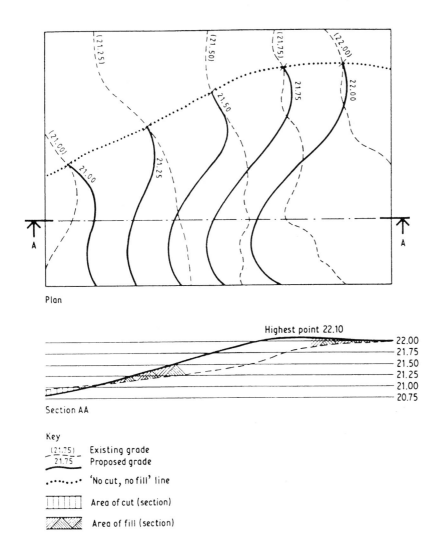

Plan

Section AA

Key

(21.75) Existing grade
21.75 Proposed grade

.......• 'No cut, no fill' line

|||||| Area of cut (section)

Area of fill (section)

Fig. 5 Grading.

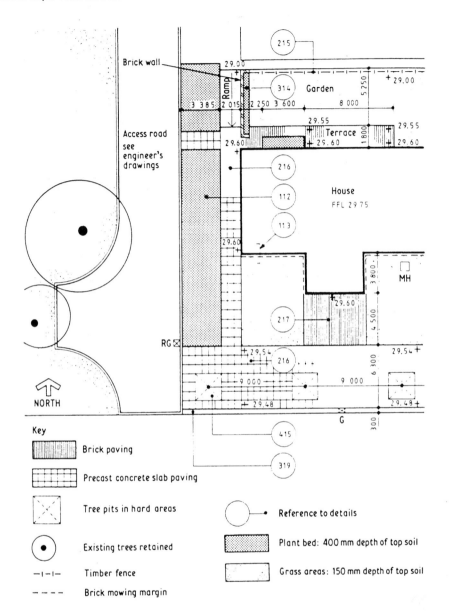

Fig. 6 Hard landscape drawing.

Note: On more complex drawings sizes can be shown on separate setting-out drawings.

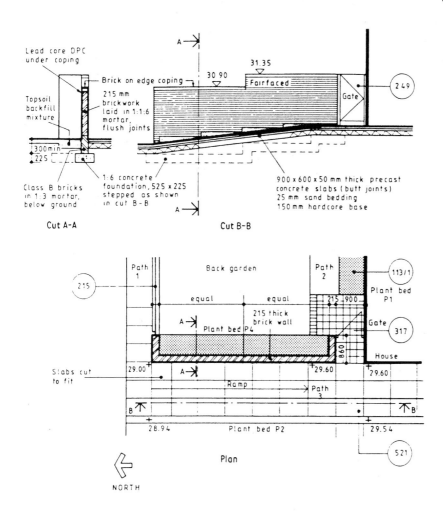

Fig. 7 Hard landscape detail.

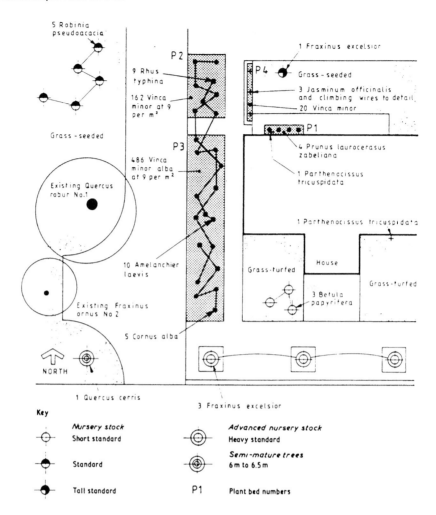

Fig. 8 Planting plan.

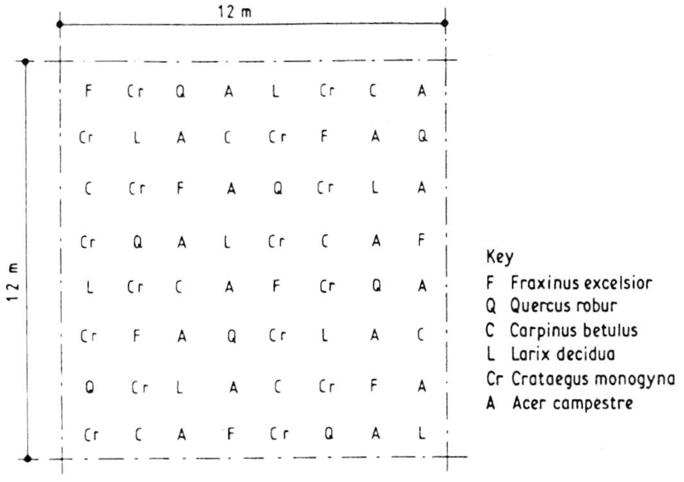

Note: This figure is based upon the use of a convention employing letters for the identification of species. However, a convention using numbers or symbols can be employed instead of letters.

Fig. 9 Layout of grid for woodland planting.

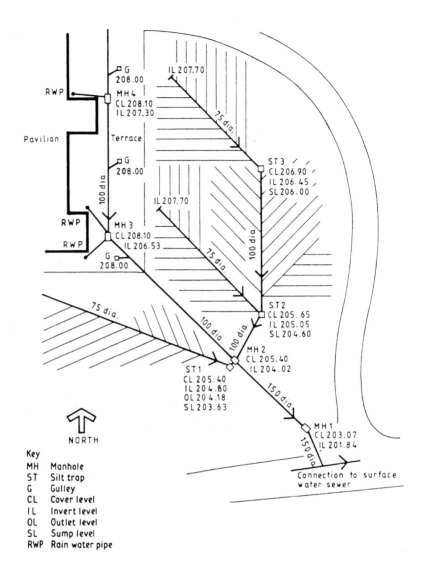

Key

MH	Manhole
ST	Silt trap
G	Gulley
CL	Cover level
IL	Invert level
OL	Outlet level
SL	Sump level
RWP	Rain water pipe

Notes:

1. An interceptor may be required if discharge is to a combined foul/surface water sewer.
2. Pipe sizes and material should be shown.
3. Show gradients to give a velocity of between 0.3 m/s and 0.6 m/s as a minimum, but on large, level sites flatter gradients may have to be used to permit a gravitational outfall to the available watercourse or sewer.

Fig. 10 Subsoil drainage.

Table 2 Graphical symbols and abbreviations

	Item	Symbol	Comments
2.1	Grass-existing	Grass - existing	
2.2	Grass-seeded	Grass - seeded	
2.3	Grass-turfed	Grass - turfed	
2.4	Grass area with bulbs, etc.	Drift	Give numbers and species on drawing or schedule
2.5	Plant bed (hard landscape drawing)		Area shaded if drawing technique allows
2.6	Plant bed (soft landscape drawing)	P4	Each plant bed numbered for identification. Ground cover and mass planting not normally shown with a dot, specify numbers and rate/m²
2.7	Subdivision of plant bed		Subdivision can be numbered
2.8	Mulch-section		
2.9	Shrubs		
2.10	Shrubs of same species		Give number and species
2.11	Trees of same species		Give number and species

cont.

Table 2 continued

Item	Symbol	Comments
2.12 Hedge existing		
existing, to be removed		
proposed (a)		
proposed (b)		Alternative when plant positions are shown
2.13 Existing trees		Spread shown to scale
2.14 Proposed trees		Refer to symbols 2.20 when size of tree is to be differentiated by symbol
2.15 Tree assessment very good	A	Give details of species, size, etc. on drawing or schedule
good	B	
fair	C	
dead/poor	D	
2.16 Existing trees to be protected temporarily during site works		Other planted areas requiring protection should be shown similarly
2.17 Existing or new trees to be protected permanently		Other planted areas requiring protection should be shown similarly
2.18 Tree pit		Drawn to scale, 'hard' areas only
2.19 Trees to be removed		Show spread to scale. Give details of size and species on drawing or schedule

cont.

Table 2 continued

2.20   Symbols for proposed trees

Basic symbol	*General symbol for transplants, nursery stock, advance stock and semi-mature trees*	*Detailed symbol when tree sizes are to be shown*	
	$+$	$+$	*Transplants and whips*
			Nursery stock
			Short standard: ss
Outline to be thicker than symbol for existing trees			Half standard: hs
			Light standard: ls
			Standard: s
			Tall standard: ts
			Selected standard: sel s
			Advanced nursery stock
			Heavy standard: hs
	Add abbreviations to indicate sizes if required		Extra heavy standard: hs
			Selected heavy standard: shs or sel. hs
			Semi-mature trees
			May be divided into:
			small: ssm
			medium: msm
			large: lsm

cont.

Table 2 continued

Item	Symbol	Comments
2.21 Climbers		
2.22 Woodland		
existing		
proposed		Draw grid to scale when practical
2.23 Levels		
existing		
plan	(25.5) +	
sect/elev	(25.5) ✗	
new		
plan	25.5 +	
sect/elev	25.5 ✗	
2.24 Existing contours + level	(25.5)	
2.25 Proposed contours + level	25.5	
2.26 Existing surfacing areas to be removed		Indicate material to be removed
2.27 Area of cut-section		
2.28 Subsoil fill-section		
2.29 Top soil fill-section		
2.30 Hardcore fill-section		

cont.

Table 2 continued

	Item	Symbol	Comments
2.31	Embankment, plan	△ △ △ △ △ △ △ △ ↑ ▽ ▽ ▽ ▽ ▽ ▽ ▽ ▽ Fall ↓	Use if contours are not shown. Arrow should extend full width of embankment
2.32	Cutting, plan	▽ ▽ ▽ ▽ ▽ ▽ ▽ ▽ ↓ △ △ △ △ △ △ △ △ Fall ↑	Use if contours are not shown. Arrow should extend full width of cutting
2.33	Large unit paving, plan, e.g. precast concrete slabs		Show exact sizes on drawings drawn to scales larger than 1:200
2.34	Small unit paving, plan, e.g. bricks		Show exact sizes on drawings drawn to scales larger than 1:200
2.35	Cobbles, plan		
2.36	Steps	1 2 3 4	Arrow should point up
2.37	Ramp/gradient	1:12	Arrow should point-up
2.38	Fall (surfaces)	1 50 fall	Arrow should point down
2.39	Fence	—I——I——I	One type on drawing
		—II——II——II	Second type on drawing
2.40	Gate (a)		
			(a) and (b) are alternatives
	(b)		

cont.

Table 2 continued

Item	Symbol	Comments
2.41 Stile		
2.42 Bench seat	B S	
2.43 Bench	B	
2.44 Bollards fixed	B	
removable	R B	
hinged	H B	
lighting bollard	L T B	
2.45 Litter bins	L B	Draw shape of bin required
2.46 Land drain mole	M	
piped		Arrow indicates fall. Indicate size, gradient and material
2.47 Open gutter		Arrow denotes direction of fall
2.48 Gulley	G	
2.49 Road gulley	R G	
2.50 Manhole	M H	
2.51 Cross-referencing	121 / 4	Drawing number (as appropriate) Detail number (as appropriate)

Table 7   Information commonly needed

Subject	Basic information
DATA COLLECTION	
Climate and microclimate	Precipitation (monthly or annual); wind rose; temperatures; dates of first and last frosts; drainage of cold air across site; frost pockets; shelter; shadow projections.
Geology and soils	Solid geology; drift geology; trial hole positions; trial hole sections (including depth of water table); nature of soil, i.e. natural or made up.
Hydrology and drainage	Water courses (including those dry in summer); lakes, ponds, etc; direction and flow of surface water; land drain systems; flood plains and temporary flooding.
Vegetation	Trees and important shrubs; Preservation Orders or covenants; ground vegetation.
Habitat factors	Type of habitat; associated flora and fauna; existing management arrangements (notes).
Land use	Purpose for which land is used at the time of survey, supplemented by notes of any factors that may affect use.
Use survey	Social and economic factors (notes); demographic factors.
Stuctures, roads and services	Identification of buildings, roads, car parks, hard standings, footpaths, paved areas, walls, fences, gates; foul and surface water drains; manholes (invert and cover levels); gas, water, electricity, telephone and TV services (overhead or underground); pipelines; notes on materials, including those suitable for reuse; underground workings.
Archaeological and local history data	Information from site or records of previous occupation or use of site.
Access and communications	Pedestrian and vehicular routes to public roads, footpaths, bus stops, railway stations, shops, schools, etc.; restrictions on access.
Pollution factors	Constraints on or off site that can affect future use, e.g. noise, chemical pollution, etc.
Legal and planning aspects	Land ownership; boundary ownership; rights of access, light, etc.; easements and way-leaves; Structure Plan designation; road improvement lines; Tree Preservation Orders; areas of outstanding natural beauty; sites of special scientific interest; conservation areas; scheduled buildings and ancient monuments; planning application.
Special data	Zones of visual influence (e.g. ISOVISTS); views into and out of site.
APPRAISAL OF SITE DATA	Evaluation and synthesis of data in survey documents, including dominant features and critical factors.
INTERPRETATION OF BRIEF	
Appraisal of client's brief	Evaluation of client's requirements to identify critical and/or contradictory factors.
Feasibility study	General zoning of the main areas to determine the suitability of site for the client.
Development sketches based on survey drawings	Reconciliation of client requirements to site conditions, including indication of the best disposition of the main elements of the proposed development.

cont.

Table 7   continued

Subject	Basic information
PRELIMINARY DESIGN STAGE	
Preliminary drawings based on survey drawings	Existing and proposed land form, buildings, paths and other hard areas, grass areas, trees, shrubs, hedges etc., site drainage and water areas. Illustrations of any special features.
FINAL DESIGN STAGE (Presentation drawings)	
Final layout	Final zoning of the main elements to form the basis of future development.
Plans	Existing and proposed contours or grid levels, buildings, paths and other hard areas, grass areas, trees, shrubs, hedges etc., site drainage, water areas and services.
Details	Plans and/or sections to show features of the overall layout to a larger scale.
PRODUCTION DRAWING STAGE	
Location plan	Location of site; means of access.
Site plan (general location)	Existing and proposed buildings and other features: paths, roads, hardstandings, etc., trees shrubs, hedges, beds etc., grass areas, water areas.
Setting out	Reference grids; main setting-out dimension.
Preparatory work	Co-ordinating grid; areas available to contractor; trees and other features to be protected; trees to be prepared for moving; trees, hedges and other plants to be removed/transplanted; materials to be salvaged for re-use; existing topsoil to be stripped and stacked for re-use; location of spoil heaps; diversion of services and drainage system; existing grass and plants to be maintained during construction.
Groundworks	Co-ordinating grid; existing and proposed contours (or grid of levels); depths of topsoil and fill required; land drainage systems.
Hard construction	Co-ordinating grid; hard surfaces; walls; fences; gates; kerbs and edgings; outline of soft landscape; reference to detail drawings; cross-reference to drawings for building and civil engineering work.
Services co-ordination	Co-ordinating grid; levels of services; surface levels; setting-out dimensions; outline of hard landscape; outline of soft landscape; details of irrigation systems.
Planting	Co-ordinating grid; grass areas (seeded and turfed); tree planting (number and species); shrubs, hedges and other planting (number and species); plant schedule (or notes on drawing); outline of hard landscape.
Construction details	Enlargement plans, sections and elevations as necessary to show details of construction.
POST CONSTRUCTION	
Record drawings	Accurate working record of the scheme as constructed, usually based on drawings amended as necessary, including record of species planted
Maintenance drawings	Thinning of initial planting; height and spread of hedges; maintenance standards of grass, trees, shrubs and other planting; location of host points; particulars of irrigation systems.

Guide to specification 4

D11 Soil stabilization: flexible materials

GENERAL GUIDANCE

This section deals with the prevention of wear on grassed areas and the prevention of erosion on shallow slopes by mesh or grid material and geotextiles.

With the increasing interest in walking, erosion is a hazard to long-distance footpaths as well as to the shorter trails to local beauty spots; this is a problem that the landscape designer may well be asked to solve. Where the paths slope, they must be divided into gentle slopes and steps by timber or concrete baulks; if the sides of the path have eroded they must be addressed before the path is renewed (refer to the section on erosion control below, or the section on retaining walls).

Earth stabilization

This is usually provided by means of plastic and wire mesh buried in the ground to prevent wear and tear on turf. Where grassed paths have worn away, one solution may be to excavate the pathway to a depth of 100–150 mm and lay a plastic/wire mesh filled with gravel (see specification below).

Erosion control on slopes

Light grass concrete blocks, plain or grass-seeded mats or thin mattress gabions are all suitable as erosion control systems on slopes that are shallow enough to retain vegetation. The permissible slope depends also on the natural inclination of the subsoil. Ready-seeded mats filled with natural or plastic fibres and nutrients are available for erosion control, and mesh-reinforced turf can be obtained that gives an instant grass surface to the slope.

Light grass concrete blocks for erosion control are usually made with interlocking nibs to prevent slippage; they are designed to have the maximum area available for grass, as they are not intended to take vehicular loading. Where the slope is steep it may be necessary to pin the blocks every metre with steel pins.

Thin flexible 'mattress' gabions filled with rock or shingle can be used for erosion control on large smooth slopes where mechanical handling is available, but they are better suited to coastal and rock control where grass cannot be established easily and where rough vegetation or exposed shingle is acceptable.

Geotextiles

Geotextiles (also known as filter fabrics) are thin woven or non-woven membranes, whose function is to separate layers of soil, aggregate or granular material from each other while still allowing free drainage of water. They range from close-woven material, which prevents very fine particles from passing through, to open mesh, which will restrict only large aggregate. Modern geotextiles are designed with specific water transmittance properties, and they can be specified to transmit × litres per second. Some typical uses are:

- to prevent fine soil particles from blocking land drains – the pipes are 'filter-wrapped' with the geotextile;
- to prevent sand or fine soil from permeating hardcore foundations to roads or paving;
- to protect the aggregate drainage layer behind retaining walls from being clogged by fine soil;
- to separate soil in planters from the aggregate under-drainage layer;
- to separate a granular surface such as bark from the foundation layer.

BRITISH STANDARDS

1377 Pt 1 : 1990	Methods of test for soils for engineering purposes: General requirements and sample preparation
1377 Pt 2 : 1990	Soil tests for civil engineering. Classification requirements
1377 Pt 4 : 1990 & AMD	Soil tests for civil engineering. Compaction related tests
1377 Pt 5 : 1990 & AMD	Soil tests for civil engineering. Compressibility, permeability, durability
1377 Pt 9 : 1990 & AMD	Soil tests for civil engineering. In situ tests
3882 : 1994	Specification for topsoil (bought and sold)
4428 : 1989 & AMD	Code of practice for general landscape operations
5930 : 1981	Code of practice for site investigations
6031 : 1981	Code of practice for earthworks
6906 Pt 4 : 1989	Methods of test for geotextiles
8002 : 1994	Code of practice for earth retaining structures

DATA

Typical erosion control mats

- 10 mm thick open texture unfilled
- 18 mm thick open texture unfilled

- 20 mm thick open texture unfilled
- 15 mm thick reinforced open texture unfilled
- 17 mm thick reinforced open texture unfilled
- 18 mm thick reinforced open texture unfilled
- 20 mm thick filled with bitumen bonded aggregate
- 10 mm thick filled with integral turf or flora, factory grown to order

Geotextile fabric

Geotextile filter fabric is made in a range of strengths to suit varying CBR percentages (see Q20 *Bases and sub-bases to roads and paving*) and water transmission capacities. The fabric should comply with BS 6906 Pts 4 and 6 for resistance to perforation, though aggregate should be placed in small quantities and not tipped in bulk.

The flow of water through geotextile fabrics is accurately calculated, and ranges from 10 to 30 l/m^2 per second. Higher rates up to 500 l/m^2 per second are obtainable with open-mesh fabrics.

Typical light grass concrete blocks

600 mm × 400 mm × 110 mm interlocking 40% grass area

Typical erosion control seed mixtures

30% Chewing's fescue	25% Perennial ryegrass
30% Creeping red fescue	30% Creeping red fescue
30% Smooth-stalked meadow grass	35% Smooth-stalked meadow grass
10% Browntop bent	10% Browntop bent

OUTLINE SPECIFICATION

Erosion control

a. *Bring bank to grade, remove all roots, stones and debris over 150 mm in any dimension, excavate fixing trenches 300 mm × 300 mm deep at top, base and edges of fabric.*

b. *Lay 20 mm thick open-texture erosion control mat with 100 mm laps, fixed with 8 × 400 mm steel pegs at 1.0 m centres, backfill fixing trenches with coarse aggregate.*

c. *Sow with low maintenance grass at 35 g/m^2, spread imported topsoil to BS 3882 General Purpose grade 25 mm thick incorporating medium-grade sedge peat at 3 kg/m^2 and Growmore fertilizer at 70 g/m^2, water lightly.*

d. *Slope to be watered regularly in dry weather until vegetation is established.*
    for grass seed mixtures   *Q30*

Alternative specification

a. *Bring bank to grade, remove all roots, stones and debris over 150 mm in any dimension.*

b. *Lay specified manufacturer's biodegradable pre-seeded erosion control mat Type XXX, fixed with 6 × 300 mm steel pegs at 1.0 m centres, water lightly in dry weather.*

c. *Slope to be watered regularly in dry weather until vegetation is established.*

Light grass concrete blocks

a. *Bring bank to grade, remove all roots, stones and debris over 150 mm in any dimension, lay 100 mm of 40 mm aggregate drainage layer, lay geotextile fabric 20 l/m²/s.*

b. *Lay specified manufacturer's grass interlocking concrete blocks Type XXX, all in accordance with manufacturer's instructions, fill three quarters full with topsoil to BS 3882 General Purpose grade incorporating medium-grade sedge peat at 3 kg/m² and Growmore fertilizer at 70 g/m².*

c. *Seed with grass seed mixture 30% Chewing's fescue, 30% creeping red fescue, 30% smooth-stalked meadow grass, 10% browntop bent at 20 g/m², and fill to top with screened topsoil as above.*

d. *Slope to be watered regularly in dry weather until vegetation is established.*

for grass concrete   Q25
for grass seed mixtures   Q30

Soil reinforcement

a. *Bring surface of subgrade to given levels and falls, remove all roots, stones and debris over 150 mm in any dimension, spread topsoil from dump on site and consolidate lightly.*

b. *Lay specified manufacturer's soil-reinforcing fabric Type XXX with 100 mm laps, spread 25 mm imported topsoil to BS 3882 General Purpose grade incorporating Growmore fertilizer at 70 g/m².*

c. *Seed with 40% Chewing's red fescue, 10% smooth-stalked meadow grass, 50% perennial ryegrass mixture at 50 g/m².*

for grass seed mixtures   Q30

Footpath

a. *Excavate footpath to 100 mm deep, bring surface of subgrade to given levels and falls, remove all roots, stones and debris over 150 mm in any dimension, dig out weak areas by hand, and backfill with approved excavated material or DoT Type 2 fill.*

b. *Lay 150 mm consolidated hardcore to given levels, blind with sand, and lay specified manufacturer's soil-reinforcing fabric Type XXX with 100 mm laps dug into ground 300 mm each side of path.*

 c. *Lay 75 mm base layer of 0–28 mm free-draining gravel, lay top layer of*
  moist graded self-binding gravel compacted to 50 mm, water, roll and
  repeat watering and rolling twice within 7 days of laying.

 d. *Watering and rolling to be repeated as necessary until completion of*
  maintenance period.

for gravel	*Q23*
for grass concrete	*Q25*
for grass seed mixtures	*Q30*

D11 Soil stabilization: earth-retaining structures

GENERAL GUIDANCE

This section covers low retaining walls, crib walling and prefabricated retaining wall units. Structural retaining walls, sea defences, waterway control, and large embankments for highways are the province of the civil or structural engineer, and are not dealt with in this book.

Prefabricated retaining wall units

This description covers precast concrete retaining wall units for holding back earth up to 2–3 m high, also heavy interlocking precast concrete units, which are handled by machine. The design of structural retaining walls is very complicated, and should not be undertaken without full understanding of the calculations involved. These are set out in BS 8002 : 1994 *Code of practice for earth retaining structures*.

 Lighter prefabricated retaining walls come in two types. The first are L-shaped panels approximately 1 m wide in varying heights from 1 m to 3.6 m high. Each unit weighs from 300 kg to over 2 tonnes, and all have to be carefully positioned by crane. Though they are used to retain earth, they present a plain vertical concrete face that the landscape designer may not find acceptable, but they are useful as enclosures for the storage of loose material such as compost and gravel, or as bunds around tanks containing flammable liquids.

 The second type are made up of shaped walling units that interlock to form strong dry-wall construction. Some of these units weigh up to 35 kg, and require mechanical handling. Unless the landscape designer is working on a large project with a civil engineer, the smaller lighter units weighing 10–15 kg, suitable for low walls and terraces, are more likely to be specified. On most of the patented systems, walls can be curved and battered to form serpentines and large circles; many have special coping units, and are available in varying colours and finishes. All are free draining, and spaces can be left between the units to allow for trailing plants.

Crib walling

The walls discussed are capable of being used to considerable heights, but unless the landscape designer is working with a civil engineer, it is advisable not to use them in walls above 2 m. Crib walls are made of timber or concrete units in lengths from about 500 mm to 2000 mm, which lock together to form the crib against a battered and drained embankment. All must be built off a firm substrate, and except for very low terrace walls most manufacturers recommend concrete footings. Whatever system the landscape designer specifies, the manufacturer's technical advice should be sought before designing the wall; most walls require reinforcement in the foundations where the substrate is liable to differential settlement. Crib walls should not be used where slopes are liable to slippage.

All crib walls make attractive terrace and garden walls when planted up, either on the flat tops of terraces or on the battering by making the final layer of backfill of fertilized topsoil to the correct depth and type for the planting specified. Properly designed and constructed timber crib walls should have a working life of up to 50 years provided the timber has been pressure impregnated after manufacture. Precast concrete units have an even longer life.

Brick and stone walls

Brick or stone retaining walls that support earth and overburden more than 1.5 m high require accurate calculation if they are to carry the pressures on them safely, and a civil engineer should be consulted where there is any doubt about the structural stability of the wall. BS 8002 gives calculations for simple retaining walls. The foundation for a retaining wall must be designed in accordance with the calculated loads of the retained earth and the weight of the wall. Larger retaining walls are usually built to a batter of 1:6, which provides better resistance to overturning. Good foundations taken well below the lower level of the wall are essential.

Many retaining-wall failures are caused by insufficient drainage of the ground behind the wall when the water pressure builds up. Traditionally, weepholes in the form of clay agricultural pipes or open joints at 900 mm intervals in both directions were provided, but it is now more common to insert a proper land drainage system behind the wall. This consists of 200 mm of coarse aggregate with a land drain at the base, often supplemented by a vertical fin drain of geotextile fabric, which is also wrapped round the pipe. Vertical retaining walls can be used for planters one and a half bricks thick and up to 900 mm high that are designed with good drainage at the back.

Revetments

A revetment is by definition a facing to a wall, but the term is used here to describe the plain banking to shallow waterways and embankments. Revetments consist of grass concrete blocks or mesh mattress gabions of galvanized metal or plastic-coated material filled with stones or soil. Gabions do not provide the smooth grassed slope attainable with grass concrete blocks, and coarser vegetation should be specified. Revetments can also be constructed with rot-proofed or naturally rot-resistant timber spiles driven into the bed of the watercourse; they can be strengthened with broken stone backing or cross timbers.

Grass concrete

Grass concrete blocks are available as light units for mild conditions (see *Erosion control*) or heavier units that can stand up to water action. These blocks are usually made with interlocking nibs to prevent movement from slippage or wave action. They can be grassed with suitable species to help stabilize the embankment, and to present a more natural appearance than plain concrete.

Gabions

Gabions are galvanized wire crates of various sizes that are delivered flat and filled on site with local broken stone or hard broken rock. At the maximum height of 2–3 m described here they are unlikely to need more than a firm level substrate as a foundation; where the bearing is softer, a row of gabions should be laid flat to project beyond the wall in the manner of a concrete foundation. As gabions are free draining, the slope cut back during construction can be backfilled without the need to drain the back of the wall.

Small gabions are mainly used at the edges of river banks or ponds, where they can be sealed with hot bitumen to make the bank waterproof, or hydroseeded to establish vegetation, but they can make attractive informal retaining walls when terraced and enhanced by careful planting.

Mattress gabions are delivered as flat packs and filled on site with pebbles, broken stone or suitably sized quarry waste; the hard fill is then topped with fertilized topsoil and seeded with indigenous grasses and plants.

BRITISH STANDARDS

882 : 1983	Aggregates from natural sources for concrete
1377 Pt 1 : 1990	Methods of test for soils for engineering purposes. General requirements and sample preparation
1377 Pt 2 : 1990	Soil tests for civil engineering. Classification requirements
1377 Pt 4 : 1990 & AMD	Soil tests for civil engineering. Compaction related tests
1377 Pt 5 : 1990 & AMD	Soil tests for civil engineering. Compressibility, permeability, durability
1377 Pt 9 : 1990 & AMD	Soil tests for civil engineering. In situ tests
3268 Pt 5 : 1977	Code of practice for the structural use of constructional timber
3882 : 1994	Specification for topsoil (bought and sold)

3921 : 1985	Clay bricks
4428 : 1989 & AMD	Code of practice for general landscape operations
4729 1990	Dimensions of bricks of special shapes and sizes
5224 1995	Masonry cement
5328 Pt 1: 1991 & AMD	Guide to specifying concrete
5328 Pt 2: 1991 & AMD	Methods for specifying concrete mixes
5628 Pt 1 : 1978 (1985)	Code of practice for use of masonry. Structural use of unreinforced masonry
5930 : 1981	Code of practice for site investigations
6031 : 1981	Code of practice for earthworks
6906 Pt 4 : 1989	Methods of test for geotextiles
8002 : 1994	Code of practice for earth retaining structures

DATA

Batter

The batter is specified as 1 horizontal unit in × vertical units.

Durability of bricks

FL   Frost resistant with low salt content
FN   Frost resistant with normal salt content
ML   Moderately frost resistant with low salt content
MN   Moderately frost resistant with normal salt content

Mortar mixes

Mortar group	Cement:lime:sand	Masonry cement:sand	Cement:sand with plasticizer
1	1:0–0.25:3		
2	1:0.5:4–4.5	1:2.5–3.5	1:3–4
3	1:1:5–6	1:4–5	1:5–6
4	1:2:8–9	1:5.5–6.5	1:7–8
5	1:3:10–12	1:6.5–7	1:8

Group 1: strong inflexible mortar
Group 5: weak but flexible

All mixes within a group are of approximately similar strength. Frost resistance increases with the use of plasticizers. Cement:lime:sand mixes give the strongest bond and greatest resistance to rain penetration. Masonry cement to BS 5224: 1995 covers a wider range of cement types and defines three strength classes. Non-air-entrained cements are included.

Table 4.1 Design of slopes in rock cuttings and embankments (by courtesy of the British Railways Board) (BS 6031 : 1981)

Type of rock	Cuttings: safe slopes (angles referred to the horizontal)	Embankments: angles of repose (angles referred to the horizontal)	Remarks
Sandstone (sedimentary), strong, massive, of considerable geological age, e.g. Old Red Sandstone (Devonian); Blue Pennant Grit, Millstone Grit (Carboniferous); Bunter Sandstone (Triassic).	Mainly vertical but may in places require cutting back to 70°	38°–42°	Very resistant to weathering although a block of stone may occasionally be loosened by frost action. Care should be taken where weak beds, e.g. shales, underlie the strong rocks, as the effects of differential weathering may lead to undermining; in such cases protection from weathering should be afforded to the weak beds.
Sandstone, weak, inferior, usually thinly bedded and more recent geological age, Hastings Beds (Lower Cretaceous); Upper Greensand (Cretaceous)	50°–70°	33°–37°	Fairly resistant to weathering, depending on the degree of hardness and the nature of the cementing material. Stone with a silicious cement resists weathering better than one with a calcareous or ferruginous cement.
Shales, e.g. Ludlow Shale (Silurian); Shales of Yoredale Series and Coal Measures (Carboniferous); Shales of Lower and Upper Lias (Jurassic).	45°–60°	34°–38°	Resists weathering to a considerable degree, though the face tends to flake in small fragments. Softening may also occur in time. Particular attention should be given to the relation between the slope of the cutting and the dip of the strata.
Marls, e.g. Keuper Marl (Triassic); Chalk Marl (Cretaceous)	55°–70°	33°–36°	Resists weathering well if care is taken with drainage of cuttings; growth of vegetation should be assisted. These rocks are liable to soften with time.
Limestone, strong, massive, of considerable geological age, e.g. Carboniferous limestone (Carboniferous); Magnesian limestone (Permian)	Mainly vertical 70°–90°	38°–42°	Resists weathering well but frost action and weathering of joints exposed in cuttings tend to loosen large blocks, which therefore have to be removed at intervals.
Limestone, weaker, including oolitic limestones, e.g. Portland Beds, Coral Rag, Lias (Jurassic)	Mainly vertical 70°–90°	38°–42°	These rocks vary considerably in their ability to resist weathering; the massive Portland Beds are usually very resistant, whilst the Coral Rag and Lias may weather badly; causing slips in overlying strata which may require apron walls of similar protective measures.
Chalk (lower, middle, upper) (Cretaceous subdivisions)	45°–80°	33°–36°	The lower chalk is generally massive, homogeneous and therefore more resistant to weathering than the middle chalk; it may thus in many cases permit a higher safe slope angle in cuttings. The upper chalk is comparatively weak and more fractured; it therefore requires a lower angle of safe slope. Frost action produces much fragmentation, and the growth of vegetation should be encouraged as an aid to binding the surface. The general effects of weathering are towards producing a 45° slope.
Igneous rocks, e.g. granite, dolerite, basalt, andesite, gabbro	80°–90°	37°–42°	Weather-resisting qualities excellent. Cuttings may be left almost vertical after removal of loose fragments. Some basalts may exfoliate to a slight extent after long periods of exposure to weather.
Metamorphic rocks, e.g. gneiss, quartzite, schist, slate	60°–90°	60°–90°	Gneiss and quartzite generally exhibit similar properties to granite or hard sandstone, weather-resisting qualities are excellent, and slopes may be left almost vertical. Gneisses are often severely contorted. Schist may vary from strong pelitic material through gradations to talc schist or mica schist which may be weak, approaching the consistency of shale. The weaker schists may weather to a considerable extent and tend to slide along the surface of schistosity. Schists are often severely contorted and may require variations in the safe slope angle in cuttings to allow for the local difference of the schistosity. Slate is generally strong fine-grained rock with excellent weathering qualities, although the effects of weathering may tend to cause sliding along the cleavage planes. Both slates and schists are subject to cleavage along planes generally at a high angle to the bedding plane and may tend to slide not only in the direction of the cleavage or schistosity but also down the dip of the original bedding planes. This is especially dangerous where the dip of either of these surfaces is in the same direction as the slope of the face of the cutting.

Class	28-day strength N/mm2	Air entrainment (X = non-air-entraining)
MC 5	5–15	yes
MC 12.5	12.5–32.5	yes
MC 12.5X	12.5–32.5	no
MC 22.5X	22.5–42.5	no

Masonry cement equals ordinary Portland cement plus a fine neutral mineral filler; it is easier to work and is more frost resistant than a cement:lime mixture. Masonry cement should never have lime added to it, and should not be used on walls subject to heavy loading.

Concrete mixes to BS 5328 Pt 1 & AMD

Application	Standard mix	Designated mix	Slump
Mass concrete fill and blinding	ST 2 (C10P)	GEN 1	75 mm
Mass concrete foundations in non-aggressive soils	ST 2 (C10P)	GEN 1	75 mm
Trench fill foundations in non-aggressive soils	ST 2 (C10P)	GEN 1	125 mm
Foundations			
in Class 2 sulphate soils	–	FND 2	75 mm
in Class 3 sulphate soils	–	FND 3	75 mm
in Class 4A sulphate soils	–	FND 4A	75 mm
in Class 4B sulphate soils	–	FND 4B	75 mm
Kerb bedding and haunching	ST 1 (C7.5P)	GEN 0	10 mm

GEN concretes with low cement content may not be suitable for good cast and direct finished surfaces.

Traditional concrete mixes

Application	Name of mix	Cement: aggregate: sand (parts by volume)
Setting posts, filling weak spots in ground, backfilling pipe trenches:	one, two, three or one to five all-in	1:2:3 or 1:5
Paving bases, bedding manholes, setting kerbs,		

Traditional concrete mixes *continued*

Application	Name of mix	Cement: aggregate: sand (parts by volume)
edgings, precast channels:	one, two, four	1:2:4
	or	or
	one to six all in	1:6
In situ concrete paving:	one, one & half, two & half	1:1.5:2.5

'All-in' is mixed aggregate as dug from the pit.

Typical precast retaining wall units

Unit (mm)	Maximum loading (earth) (kg/m³)
1000 high × 1000 wide	1600
1500 high × 1000 wide	1600
2000 high × 1000 wide	1600
2400 high × 900 wide	1600
3000 high × 1000 wide	1600
3600 high × 900 wide	1600

Typical gabions

All dimensions in mm.

Box gabions
1000 deep × 1000 high × 2000 wide
1000 deep × 500 high × 2000 wide
1000 deep × 1000 high × 3000 wide
1000 deep × 500 high × 3000 wide
1000 deep × 1000 high × 4000 wide
1000 deep × 500 high × 4000 wide
1000 deep × 1000 high × 1500 wide

Mattress gabions
2000 deep × 170 high × 6000 wide
2000 deep × 230 high × 6000 wide
2000 deep × 300 high × 6000 wide

Typical grass concrete blocks

500 mm × 300 mm × 110 mm thick with small slots for vegetation
600 mm × 400 mm × 125 mm thick heavy duty with 56% grassed area

Typical seed mixtures for grass concrete

30% Chewing's fescue	25% Perennial ryegrass
30% Creeping red fescue	30% Creeping red fescue
30% Smooth-stalked meadow grass	35% Smooth-stalked meadow grass
10% Browntop bent	10% Browntop bent

Suggested planting for crib walling and gabions

Clematis spp.	*Lonicera* spp.
Cotoneaster horizontalis	*Parthenocissus* spp.
Cytisus praecox	*Potentilla* spp.
Genista spp.	*Rosmarinus officinalis*
Hedera spp.	*Thymus vulgaris*
Jasminum nudiflorum	*Ulex europaeus*
Juniperus spp.	*Vinca* spp.
Lavandula spp.	*Wisteria sinensis*

BS 3882 : 1994 topsoil quality

Topsoil grade	Properties
Premium	Natural topsoil, high fertility, loamy texture, good soil structure, suitable for intensive cultivation
General Purpose	Natural or manufactured topsoil of lesser quality than Premium, suitable for agriculture or amenity landscape, may need fertilizer or soil structure improvement.
Economy	Selected subsoil, natural mineral deposit such as river silt or greensand. The grade comprises two subgrades: 'low clay' and 'high clay', which is more liable to compaction in handling. This grade is suitable for low-production agricultural land and amenity woodland or conservation planting areas.

OUTLINE SPECIFICATION

Excavation generally

Excavate topsoil and stockpile on site where shown on drawing, excavate subsoil and dump on site where shown on drawing, bring ground to given slope, excavate for foundation to wall.
    for excavation   *D20*

Drainage generally

a. *Excavate pipe trench as shown on drawing at back of wall.*
b. *Lay filter-wrapped 100 mm perforated plastic land drain to BS 4962 bedded on 100 mm DoT Type 2 bedding to back of wall at foundation level.*

c. *Lay filter fabric at back of wall, lay aggregate drainage backing of 200 mm of 40–60 mm aggregate or ballast between the wall and the filter fabric. Aggregate to be to BS 63 Pt 1 or BS 882.*

for drainage   R13

Proprietary precast concrete crib walling

a. *Lay foundation concrete FND3 to crib manufacturer's design, supply and construct specified manufacturer's precast concrete crib walling units Type XXX to form battered crib wall 2400 mm high, inclination 1 in 1.5, laid in accordance with manufacturer's instructions.*
b. *Backfill with approved excavated material compacted as the work proceeds, infill cribs with clean broken rock well graded up to 100 mm.*

for concrete mixes   E10

Proprietary timber crib wall

a. *Lay foundation concrete FND2 to crib manufacturer's design, supply and construct specified manufacturer's CCA-treated softwood crib walling units Type XXX to form battered crib wall 1200 mm high, inclination 1:2, laid all in accordance with manufacturer's instructions.*
b. *Backfill with approved excavated material compacted as the work proceeds, infill cribs with approved topsoil taken from dump.*
c. *Plant with specified plant material. Water and weed regularly until planting material is established.*

for concrete mixes   E10
for planting   Q31

Proprietary block retaining wall

a. *Lay foundation concrete FND3 600 mm × 300 mm deep, supply and construct specified manufacturer's precast concrete blocks and coping Type XXX to form battered wall 1800 mm high, inclination 1 in 3, laid all in accordance with manufacturer's instructions.*
b. *Backfill with approved excavated material compacted as the work proceeds.*

for concrete mixes   E10

Precast concrete retaining-wall units

a. *Excavate for wall, supply and erect specified manufacturer's precast concrete units Type XXX 2400 mm high × 900 mm wide, constructed all in accordance with manufacturer's instructions.*
b. *Backfill with approved excavated material compacted as the work proceeds.*

for excavation   D20
for concrete mixes   E10

Precast concrete dwarf retaining wall

a. *Lay foundation concrete FND3 600 mm × 200 mm deep, supply and erect specified manufacturer's precast concrete L-shaped units 400 mm × 600 mm × 600 mm high Type XXX to form dwarf retaining wall 600 mm high, constructed all in accordance with manufacturer's instructions.*
b. *Backfill with topsoil to BS 3882 Premium grade.*

   for concrete mixes   *E10*

Brick retaining wall

Construct one and a half brick thick wall 1100 mm high above finished ground level on foundation 600 mm × 300 mm deep of FND3 concrete, in Class B engineering brick in English garden wall bond, laid in Group 3 cement:lime:sand mortar with flush joints struck as the work proceeds, lateral movement joints at maximum intervals of 15 m, width of joint to be 15 mm wide, coping of bull-nosed engineering brick-on-edge laid in Group 3 masonry cement:sand 1: 4 mortar.

   for concrete mixes   *E10*
   for brickwork   *F10*

Gabions

a. *Excavate to given levels, supply and lay three courses staggered galvanized wire three-cell box gabions 3000 mm wide × 1000 mm deep × 500 mm high, set back 600 mm each course, fill by hand with clean broken stone 150–200 mm size tightly packed, wire up gabions in accordance with manufacturer's instructions.*
b. *Backfill slope with excavated material. Maximum permitted deviation from straight face ±50 mm.*

   for gravel and shingle   *Q23*

Grass concrete block revetment

a. *Construct bank revetment, bring bank to grade, lay regulating layer of quarry scalpings, lay filter membrane, lay 100 mm drainage layer of broken stone or approved hardcore 40–60 mm size, blind with sharp sand.*
b. *Lay specified manufacturer's interlocking grass concrete blocks 600 mm × 400 mm × 125 mm 56% grass area Type XXX on 25 mm bedding sand to BS 882 Grade C, complete with toe beams and precast concrete kerbs to edges.*
c. *Fill with topsoil to BS 3882 General Purpose grade, incorporating peat at 3 kg/m² and Growmore fertilizer at 70 g/m², seed with dwarf rye-based grass seed mix.*

As above but blocks filled with 20–50 mm shingle.
for kerbs	*Q10*
for gravel and shingle	*Q23*
for grass concrete	*Q25*
for grass seeding	*Q30*

D20 Excavation, filling and groundwork

GENERAL GUIDANCE

This section deals with the earth-moving work necessary to prepare the site for hard and soft landscaping work. Soil stabilization is covered in section D11.

Depositing soil

Excavated soil bulks up to at least two or three times its original bulk, and the removal of vast quantities of soil can be a very expensive business.

The landscape designer should make provision for its use on site wherever possible, and careful initial thought over land forming should make this possible in many schemes. Whether it is deposited on site in spoil heaps or spread directly over the site, it is much cheaper than having it taken to a tip. Comparative current costs suggest that it costs eight times as much to deposit soil to a tip 13 km away as it does to stack it in site spoil heaps 50 m from the work, or three times as much if the spoil heap is 200 m away. Also, cost of depositing material in spoil heaps 200 m from the work is just over two thirds that of spreading it 500 m away; this will produce an obvious saving on large-scale work.

Good quality topsoil

When this is required for reuse it should be stacked at a point marked both on the drawing and the site. While it is stacked, weeds must be controlled by the use of either selective or total weedkillers, and the soil must not be reused until the active ingredient is negated. The choice is at the landscape designer's discretion, and will vary according to season; the use of herbicides must also be discussed with the client, who may not wish them to be used. Site traffic must not be allowed to compact areas that will be soft landscaped.

Turf

Existing turf may be either retained on site or stored elsewhere for future use, though it will not keep for long periods, even when carefully watered and sheltered during hot dry or frosty weather. If it is in poor condition it may be ploughed in during cultivation operations.

Other excavated material

Other non-toxic excavated material can be used as infill on site to make up levels before laying a finished material; this will, as stated above, either be deposited in marked spoil heaps or spread directly on site where required. Where there are indications that the soil may be contaminated, it may be necessary to employ expert advice on the type of contaminants present.

Unstable ground

If freshly excavated ground shows signs of slipping onto a road or nearby building, the contractor must seek advice from the landscape designer for a satisfactory method of emergency retention, as for example the use of temporary sheet piling.

Foundations

Where poor soil conditions exist, and foundation trenches have to be dug below the required depth for the bottom of the concrete, weak patches must be dug out by hand and backfilled with well-rammed granular fill.

All foundations must be backfilled to suit the conditions specified for the area at the foot of the wall, whether hard or soft landscape; this will normally require the soil to be well rammed to the specified depth, ready to receive such items as paving or mowing stones.

For the maximum wall height of 2.4 m for free-standing walls described in this handbook, the minimum excavated trench depth should be 500 mm. The trench should be cut with clean sides and level bottom ready to receive the concrete. As mentioned above, where the ground is unstable or full of rubble from old buildings, trenches may have to be made much deeper. This will incur an extra, and may require temporary timber or sheet steel supports to support the excavations (these are described and illustrated in BS 6031); under normal conditions the shuttering may be struck (removed) after half a day in warm weather or one day in cold weather (minimum 3 °C).

Where information is not available at the tender stage from an accurate soil and strata survey, assumed types may be given. Similarly, where the water-table is thought to be stable over the whole site, give an assumed average. Where the site is subject to tidal water, both high and low levels should be given. On sites where large quantities of excavated material are likely to be waterlogged it may be necessary to allow for a lagoon to take the excess run-off water; the position of this should be marked on the drawing.

Services

The approximate position of existing services should be shown on the drawing, but these must be checked by the contractor. He is required to inform the relevant authority of any exposed or damaged services. The contractor must inform the landscape designer of any other unplotted services, and seek his advice on dealing with them.

Rubbish

All areas to be landscaped should be kept clear of builders' rubbish, oil and diesel storage drums and so forth, and any badly compacted areas should be ripped or dug over by hand.

Pesticides

Pesticides are controlled by legislation, and many of the chemicals likely to be used can be applied only by a certified operator, who must comply with the Control of Pesticides Regulations and relevant MAFF codes. Herbicides must also be used with great care to prevent damage to surrounding vegetation. All should be specified as leniently as possible so as to reduce the risk of contaminating adjoining land. Where they are used, all empty containers and unused chemicals, including dilutions, must be safely disposed of in accordance with the manufacturer's instructions; none must be allowed to spill onto the surrounding area or watercourses.

Do not apply any of these if the client intends to run a totally organic enterprise or is hoping for a clean bill of health from the Soil Association. Before applying them on nature conservation areas discuss the chemical to be used with the client; many are inimical to the welfare of insects, fish and so forth.

BRITISH STANDARDS

1377 Pt 1 : 1990 & AMD	Soil tests for civil engineering. Sample preparation
1377 Pt 2 : 1990	Soil tests for civil engineering. Classification requirements
1377 Pt 4 : 1990 & AMD	Soil tests for civil engineering. Compaction related tests
1377 Pt 5 : 1990 & AMD	Soil tests for civil engineering. Compressibility, permeability, durability
1377 Pt 9 : 1990 & AMD	Soil tests for civil engineering. In situ tests
3882 : 1994	Specification for topsoil (exported/imported)
4428 : 1989 & AMD	Code of practice for general landscape operations
5930 : 1981	Code of practice for site investigations for civil engineering operations
6187 : 1982	Code of practice for demolition
8000 Pt 1 : 1989	Workmanship on building sites. Part 1 Excavation and filling

DATA

Topsoil quality to BS 3882 : 1994

Topsoil grade	Properties
Premium	Natural topsoil, high fertility, loamy texture, good soil structure, suitable for intensive cultivation
General Purpose	Natural or manufactured topsoil of lesser quality than Premium, suitable for agriculture or amenity landscaping; may need fertilizer or soil structure improvement
Economy	Selected subsoil, natural mineral deposit such as river silt or greensand. The grade comprises two subgrades: 'low clay' and 'high clay', which is more liable to compaction in handling. This grade is suitable for low-production agricultural land and amenity woodland or conservation planting areas.

Excavation and filling BS 8000 Pt 1 : 1989 Workmanship on building sites

The main recommendations in this Standard relevant to landscape work are as follows:

- The location of all underground structures and services must be determined and marked before excavation.
- Existing services and land drainage must be maintained during excavation.
- Vegetation and roots must be removed with minimum disturbance to topsoil.
- Topsoil must be stripped to not less than 150 mm.
- Spoil heaps of topsoil must be not more than 1500 mm high.
- Prevent adulteration of topsoil by other materials, oil or chemicals.
- Prevent compaction of topsoil dumps.
- Spoil must be dumped to authorized sites only.
- Support to buildings and structures adjoining excavations must be provided where necessary.
- Excavations must be kept clear of water.
- The bottom 50 mm of trenches must be excavated immediately before laying concrete.
- Trenches must be adequately supported.
- Filling that supports a structure must be compacted in layers not more than one third thicker than the largest material in the fill.
- Hardcore that supports a structure must be compacted in layers not exceeding 150 mm.
- Formation levels for soft landscape must be consolidated (not compacted) to:

- 100 mm below finished levels for grassed areas;
- 250 mm below finished levels for herbaceous-planted areas;
- 400 mm below finished levels for shrub-planted areas and hedges;
- 600 mm below finished levels for a distance of 2.0 m round trees.

Soils: angle of repose

Earth,	loose and dry	35–40° from horizontal
	loose and naturally moist	45°
	loose and saturated	27–30°
	consolidated and dry	42°
	consolidated and naturally moist	37
Loam,	loose and dry	40–45°
	loose and saturated	20–25°
Gravel,	medium coarse and dry	30–45°
	medium coarse and wet	25–30°
Sandy gravel		35–45°
Sand,	compact	35–50°
	loose and dry	30–35°
	saturated	25°
Shale (blaes)	filling	30–35°
Broken brick or broken rock		35–45°
Clay,	loose and wet	20–25°
	solid naturally moist	70°

OUTLINE SPECIFICATION

Existing services

Locate and mark with fluorescent markers 900 mm high all underground service runs and drainage runs on site. Locate and mark as above all underground structures. Notify local authority and statutory undertakers where necessary to divert or stop off existing service runs. All stopping off to be approved by the relevant authority before backfilling.

Existing structures

Retain all structures as shown on drawing, mark clearly with weather-proof markers on site. No plant to be used and no excavation, stored materials, or contamination permitted within 3.0 m of the structures without approval.

Existing hard landscape materials

Lift by hand and store on site for reuse hard landscape materials (e.g. bricks, paving flags, setts) as shown on drawings.

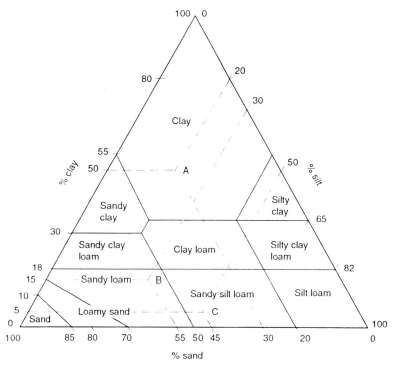

Fig. 4.1 Textural classification (limiting percentage of sand, silt and clay sized particles for the mineral texture class).
Note: Examples of textural classification:
Soil A with 30% sand, 20% silt and 50% clay is in the 'Clay' textural class.
Soil B with 55% sand, 30% silt and 15% clay is in the 'Sandy loam' textural class.
Soil C with 45% sand, 50% silt and 5% clay is in the 'Sandy silt loam' textural class.
(BS 3882 : 1984)

Existing turf

Remove turf where shown on drawings in turves not greater than 300 mm × 900 mm and not less than 75 mm thick to stack on site, water regularly in dry weather and protect against frost or hot sun. Turf not so indicated to be ploughed in.
or
Retain all turf as shown on drawings, erect temporary protective fencing, water and treat with weedkiller and turf fertilizer regularly until the completion of the works.

for turfing   Q30

Existing vegetation

Retain all vegetation as shown on drawings, erect temporary protective fencing, water and weed regularly until the completion of the works. Trees to be protected at limit of branch spread, and no excavation, stored materials, or contamination permitted within protective fencing.

Take about 10 ml of soil, wet up gradually while kneading between fingers and thumb until there is enough moisture to hold the soil together and for the soil to exhibit its maximum cohesion.

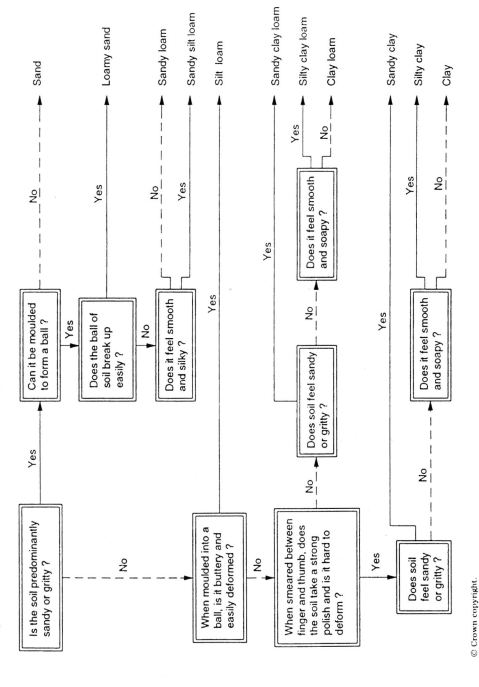

Fig. 4.2 Identification by hand texture. (BS 3882 : 1984)

Clearing vegetation

Clear existing site vegetation from areas shown on drawings, soil distur-bance to be kept to minimum. Fell trees as shown on drawings, grub up roots and remove debris, soil disturbance to be kept to minimum. The contractor is responsible for protection of site personnel and the public during felling.

Clearing hard landscape materials

Break up and remove to temporary storage site concrete or road bases where shown on drawings, and break up and remove to temporary storage brick, stone and concrete from existing building structures where shown on drawings. These materials not to be used for hardcore or backfill without prior approval. Any material to be used for hardcore must be crushed by hand or machine to 20–75 mm sizes and washed. All other material to be removed from site to approved tip not exceeding 13 km from site.

Disposal of debris

All debris arising from demolition and site clearance to be removed to authorized tip not exceeding 13 km from site. Site burning of refuse not permitted.

Excavation for foundations

Excavate trenches for foundations to depth and width shown on drawings, excavations to be kept clear of water and supported where over 750 mm deep or in soft ground, bottom 50 mm of trenches to be excavated immedi-ately before laying concrete. Weak spots in trenches to be dug out by hand and backfilled with rammed DoT Type 1 granular material. Maximum permitted deviation in given levels 50 mm, any accidental over-excavation to be cleaned by hand and backfilled with well-rammed DoT Type 1 gran-ular fill to given level.

Excavation of topsoil for levels

Excavate 200 mm topsoil from areas shown on drawings and remove to spoil heap on site not exceeding 1500 mm high; treat topsoil with agreed weedkiller and turn topsoil heaps at 3 week intervals, topsoil to be pro-tected from contamination by other materials, chemicals, and run-off. No site traffic permitted on areas shown on drawings as soft landscape.
or
Excavate 200 mm topsoil from areas shown on drawings. Spread on site to specified thickness on areas shown on drawings in layers not exceeding 150 mm.

Excavation of subsoil for levels or foundations

Excavate clean subsoil to levels given and remove to spoil heap on site not exceeding 1500 mm high; treat with agreed weedkiller if necessary, subsoil to be protected from contamination by other materials, chemicals, and run-off. No site traffic permitted on areas shown on drawings as soft landscape.
or
Excavate subsoil to levels given and remove from site to approved tip not exceeding 13 km from site.

Spreading topsoil

Spread approved imported topsoil to BS 3882 General Purpose grade on site to specified levels on areas shown on drawings in layers not exceeding 150 mm.
or
Spread topsoil from spoil heap on site in layers not exceeding 150 mm to bring surface to specified levels.

Cultivation

Where existing topsoil is less than 250 mm thick, rip subsoil using approved subsoiling machine where shown on drawings to a depth of 600 mm below topsoil at 600 mm centres, all tree roots and debris over 150 mm × 150 mm to be removed. Plough uncultivated ground where shown on drawings to a depth of 300 mm, harrow, remove debris and stones over 25 mm and roll lightly. Lightly consolidate previously spread topsoil to the following thicknesses, finished and settled levels to be as shown on drawings:

- *100 mm thick below finished levels for grassed areas*
- *250 mm thick below finished levels for herbaceous-planted areas*
- *400 mm thick below finished levels for shrub-planted areas*
- *600 mm thick below finished levels for a distance of 2.0 m round trees.*

for seeding and turfing   *Q30*
for planting   *Q31*

Chemical treatment generally

Agree with contractor the horticultural chemicals to be used on specified areas, including topsoil dumps, all chemicals to be applied in accordance with the manufacturer's instructions and in compliance with the current Control of Pesticides Regulations. Chemicals to be applied by certified operators only.

E10 *In situ* concrete

GENERAL GUIDANCE

This section deals with concrete foundations to walls, roadways, and paving. The use of designed concrete structures is outside the scope of this book. If the landscape designer is required to build walls higher than 2 m, or is in any doubt as to the stability of the ground on which walls or other load-bearing structures are to be built, the advice of a structural or civil engineer must be sought.

Solid concrete roads have mostly been replaced for urban paved areas and pedestrian routes by the 'interlocking block flexible pavement'. Highways constructed purely of concrete have been superseded by macadam, as the concrete is more subject to frost action and is much noisier than macadam.

Ordinary concrete can be used for roadways, but concrete made with a blend of Portland cement and PFA (pulverized fuel ash) gives greater density, less permeability, and greater resistance to chemical attack. The wearing surface of the roadway is safer if it is made with air-entrained concrete, which is not so strong but reduces permeability.

Concrete foundations

These may be either strip foundations under walls or slab foundations under roads or paving. The standard size for foundations for walls under 2 metres high is 150 mm deep, with a width 100 mm wider than the wall on either side, but in soft or made-up ground this should be increased to 150 mm wide. Where the ground is unstable or unsuitable for the normal 150 mm deep foundation, lightly reinforced foundations can be made with one of the standard brick-reinforcing meshes; these will enable the foundation to take up minor differential settlement in the soil. The reinforcement should be placed centrally in the concrete and given a minimum of 25 mm cover at the edges. Slab foundations under pedestrian paving may be as little as 100 mm thick, while for light traffic 150 mm is better. The strength of any wall, road or paving depends almost entirely on the strength of the sub-base and of the substrate below it; if these are properly constructed the concrete can be designed to take only its fair share of the loading.

Where the wall is to be used as a backdrop for shrubs or herbaceous borders it may be necessary to make the foundations a little deeper to prevent them being damaged by cultivation operations, and to allow a sufficient depth of soil for planting. Special detailing is required where walls have to be built near or to either side of existing trees.

Where steps have to be formed in the wall or paving, and therefore in the foundation, the foundation must overlap 300 mm at the change of level.

Concrete grades

Concrete is classified in BS 5328 Parts 1 and 2 by grades defining its strength at 28 days after laying; this strength is expressed in N/mm^2. The exposure to which the concrete is subjected affects the grade to be specified.

Non-structural concrete for external works in ordinary conditions is not required to comply with a grade specification. There are four types of concrete specification: Designed mix, Prescribed mix, Standard mix, and Designated mix. Standard and Designated mixes are most suitable for hard landscape work

Designed mix

The purchaser specifies the performance required of the concrete; the producer is then responsible for selecting an appropriate mix. Strength testing is essential.

Prescribed mix

The purchaser specifies the constituents of the concrete and is responsible for ensuring that the concrete will meet his performance requirements. The mix proportion is essential.

Standard mix

The mix is specified from the list given in BS 5328 Pt 2 : 1991 s.4, and is made with a restricted range of materials. The mix proportion is essential. The specification must include the proposed use of the concrete, and it must also include:

- the standard mix reference;
- the specified type of cement;
- the required type and maximum size of aggregate;
- the workability (slump) of the mix;
- quality assurance will be required.

These clauses can be chosen only from the standard lists given in Tables 4 and 5 in BS 5328 Pt 2 : 1991.

Designated mix

The mix is specified in BS 5328 Pt 2 : 1991 s.5, and requires the producer to hold current product conformity certification with quality approval to BS 5750 Pt 1 (EN 29001). Quality assurance is essential, and should be checked. These mixes cannot be modified by the specifier.

Reinforced concrete

Landscape contracts do not normally include designed structural reinforcement, but bars, wall mesh, or light steel mesh can be used to strengthen concrete over doubtful ground or where changes of level may stress the slab or foundation. Reinforcement may be by means of steel bars laid in both directions and wired together, but it is easier to use prefabricated steel mesh laid halfway through the thickness of the slab. Highway reinforcement calculation is a specialist job, but a typical reinforcement for light paving would be steel mesh to BS 4483, 200 mm × 200 mm square, either 2.22 kg/m^2 or 3.02 kg/m^2.

Brick walls may be reinforced with expanded steel mesh, or with a narrow steel grid specifically designed for the purpose laid in the joints.

Movement joints

Any change in thickness, material or direction requires a movement joint. Correctly placed joints can help to control cracking, because cracks will develop along weak lines such as joints rather than in the middle of the slab.

Longitudinal joints should be at a maximum width of 4.2 m for unreinforced concrete, and a maximum width of 6.0 m for reinforced concrete with an offset between one layer and the next of 300 mm; they are formed in the wet concrete.

Transverse joints in unreinforced slabs should be at spacings giving a longitudinal:transverse ratio no more than 2:1.

Protection of concrete

All newly laid *in situ* concrete must be protected from frost, rain, hot sun and drying winds. Any *in situ* concrete that is to have an exposed surface, or which is to receive a decorative finish, must also be protected from indentations, dirt or staining. This means that the contractor must not store oil or paint or any other drum on the surface, and if they are being used next to these surfaces, the whole must be properly protected.

All concrete used for bases and pavements must be left to cure for at least 10 days, and the final loading must be not be placed on it until full structural strength has been reached. Striking formwork can take place earlier, because this allows the curing to develop faster, but only foot traffic should be allowed on the concrete at this stage. Rapid-hardening concrete can be used to strike formwork earlier, but this is normally used where formwork would obstruct other building operations.

BRITISH STANDARDS

882 : 1992 & AMD	Aggregates from natural sources for concrete
1881 : Pts 101–107, 114, 116	Methods of sampling fresh concrete on site
3892 Pt 2 : 1993	PFA for use in grouts and for miscellaneous uses in concrete
5075 Pt 1 : 1982 & AMD	Concrete admixtures: accelerating, retarding, and water reducing
5075 Pt 2 : 1982 & AMD	Concrete admixtures: air-entraining admixtures

5328 Pt 1 : 1991 & AMD	Guide to specifying concrete
5328 Pt 2 : 1991 & AMD	Methods for specifying concrete
8110 Pt 1 : 1985	Code of practice for design and construction (buildings and structures)

DATA

Concrete roadway bases

Roadway layer	Concrete mix to BS 5328
Sub-base or road base:	
wet lean concrete 1	ST 1 (C7.5, C7.5P)
wet lean concrete 2	ST 2 (C10, C10P)
wet lean concrete 3	ST 3 (C15, C15P)
wet lean concrete 4	ST 4 (C20, C20P)
Road base:	
continuously reinforced	C40
Surface slab (wearing course):	
unreinforced	C40
jointed reinforced	C40
continuously reinforced	C40

A waterproof membrane must be laid on the sub-base to prevent water from draining out of the concrete road base before it has set.

Exposure classification for structural concrete

Condition Type of exposure	Minimum grade		28 day strength (N/mm²)
Mild			
= surfaces protected against weather	Unreinforced	C20	20
or aggressive conditions	Reinforced	C30	30
Moderate			
= exposed but sheltered from severe rain	Unreinforced	C30	30
or freezing when wet	Reinforced	C35	35
= concrete under non-aggressive water			
= concrete in contact with non-aggressive soil			
= concrete subject to condensation			
Severe			
= exposed to severe rain, alternate	Unreinforced	C35	35
wetting and drying, occasional freezing,	Reinforced	C40	40
severe condensation			

Exposure classification for structural concrete *continued*

Condition Type of exposure	Minimum grade		28-day strength (N/mm²)
Very severe = occasionally exposed to seawater or de-icing salts	Unreinforced Reinforced	C40 C40	40 40
Most severe = frequently exposed to seawater or de-icing salts, or in tidal zone	Unreinforced Reinforced	C45 C50	45 50
Abrasive = exposed to abrasive action by machinery or grit	Unreinforced Reinforced	C45 C45	45 45

Non-structural concrete

C7.5 concrete has a strength of 7.5 N/mm².
C10 concrete has a strength of 10.0 N/mm².
C15 concrete has a strength of 15.0 N/mm².

Concrete mixes to BS 5328 : Pt 1 & AMD

Application	Standard mix	Designated mix	Slump
Mass concrete fill and blinding	ST 2 (C10P)	GEN 1	75 mm
Mass concrete foundations in non-aggressive soils	ST 2 (C10P)	GEN 1	75 mm
Trench fill foundations in non-aggressive soils	ST 2 (C10P)	GEN 1	125 mm
Foundations in Class 2 sulphate soils	–	FND 2	75 mm
in Class 3 sulphate soils	–	FND 3	75 mm
in Class 4A sulphate soils	–	FND 4A	75 mm
in Class 4B sulphate soils	–	FND 4B	75 mm
Kerb bedding and haunching	ST 1 (C7.5P)	GEN 0	10 mm

GEN concretes with low cement content may not be suitable for good cast and direct finished surfaces.

Traditional concrete mixes

Application	Name of mix	Cement:aggregate:sand (parts by volume)
Setting posts, filling weak spots in ground, backfilling pipe trenches	one, two, three	1:2:3
	or	or
	one to five all-in	1:5
Paving bases, bedding manholes, setting kerbs, edgings, precast channels	one, two, four	1:2:4
	or	or
	one to six all in	1:6
In situ concrete paving:	one, one & half, two & half	1:1.5:2.5

'All-in' is mixed aggregate as dug from the pit.

Curing concrete

Hot weather and drying winds can cause differential shrinkage in concrete, resulting in cracking.

Use impervious sheeting laid immediately after placing concrete.

Curing sheet should be left in position for 10 days after placing concrete.

OUTLINE SPECIFICATION

Foundations

a. After excavation as specified in D20, dig out weak spots in the substrate and fill with well-rammed granular fill.

b. Pour FND2 concrete foundations to given levels as shown on drawings in trench previously excavated, 300 mm wide × 150 mm deep, concrete to be poured not higher than 600 mm from bottom of trench, vibrate concrete lightly.

c. Cover with impervious sheeting for 10 days.

d. Variations in level not to exceed ±5 mm from given levels. Movement joints to be maximum distance apart of 4.2 m.

for excavation   D20

Slabs

a. After excavation to levels as specified in D20, dig out weak spots in the substrate and fill with well-rammed granular fill.

    b. Lay 150 mm hardcore and compact, blind with sand.

    c. Lay concrete slab 150 mm thick GEN4 concrete to given levels as shown on drawings in formwork specified elsewhere, to falls shown on drawing, concrete to be poured not higher than 600 mm from bottom of slab, tamp concrete lightly, excess laitance not permitted.

    d. Cover with impervious sheeting for 10 days, strike formwork in 4 days in warm weather (7 days in cold weather). No material to be stored on the finished surface, and no site traffic permitted after completion.

    e. Maximum deviation from finished level ±3 mm for self-finished surface to slab; ±5 mm for applied finish to slab. Movement joints to be maximum distance apart of 4.2 m.

for excavation	*D20*
for formwork	*E20*

Residential driveway

    a. After excavation to levels as specified in D20, dig out weak spots in the substrate and fill with well-rammed granular fill.

    b. Lay 100 mm hardcore and compact, lay concrete slab 100 mm thick Designated mix PAV1 (concrete 150 mm thick for poor substrate laid on 100 mm hardcore using designated mix PAV1) with minimum cement content 300 kg/m³ and 5.5% entrained air, 75 mm slump.

    c. Cover with plastic sheeting or wet hessian and cure for 10 days. No material to be stored on the finished surface, and no site traffic permitted after completion.

    d. Maximum deviation from finished level ±5 mm for self-finished surface to slab; ±7.5 mm for applied finish to slab. Movement joints to be maximum distance apart of 4.2 m.

for formwork	*E20*
for trench excavation	*D20*

E20 Formwork for *in situ* concrete

GENERAL GUIDANCE

This section covers the metal or timber formwork used for retaining concrete during construction.

    Formwork (or shuttering) is the responsibility of the contractor, not the landscape designer, but the landscape designer should ensure that the type of formwork used is capable of producing the finish required. Joints in formwork can be used decoratively, and in any case they should be designed to fit in with the structural elements of the design. Formwork must be designed so that the minimum cover to reinforcement is always maintained. Where concrete is poured against the natural ground, at least 75 mm of cover to reinforcement should be maintained. Formwork must be substantial enough to hold wet concrete and the stress of pouring and vibration without deformation.

Controlled permeability linings to formwork can be used to produce a harder and better-looking finish by allowing the surplus air and water in the concrete to drain away at a controlled rate.

Concrete road bases, roads and paving can be laid in steel road formers, which are heavy prefabricated steel forms bolted together; they are made to standard highway curves and angles.

BRITISH STANDARDS

8110 Pt 1 : 1985   Code of practice for design and construction (buildings and structures)

DATA

Finishes to concrete

Smooth finished surface
Use plywood formwork with panel joints designed to produce regular pattern.

Fine smooth finished surface
Use resin-film-faced plywood panel with joints designed to produce regular pattern.

Rough board finished surface
Use softwood boards arranged in regular pattern, free from major shakes, knotholes and damage. Formwork to be fixed with close butted joints.

Patterned finished surface
Use plywood panels with battens screwed to panels to form ribbing, or proprietary shaped steel formwork.

Blasted or tooled finished surface
Even small blemishes in the formwork will show up after treatment; use smooth plywood panels with the joints covered with tape. Construction joints and other marks cannot be covered by subsequent surface treatment of the concrete, so the formwork must be constructed to a high standard. Sample panels may be specified.

OUTLINE SPECIFICATION

Formwork to planter

Construct formwork to in situ *concrete planter with softwood boards with maximum moisture content of 25%, 900 mm long × 150 mm wide free from major defects with close-butted joints; end joints to be in line vertically. Formwork to be treated with release agent evenly applied without ponding. No damage must accrue to formwork during pouring of concrete. Strike formwork after agreed period. Deviations from the finished surface to be not more than 3 mm.*

for concrete   *E10*

F10 Brick and block walling

GENERAL GUIDANCE

This section describes non-structural walls up to 2.4 m high. Bricks for landscape works are covered by three main groups: commons, facings, and engineering bricks. Materials used in brickmaking are clay or calcium silicate and occasionally concrete.

Clay bricks are most commonly used, and they can produce attractive textural and colour variations. Calcium silicate bricks produce a more uniform texture, though of a satisfactory strength and quality. Concrete bricks are rarely used in landscape work, and must not be used in ground likely to cause sulphate attack to them. Where concrete bricks are called for in these conditions, they must be manufactured for this specific purpose.

Fig. 4.3 The Brunswick range of clay facing bricks.
(Ibstock Building Products Ltd)

Selecting bricks

Before selecting the bricks for a landscape wall the landscape designer should consider what it is to be used for, and where it is to be built. Points to consider are:

- type of soil, sulphate level, level of water-table and waterlogged soil;
- salt spray, atmospheric pollution, frost, exposure to wind, rain penetration;
- any local problems with vandalism or graffiti;
- any aesthetic requirements, the need to blend with existing buildings or landscape;
- any conservation area controls imposed by the local planning authority.

Equally important is the safety aspect, and the injury that badly positioned or constructed walls may cause if they collapse. For this reason, two half-brick skins joined together with wall ties should not be specified unless specifically requested in writing by the client, and the client should then be advised that these walls are at risk if the wall ties are incorrectly placed. If they have to be built, rigorous supervision is required to ensure that the specified number and type of wall ties are correctly built in as the work proceeds.

Brick bonds

The main purpose of bonding brickwork is to achieve the maximum strength and stability compatible with the minimum of cutting. Bonds are also designed to give different textures to a wall, such as evenness, vertical or horizontal emphasis.

The calculations for landscape walls take no account of the bond used as a factor in determining strength, but where there is likely to be considerable stress on the wall it is better to use either English or double Flemish bond rather than one of the garden wall bonds.

Half-brick or one-brick walls should not be specified to be 'fairfaced both sides'; it is impossible for the former, and difficult for the latter. Fairface work is expensive, and the landscape designer may be able to specify it on one side of the wall where the other side is invisible, as would be the case for brick planters. Where truly accurate work is specified to both sides of a one-and-a-half-brick wall, the landscape designer must add a clause to the specification forbidding 'over-hand' working (that is, where the bricklayer works from one side of the wall only and consequently cannot see the other side of the wall as it is being built), because for all but the lowest walls scaffolding or trestles will be required both sides, and this will form part of the extra cost.

Quoins, junctions, piers and pilasters should always be of generous proportions, as they are particularly vulnerable to damage or vandalism. The landscape designer should also inspect them carefully during construction, because they are important to the whole stability of the wall.

Descriptions of the more common bonds are given below; for more detailed descriptions of these and other bonds, refer to books on hard landscape design.

Half-brick walls

This bond is most frequently used on low walls in conjunction with one-brick piers, which help to give them greater stability. Where an open wall is acceptable, panels of honeycomb bond may be specified, but they must be surrounded by one-brick walling. Honeycomb bond is built by spacing the stretchers 50 mm or so apart, so that light can pass through the wall.

One-brick or thicker walls

The main structural wall bonds used in building work are double Flemish bond and English bond. These are used for walls subject to structural stresses in walls one, one and a half, and two bricks thick.

English bond

This is the strongest unreinforced brick bond, because there are no straight vertical joints in the wall. It is built with alternate courses of headers and stretchers, and gives a regular pattern with very even texture.

Double Flemish bond

This bond is built with alternate headers and stretchers in each course, with a header centrally over each stretcher; it is not as strong as English bond, but can be more decorative when, for example, darker headers are used to provide specific patterns. When this work is specified, the landscape designer must show these on the detailed drawings, and allow for the extra cost involved in selecting the bricks and working them into the required pattern.

Single Flemish bond

This can be used only in one-and-a-half-brick or thicker walls. It is a useful bond where Flemish bond is used on the face with handmade or good quality machine-made bricks and less expensive though compatible bricks are used at the back for the English bond. Where specified, an extra will be required in the cost of labour.

Garden wall bonds

These bonds contain a high proportion of stretchers, and comparatively little brick cutting is required. This makes them economical to build, and gives them a horizontal character. It also produces long straight joints in the centre, and thus reduces their strength even when built with substantial piers at frequent intervals. They are not recommended for walls likely to be under stress or subject to impact from vehicle overrun, or where the wall acts as a barrier between two differing levels.

English garden wall bond

This bond has one course of headers to three courses of stretchers. Unless appearance is of little importance, the landscape designer should not specify this unless at least seven courses are visible above ground.

Flemish garden wall bond

Each course consists of three stretchers and one header, staggered so that the headers form vertical lines. This is simple to lay, and the minimum number of headers used enables acceptable fairface work to be done on both sides of a one-brick-thick wall.

Rat-trap bond

This is similar to Flemish garden wall bond, but the bricks are laid on edge, making each course 112.5 mm high. There is a central gap between the stretchers, which are faced on each side, thorough bonding being dependent on the headers. It is an economic wall of little strength, and the landscape designer should not specify this bond in unprotected public areas, or where there are vehicles.

Specials

To save cutting bricks on site and to produce perfect half-bats and other shapes, 'standard specials' are made by brickworks to match most of their bricks; on large contracts 'special specials' will be made to order.

The most commonly used 'standard specials' are the three types that can be cut from a standard brick:

- The **snapped header** or half-bat is equal to a standard brick cut across the centre. It is advisable to specify these for bonds requiring many half-bats, to save wasteful cutting on site.
- The **king closer**, available right or left handed, is equal to cutting one corner from the centre of one long and one short face to form a splayed angle. Specify these for complicated pier and corner bonds to avoid straight vertical joints in the wall.
- The **queen closer** is equal to a brick split lengthways down the centre to give two narrow half-bricks; it is used at the end of a run of quarter-bond brickwork to bring the whole course to bond.
- External and internal angled **corner bricks** are made in 30°, 45° and 60° angles. They are used to form angles in walls, planters and so forth.
- **Bull-nosed** and **plinth bricks** are used for string courses, plinths and copings.

Damp-proof courses

These are intended to protect the wall from rising or descending damp, which can result in the growth of algae, efflorescence, and deterioration of

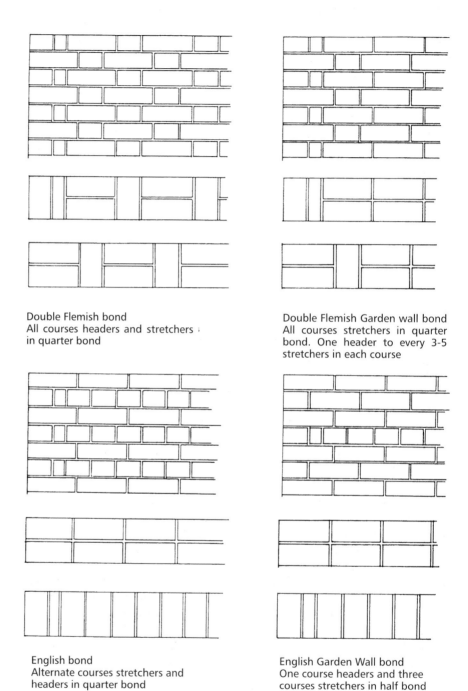

Double Flemish bond
All courses headers and stretchers
in quarter bond

Double Flemish Garden wall bond
All courses stretchers in quarter
bond. One header to every 3-5
stretchers in each course

English bond
Alternate courses stretchers and
headers in quarter bond

English Garden Wall bond
One course headers and three
courses stretchers in half bond

Fig. 4.4 Brick landscape walls one brick thick.

the mortar. DPCs are placed at 150 mm above finished ground level and under copings that are not waterproof. Where DPCs are used, it is advisable to use either engineering DPC1 or DPC2 bricks, or high-bond membrane DPCs in order to prevent the wall or coping from slipping at the DPC level. When dampness in the wall is of no import, the DPCs may be omitted.

Movement joints

All brickwork moves to a greater or lesser extent. There are two kinds of movement that must be considered by the landscape designer: lateral movement caused by the contraction and expansion of the brickwork itself due to temperature and weather conditions, and the vertical movement caused by differential settlement between one part of the wall and the other.

Neither structural brickwork nor its movement and expansion joints are discussed here, and if the landscape designer needs such walls he or she is advised to employ a structural or civil engineer. Traditionally, landscape walls were built without movement joints, because they had substantial footings, and softer bricks laid in lime mortar, and were often thicker than modern walls. Where minor cracking and settlement of landscape walls is of no consequence, movement joints may be omitted.

Lateral movement joints

Lateral movement joints should be provided at maximum intervals of 15 m in long walls, and the width of the movement joint should be 1 mm per metre of wall. Obviously these joints should be made at points where they can best be accommodated in the design: at junctions, or where they can be made a feature of the design. Lateral expansion and contraction joints must continue right to the top of the wall, including any coping or cappings of brick, slate, concrete or reconstructed stone. They are not continued below the wall into the foundations, because these will be designed to take into account the type of subsoil and the weight of the wall, and a break in this will cause differential settlement in the wall.

Joints can be made by leaving 25 mm open gaps between two sections of wall strengthened by piers, by overlapping two sections of wall, or by interlocking the two sections, either with a half-and-half joint or a splayed joint.

Gaps may be filled by filling to the centre with a compressible material and sealing the exterior faces to a depth of 10 mm to prevent water or frost penetration. The landscape designer should contact the manufacturer of the chosen sealant for their approved specification and method of application.

Straight-through sealants are either gaskets in neoprene rubber or similar material; these must be specified to be built into the wall as the work proceeds. Gun-applied soft mastics applied after the wall has been completed can be specified, but it requires careful supervision to check that the sealant has penetrated right through the wall. Here also the chosen manufacturer should be asked to provide a suitable specification and method of application for the situation and material to be used.

Settlement joints

Where landscape walls, long planters and the like are built on made-up or unstable ground, reinforced beam foundations should be used.

Even very low walls will be stressed by a change in ground conditions, and a vertical break should be provided both in the wall itself and in the foundations, thus allowing the two sections to slide vertically against each other should differential settlement take place.

BRITISH STANDARDS

12 : 1996	Portland cement
187 : 1978 & AMD	Calcium silicate (flint/lime and sand/lime) bricks
743 : 1970 & AMD	Materials for damp-proof courses (Appendix C superseded by BS 8215 : 1991)
890 : 1995	Building limes
1200 : 1976 & AMD	Building sands from natural sources. Sands for mortar for brickwork
3921 : 1985 & AMD	Clay bricks and blocks
4027: 1996	Sulphate-resisting Portland cement
4721 : 1981 (1986) & AMD	Ready mixed building mortars
4729 : 1990	Shapes and dimensions of special bricks
4887 Pt 1 : 1986	Mortar admixtures. Air-entraining (plasticizing) admixtures
4887 Pt 2 : 1987	Mortar admixtures. Set retarding admixtures
5224 : 1995	Masonry cement
5628 Pt 1 : 1992	Code of practice for masonry: structural use of unreinforced masonry
5628 Pt 3 1985 & AMD	Code of practice for masonry, materials and components, design and workmanship
6398 : 1983	Bitumen damp courses for masonry
6477 : 1992	Water repellents for masonry surfaces
6649 : 1985	Clay and calcium silicate modular bricks
7583 : 1996	Portland limestone cement
8000 Pt 3 : 1989	Code of practice for masonry (brick and block)
8104 : 1992 & AMD	Code of practice for assessing exposure of walls to wind-driven rain

DATA

Durability of bricks

FL   Frost resistant with low salt content
FN   Frost resistant with normal salt content
ML   Moderately frost resistant with low salt content

MN   Moderately frost resistant with normal salt content
OL   Not frost resistant with low salt content
ON   Not frost resistant with normal salt content

Limits for the soluble salts content of all bricks have now been specified. The liability to efflorescence test has accordingly been deleted.

The sum of the content of sodium, potassium and magnesium shall not exceed 0.25%.

Sulphate shall not exceed 1.6%.

Test methods for salts content are given.

Choice of bricks for given locations

Position	Brick type
Brickwork at ground level	
well-drained ground	FL FN ML MN
badly drained ground	FL FN
Brickwork below ground	FL FN ML MN
Damp-proof courses	FL FN ML MN, or engineering
Copings to walls	FL FN ML MN, or engineering
Brick on edge copings	FL FN, or engineering
Earth retaining walls with	
waterproofed retaining face	FL FN ML MN, or engineering
Manholes, silt pits	FL FN, or engineering

Calcium silicate bricks

Type	Strength (N/mm^2)	Location
Class 2 crushing strength	14.0	Not suitable for walls
Class 3	20.5	Walls above DPC
Class 4	27.5	Cappings and copings
Class 5	34.5	Retaining walls
Class 6	41.5	Walls below ground
Class 7	48.5	Walls below ground

The Class 7 calcium silicate bricks are therefore equal in strength to Class B engineering bricks. Calcium silicate bricks are not suitable for DPCs.

Concrete bricks

BS 6073 gives the following work sizes for concrete bricks:

290 mm × 90 mm × 90 mm
190 mm × 90 mm × 90 mm
190 mm × 90 mm × 65 mm
215 mm × 102.5 mm × 65 mm

Brickwork dimensions

Horizontal		Vertical	
No of bricks	Dimensions (mm)	No of courses	Dimensions (mm)
½	112.5	1	75
1	225	2	150
1½	337.5	3	225
2	450	4	300
2½	562.5	5	375
3	675	6	450
3½	787.5	7	525
4	900	8	600
4½	1012.5	9	675
5	1125	10	750
5½	1237.5	11	825
6	1350	12	900
6½	1462.5	13	975
7	1575	14	1050
7½	1687.5	15	1125
8	1800	16	1200
8½	1912.5	17	1275
9	2025	18	1350
9½	2137.5	19	1425
10	2250	20	1500
20	4500	24	1800
30	6750	28	2100
40	9000	32	2400
50	11260	36	2700
60	13510	40	3000

Mortar mixes

Mortar group	Cement:lime:sand	Masonry cement:sand	Cement:sand with plasticizer
1	1:0–0.25:3		
2	1:0.5:4–4.5	1:2.5–3.5	1:3–4
3	1:1:5–6	1:4–5	1:5–6
4	1:2:8–9	1:5.5–6.5	1:7–8
5	1:3:10–12	1:6.5–7	1:8

As the grades progress from 1 to 5 they become weaker, but have an increasing ability to accommodate movement caused by settlement or shrinkage.

For example, Grade 1 produces a strong mortar that is inflexible, particularly when no lime is added; conversely, Grade 5 produces a weak but flexible mix.

All mixes within a group are of approximately similar strength. Frost resistance increases with the use of plasticizers. Cement:lime:sand mixes give the strongest bond and greatest resistance to rain penetration.

Masonry cement to BS 5224: 1995 covers a wider range of cement types and defines three strength classes. Non-air-entrained cements are included.

Class	28-day strength N/mm^2	Air entrainment (X = non-air-training)
MC 5	5-15	yes
MC 12.5	12.5-32.5	yes
MC 12.5X	12.5-32.5	no
MC 22.5X	22.5-42.5	no

Masonry cement equals ordinary Portland cement plus a fine neutral mineral filler; it is easier to work and is more frost resistant than a cement:lime mixture. Masonry cement should never have lime added to it, and should not be used on walls subject to heavy loading.

Wind zones

Wind zone	Ratio between maximum permitted wall height to wall thickness
1	8.5
2	7.5
3	6.5
4	6.0

Refer to wind zone map in Fig. 4.5.

OUTLINE SPECIFICATION

Brick wall

a. *Construct one-brick-thick wall 1500 mm high above finished ground level on foundations specified elsewhere with one-and-a-half brick piers at 3.0 m centres, all in type FL rough stocks in English garden wall bond, laid in Group 3 cement:lime:sand (1:1:6) mortar with flush joints struck as the work proceeds.*

b. *Lateral movement joints at maximum intervals of 15 m, joint to be 15 mm wide.*

c. *Construct DPC 150 mm above ground of two courses engineering brick in Group 2 cement:sand (1:3) mortar.*

Fig. 4.5 Windspeed zones. (Based on BS 5628 Pt 3)

d. *Construct coping of bull-nosed engineering brick Class B on edge laid in Group 2 cement:sand (1:3) mortar.*

for excavation for foundations	*D20*
for concrete foundations to walls	*E10*
for damp-proof courses	*F30*
for precast concrete copings	*F31*

Planter

a. *Construct planter 750 mm high above finished ground level, on foundation specified elsewhere, one brick thick of type FL rough stocks laid in Flemish garden wall bond in Group 3 cement:lime:sand (1:1:6) mortar, joints raked and pointed in coloured mortar, joints left open for weepholes at 1.0 m centres 150 mm above ground level.*

b. *Coping of precast concrete specified elsewhere on DPC of two courses slates laid in 1:3 Group 2 cement:sand mortar.*

for excavation for foundations	*D20*
for concrete foundations to walls	*E10*
for damp-proof courses	*F30*
for precast concrete copings	*F31*

Pier

Construct isolated pier three bricks square in type FL rough stocks laid in Flemish bond in Group 3 cement:lime:sand (1:1:6) mortar, joints raked and pointed in coloured mortar, on foundation specified elsewhere, with special capping bricks to match laid in 1:3 Group 2 cement:sand mortar DPC of two courses slates laid in 1:3 Group 1 cement:sand mortar.

for excavation for foundations	*D20*
for concrete foundations to walls	*E10*
for damp-proof courses	*F30*
for precast concrete copings	*F31*

F20 Natural stone walling

GENERAL GUIDANCE

This section covers natural undressed or dressed stone, including drystone walling.

In traditional stone areas, the stone used to be quarried very locally, and even now stone is the preferred option when economically viable. Stone walls were widely used for boundary and garden walls, and are still required in National Parks, AONBs and some urban conservation areas. Before finalizing any design drawings, the landscape designer would be well advised to seek local advice firstly from the local stone yard, and secondly from a master mason. Basic information on different types of stone is given below, but care should be taken that a suitable stone is specified that is suited both to the climate and to the situation in which it is to be used. Granite, because it is so dense and impervious to most pollutants, survives in London and other large cities; Portland stone is also a good survivor under these conditions.

Nearly all natural stone will change colour by weathering, and some kinds are unable to withstand frost; if possible, a site should be inspected where the stone of choice has been exposed to similar conditions and weathered for several years before finally specifying it. Where two different types of stone are to be used, or where the stonework is to be adjacent to brickwork, chemical action or staining can occur.

Fig. 4.6 Dry stone walling with opening.
(Stone supplied by CED Ltd)

Fig. 4.7 Knapped chalk flints. An old tradition, now revived.
(Stone supplied by CED Ltd)

BRITISH STANDARDS

PD 6472 : 1974	Guide to specifying the quality of building mortars
12 : 1996	Portland cement
890 : 1996	Building limes
1200 : 1976 & AMD	Building sands from natural sources
4027 : 1996	Sulphate-resisting Portland cement
4721 : 1981 (1986) & AMD	Ready mixed building mortars
4887 Pt 1 : 1986	Mortar admixtures. Air-entraining (plasticizing) admixtures
4887 Pt 2 : 1987	Mortar admixtures. Set retarding admixtures
5224 : 1995	Masonry cement
5390 : 1976 (1984) & AMD	Code of practice for stone masonry
6577 : 1985	Mastic asphalt for building (natural rock asphalt aggregate)
7583 : 1996	Portland limestone cement

DATA

Types of stone

Useful information about the types of stone available is contained in the *Natural Stone Directory*, published by Ealing Publications at £25.00. This is a new edition, which covers all British quarries and also lists importers of foreign stone. It contains coloured illustrations, and gives the characteristics of the stone abstracted from each quarry.

Portland stone	Grey-white, hard limestone, smooth with shell flecking
York stone	More often used for paving, yellow-buff, shows laminations clearly
Cotswold stone	Grey-buff limestone in smaller sizes than Portland, splits easily
Bath stone	Creamy-buff, softer, smooth textured, not strongly marked
Granite	Pink or grey, sparkling, unlaminated igneous stone, hard to cut, and virtually indestructible

Types of walling

There are six types of walling likely to be met in landscape work:

- uncoursed random rubble;
- random rubble brought to courses;
- uncoursed squared rubble;
- snecked squared rubble;
- coursed squared rubble;
- ashlar.

Mortars

Mortars for natural stonework vary according to the type of stone; as a guide the following mixes are given:

- Granite and other igneous rocks require a strong 1:3 cement:sand mortar, but being non-absorbent granite needs less than the normal amount of water used in the mix.
- Hard sandstones: use 1:1:6 cement:lime:sand mortar.
- Soft sandstones: use 1:3:12 cement:lime:sand.
- Copings, plinths and other ornamental features: use 1:2:9 cement:lime: sand.
- Very soft absorbent stone: use 1:2.5 hydraulic lime:sand.

Finishes to stone

There are three main finishes to stone walls:

- uncut or quarry-pitched stone;
- roughly squared stone shaped on site;
- ashlar, or carefully dressed stone blocks cut to accurate sizes.

Natural stone ashlar may be 'dressed', or face finished, in several ways:

- sawn: the face as left by the stone saw, which does not make the best of the natural stone appearance, but is cheap;
- dragged: the stone is brought to a true face with a rough grinder;
- tooled: various tools are used across the face of the stone to produce regular or irregular grooving of varying depths;
- rubbed: the stone is ground to a perfectly smooth regular face, which shows off the nature of the grain;
- polished stone: used for internal stonework, and is therefore unlikely to be used for landscape walls except in atria or covered courtyards. The polishing process involves bringing the stone to a reflective finish.

OUTLINE SPECIFICATION

Drystone wall

Remove topsoil down to firm subsoil and construct Cotswold drystone wall 1400 mm high and 700 mm wide at base on foundation stones 900 mm wide set below ground level on firm subsoil (or hardcore or stone waste 150 mm deep) complete with covers, copes laid vertically, all in accordance with the Drystone Walling Association technical specifications.

Squared rubble wall

a. Construct squared rubble wall 900 mm high × 250 mm thick of limestone from specified quarry, laid on GEN4 concrete foundation

400 mm wide × 200 mm thick, bedded and jointed in 1:3:12 white cement:lime:sand mortar, all stones to be free of quarry sap and not laid surbedded.

b. Lay coping of cast stone 300 mm wide × 100 mm thick twice weathered laid in 1:1:9 cement:lime:sand mortar.

Ashlar stone wall

a. Construct wall 1100 mm high × 300 mm thick of Guiting Stone limestone in random sizes with dragged finish on both faces, laid on GEN4 concrete foundation 450 mm wide × 200 mm thick, bedded and jointed in 1:3:12 white cement:lime:sand mortar, all stones to be free of quarry sap and not surbedded.

b. Lay coping of Portland Roach stone rubbed finish 290 mm × 125 mm thick twice throated and once weathered laid in 1:1:9 cement:lime:sand mortar on DPC of two courses slates bedded in 1:3 cement:sand mortar.

for excavation	D20
for foundations	E10

F22 Cast stone

GENERAL GUIDANCE

This section covers precast artificial stone walling units used for small landscape walls. It does not deal with the large structural units used for retaining walls, as these are the province of the civil engineer.

These are manufactured from natural aggregates mixed with ordinary Portland cement together with colouring pigments. They are moulded to give various traditional masonry walling types, or hydraulically pressed and split or pitch faced to give different effects. Finishes available include ashlar, rough dressed, and squared and pitched. Coloured mortar may be used, but contrasting colours can radically change the character of the wall; in most cases it is advisable to use both reconstructed stone colour and mortar to match that of walls within the district, and where possible sites where these have been used should be visited in order to see how they have weathered over a few years.

In order to give an authentic appearance to 100 mm walls some manufacturers produce 'jigsaw'-shaped pieces that fit snugly together. All through-joints should be 10 mm thick and faced up to a wider finish; those within the 'jigsaw' pieces have faced joints only. Flush or bucket-handle joints are recommended in exposed areas. The correct mortar to suit the conditions in which the wall is to be used must be specified: the greater amount of lime in the mix gives less strength, but greater ability to accommodate movement

Fig. 4.8 Five patterns of Chilstone reconstituted stone balustrades including Renaissance, Georgian and Regency patterns. (Chilstone)

Fig. 4.9 Chilstone reconstituted stone polo balustrade illustrated with serpentine seat and southwood finial. (Chilstone)

Fig. 4.10 Haddonstone's standard balustrading is available in six styles with 26 standard balustrades from 311 mm to 779 mm in height.

from moisture and temperature changes. Movement joints are normally required as for brickwork. Walls can be dressed with brick or a different type of masonry finish, or even backed with brick. Pier caps, copings, balusters and ornamental cappings are available for formal walling.

BRITISH STANDARDS

12 : 1996	Portland cement
890 : 1996	Building limes
1200 : 1976 & AMD	Building sands from natural sources
1217 : 1986	Cast stone
4027 : 1996	Sulphate-resisting Portland cement
4721 : 1981 (1986) & AMD	Ready mixed building mortars
4887 Pt 1 1986	Mortar admixtures. Air-entraining (plasticizing) admixtures
4887 Pt 2 1987	Mortar admixtures. Set retarding admixtures
5224 : 1995	Masonry cement
5390 : 1976 (1984) & AMD	Code of practice for stone masonry
6457 : 1984	Reconstructed stone masonry units
7583 : 1996	Portland limestone cement (A type of cement with up to 20% of specified limestone. Limestone cement can be used as ordinary Portland cement in designed mixes)

DATA

The British Standards give the materials to be used for cast and reconstructed stone, but do not give sizes of units.

Typical sizes of precast stone are:

Simulated squared rubble	Simulated tooled ashlar
200, 225, 250, 275, 300, 325 mm long	800, 300, 450 mm long
75, 100, 125, and 150 mm high	75, 103, 150, and 225 mm high
100 mm bed thickness	100 mm bed thickness

Masonry components available include copings, parapets, decorative urns and balls, corbels, finials, and kickers.

Mortar mixes

Mortar group	Cement:lime:sand	Masonry cement: sand	Cement:sand with plasticizer
1	1:0–0.25:3		
2	1:0.5:4–4.5	1:2.5–3.5	1:3–4
3	1:1:5–6	1:4–5	1:5–6
4	1:2:8–9	1:5.5–6.5	1:7–8
5	1:3:10–12	1:6.5–7	1:8

Group 1: strong inflexible mortar
Group 5: weak but flexible.

All mixes within a group are of approximately similar strength. Frost resistance increases with the use of plasticizers. Cement:lime:sand mixes give the strongest bond and greatest resistance to rain penetration.

Masonry cement to BS 5224: 1995 covers a wider range of cement types and defines three strength classes. Non-air-entrained cements are included.

Class	28 day strength N/mm2	Air entrainment (X = non-air-entraining)
MC 5	5–15	yes
MC 12.5	12.5–32.5	yes
MC 12.5X	12.5–32.5	no
MC 22.5X	22.5–42.5	no

Masonry cement equals ordinary Portland cement plus a fine neutral mineral filler; it is easier to work and is more frost resistant than a cement:lime mixture. Masonry cement should never have lime added to it, and should not be used on walls subject to heavy loading.

OUTLINE SPECIFICATION

Cast stone squared rubble wall

a. *Construct squared rubble wall 900 mm high × 250 mm thick of cast stone to comply with BS 6457 by specified manufacturer to simulate Cotswold stone, laid on GEN4 concrete foundation 400 mm wide × 200 mm thick, bedded and jointed in Group 5 white cement:lime:sand mortar 1:3:12.*

b. *Coping of cast stone 300 mm wide × 100 mm thick twice weathered laid in Group 4 masonry cement:sand mortar 1:5:6.*

for excavation	*D20*
for foundations	*E10*

Ashlar cast stone wall

a. *Construct wall 1100 mm high × 300 mm thick of cast stone to BS 6457 to simulate Portland stone limestone in random sizes with dragged finish on both faces laid on GEN4 concrete foundation 450 mm wide × 200 mm thick, bedded and jointed in 1:3:12 white cement:lime:sand mortar.*

b. *Coping of same stone rubbed finish 290 mm × 125 mm thick twice throated and once weathered laid in 1:1:9 cement:lime:sand mortar on DPC of two courses slates bedded in 1:3 cement:sand mortar.*

for excavation	*D20*
for foundations	*E10*

F30/F31 Copings and accessories for brick/block walling

GENERAL GUIDANCE

This section covers damp-proof courses, metal and stone copings, and fixings for masonry walls.

Cappings and copings

All masonry walls made from bricks, precast concrete blocks, stone, or even drystone walls, require a waterproof and frostproof capping. Some copings may also require a steep pitched top where it could be considered a potential danger in allowing children to walk along the top of the wall, as for example where a retaining wall is low on one side and high on the other. Where vandals take a delight in removing copings it may be advisable to cap walls under 2000 mm high with a flush-sided coping that will not provide leverage.

BS 5642 Pt 2 covers precast concrete, cast stone, clayware, slate and natural stone copings, and specifies two types: the clip type, which sits over the wall, and the flat-bottomed type, which sits on the wall.

Precast concrete copings clip types must have a 65 mm high outer vertical edges, and flat bottomed types 50 mm. Both require at least 45 mm overhang on either side of the wall, with a throating 12 mm wide × 8 mm deep; throats must be 15 mm from the outer edge. This helps to shed water and snow from the wall beneath. They are available in various profiles and textures, and are manufactured in straight lengths together with end and corner units to match, and in single large units for piers. Although more expensive, special units can be manufactured to order.

Damp-proof courses

These are intended to protect the wall from descending damp, which can result in the growth of algae, efflorescence, and deterioration of the mortar. They are used under copings that are not waterproof, and it is advisable to use either engineering DPC1 or DPC2 bricks to BS 3921 Table 4, or high-bond membrane DPCs in order to prevent the coping from slipping at the DPC level. When dampness in the wall is of no import, the DPCs may be omitted. High-bond DPCs are inserted in the top joint under the coping or in the masonry joint below it; where no DPC is used the waterproof coping is laid in the full 10 mm standard mortar bed. In both cases all joints between the coping units must be carefully filled. Cappings and copings are bedded in 1:¼:3 cement:lime:sand mortar.

Fig. 4.11 Precast concrete dock edge. (Shap Concrete Products)

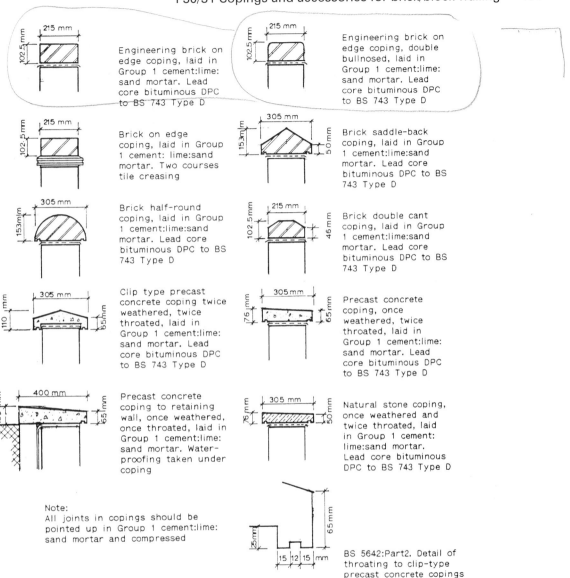

Fig. 4.12 Copings to walls.
Note: All joints in copings should be pointed up in group 1 cement:lime:sand mortar and compressed.

Stone copings

Cast stone and natural stone copings have very similar specifications in the British Standard. Natural stone finishes can be either sawn or rubbed.

Where slate copings are used, type A must be specified for areas of high pollution, while type B can be used in areas of moderate to slight pollution. Where the flat-bottomed type is specified, it must have a serrated base. In either case the outer vertical height is 19 mm and the throat is 10 mm wide; they are available with either a sawn or rubbed finish. Slate copings do not require a DPC.

Metal copings

Metal copings are not covered by a British Standard, because they are usually made to order, either as preformed profiles or as sheet material formed on site. They can be made in copper, zinc, aluminium, lead or stainless steel to match the thickness of the wall. Stainless steel and aluminium sections are preformed, while copper, zinc and lead sheet are usually formed on site.

Metal copings are not suitable for capping walls where heavy use or exposed conditions are likely. The manufacturer will provide the correct profile for the wall, including all necessary clips and fixings.

Cramps and dowels

Where cramps and dowels are required the British Standard specifies copper, copper alloy or austenitic stainless steel 18/8 cramps and dowels.

BRITISH STANDARDS

5224 : 1995	Masonry cement
5628 Pt 3 : 1985	Use of masonry, Part 3, materials and components, design and workmanship
5642 Pt 2 : 1983	Copings of precast concrete, cast stone, clayware, slate and natural stone
6398 : 1983	Bitumen damp courses for masonry
6477 : 1992	Water repellents for masonry surfaces
8000 Pt 3 : 1989	Code of practice for masonry (brick and block)

DATA

DPCs to copings

- High-bond bituminous felt; types D and E.
- Lead coated with bituminous paint.
- Engineering brick copings bedded in 1:3 cement:sand mortar.
- Preformed copper, aluminium or stainless steel copings need no DPC.
- Two courses of well-burnt clay tiles lapped and tilted to fall, in 1:3 cement:sand mortar.

Mortar mixes

Mortar group	Cement:lime:sand	Masonry cement: sand	Cement: sand with plasticizer
1	1:0–0.25:3		
2	1:0.5:4–4.5	1:2.5–3.5	1:3–4
3	1:1:5–6	1:4–5	1:5–6
4	1:2:8–9	1:5.5–6.5	1:7–8
5	1:3:10–12	1:6.5–7	1:8

Group 1: strong inflexible mortar
Group 5: weak but flexible.

All mixes within a group are of approximately similar strength. Frost resistance increases with the use of plasticizers. Cement:lime:sand mixes give the strongest bond and greatest resistance to rain penetration. Masonry cement to BS 5224: 1995 covers a wider range of cement types and defines three strength classes. Non-air-entrained cements are included.

Class	28 day strength N/mm2	Air entrainment (X = non-air-entraining)
MC 5	5-15	yes
MC 12.5	12.5-32.5	yes
MC 12.5X	12.5-32.5	no
MC 22.5X	22.5-42.5	no

Masonry cement equals ordinary Portland cement plus a fine neutral mineral filler; it is easier to work and is more frost resistant than a cement:lime mixture. Masonry cement should never have lime added to it, and should not be used on walls subject to heavy loading.

OUTLINE SPECIFICATION

Precast concrete coping

Fix precast clip-type concrete coping to wall previously constructed, minimum concrete grade to be C50, 45 mm overhang both sides with throating 12 mm × 8 mm deep 15 mm from outer edge both sides, twice weathered with 65 mm vertical face, bedded in 1:¼:3 cement:lime:sand mortar with high-bond bituminous felt DPC type D or E. Joints between coping units to be filled with Group 1 masonry cement:lime:sand mortar.

for brick and block walls	*F10*
for natural stone walls	*F20*
for precast stone walls	*F22*

J Waterproofing

GENERAL GUIDANCE

This section includes water repellents for masonry surfaces, waterproof rendering, mastic asphalt tanking, and liquid-applied tanking.

Many waterproofing agents are proprietary brands not covered by a British Standard, and unless the landscape designer knows the product well from experience, he or she should check with the manufacturer that it has a British Board of Agrément (BBA) or similar quality use certificate.

Water repellents

These products are clear transparent liquids brushed onto the surface; even on walls in good condition they do not have a very long life, and need regular maintenance. Water-repellent treatment is used mainly for protecting fine masonry from pollution and other chemical attack, and it is often used in conservation work, though there can be problems if the brick or stone is not physically or chemically stable before treatment. Suitable products are covered in BS 6477 : 1992, but the manufacturer's recommendations should always be followed carefully.

Waterproof rendering

Properly made cement rendering on a backing of good concrete is waterproof in itself, and most leakage is due to fine cracking caused by poor construction. Waterproof rendering is intended to prevent rainwater from entering the wall, and not to provide a completely water-containing surface, because it is a rigid material, and cannot give with the movement of the structure in the same way as asphalt. Waterproof rendering is composed of ordinary Portland cement, lime and sand mortar with a waterproofing agent added. It is usually put on in two coats, and care must be taken to ensure that the wall or slab is clean, dust and oil free, and rough enough to form a good key for the rendering; if necessary it must be hacked or gritblasted, and this is essential on old walls. Surfaces that are inclined to spall, crack or dust up are not suitable for rendering.

Mastic asphalt

Mastic asphalt can be either natural rock asphalt to BS 6577 or limestone aggregate asphalt to BS 6925. Both perform in much the same way, but the former is about twice as expensive, and is not likely to be specified by the landscape designer unless requested by the client. The wall or slab should be quite clean and free from dust, oil, cement or other material that will prevent adhesion of the asphalt. Prior to mastic asphalt tanking, the landscape designer should specify a primer recommended by the mastic asphalt manufacturer in order to ensure a good bond with the surface; this is normally proprietary bitumen or rubber emulsion. Recessed horizontal joints to new brickwork should be specified in order to provide a key for the mastic asphalt, while for existing brickwork the joints should be raked out and thoroughly cleaned to provide a key. All surfaces that are to have mastic asphalt coatings should be fully cured, and free from injurious chemicals. All finished work must be protected as soon as possible, either by a 50 mm cement and sand screed for horizontal work, or by a half-brick wall for vertical work, because even slight damage to a waterproof coating will be liable to cause leaks.

Liquid-applied waterproofing

This is not so effective as tanking with mastic asphalt, and is suitable only where no water pressure is likely to develop; it should not be relied on to give an absolutely waterproof structure. The materials are all proprietary, and the manufacturer's recommendations should be closely followed, as the timing of applying coats is important. The surface should be clean, smooth, and free from injurious chemicals; concrete or rendering must be fully cured.

BRITISH STANDARDS

6477 : 1992	Water repellents for masonry surfaces
6925 : 1988	Mastic asphalt for building and civil engineering (limestone aggregate)
6949 : 1991	Bitumen based coatings for cold application, not for potable water
8000 Pt 4 : 1989	Code of practice for waterproofing
8000 Pt 10 : 1989	Code of practice for plastering and rendering

DATA

Water repellents

BS 6477 : 1992 classification of water repellents for masonry surfaces by substrate on which they are to be used:

- Group 1 siliceous: clay bricks, sandstone, mature cement or cement/lime render, fibre reinforced cement.
- Group 2 calcareous: limestone, cast stone (check with supplier)
- Group 3 fresh alkaline material: new work or repairs to painted or rendered brickwork (check with supplier)
- Group 4 calcium silicate brickwork: sandlime or flintlime bricks

Note: Some water repellents are suitable for use on more than one group of substrate. All should be capable of being applied by a single generous coat.

Identification marking
Includes BS 6477, group or groups to which it can be applied, instructions and precautions.

Mastic asphalt

BS 6577 : 1985 *Mastic asphalt for building* (natural rock asphalt aggregate)
    Use for roofs Type R1162, for tanking and DPCs Type T1418.
BS 6925: 1988 *Mastic asphalt for building and civil engineering* (limestone aggregate)
    Use for roofs Type R988, for tanking and DPCs Type T1097.

Approximate weight of mastic asphalt

1 mm thick mastic asphalt weighs 2.4 kg/m^2
e.g. 10 mm = 24 kg/m^2
        13 mm = 31
        20 mm = 48
        25 mm = 60
        30 mm = 72

Tanking

Roof garden tanking

Horizontal surfaces:	30 mm thick three-coat work
Vertical surfaces:	300 mm or less high 13 mm thick two-coat work
	over 300 mm high 20 mm thick three-coat work

Planter tanking

Horizontal surfaces:	20 mm thick two-coat work
Vertical surfaces:	13 mm thick two-coat work

Key vertical surfaces and apply a primer recommended by the mastic asphalt supplier before laying; this is normally either a proprietary bitumen or rubber emulsion. All surfaces to be protected as soon as possible after laying.

Typical proprietary waterproofing

Waterproof rendering:	clear, black or cementitious
Mastic asphalt tanking:	to BS 6925 type T 1097
Liquid-applied tanking:	liquid asphaltic composition
	pitch epoxide coating
	pitch-based tanking
Flexible sheet tanking:	bituthene 1.5 mm thick

OUTLINE SPECIFICATION

Mastic asphalt tanking to flat roofs

a. *Clean off surfaces and remove all dust, oil and chemical stains, grit blast or hack surface if necessary to form key for tanking. Ensure that the surface is completely dry before laying mastic asphalt.*

b. *Lay felt isolating membrane.*

c. *Apply manufacturer's approved primer to surface, lay 30 mm thick three-coat work mastic asphalt to BS 6925 : 1988 Type R 988.*

d. *Provide temporary protection to surface unless screed is to be laid immediately afterwards.*

for brickwork	*F10*
for concrete	*E10*

Mastic asphalt tanking to planters

a. *Clean off surfaces and remove all dust, oil and chemical stains, grit blast or hack surface if necessary to form key for tanking. Ensure that the surface is completely dry before laying mastic asphalt.*

b. *Apply manufacturer's approved primer to surface, lay 20 mm two-coat work mastic asphalt to BS 6925 Type 1097 to base and lay 13 mm thick two-coat work to walls.*

c. *Provide temporary protection to surface unless screed is to be laid immediately afterwards.*

for brickwork	*F10*
for concrete	*E10*

Liquid-applied waterproofing

a. *Clean off all dust, oil, chemical stains and ensure that the surface is smooth and dry before applying the compound.*

b. *Apply two coats proprietary waterproofing compound in accordance with the manufacturer's instructions.*

c. *Provide temporary protection to surface unless permanent protection is to be constructed immediately afterwards.*

Waterproof rendering

a. *Rake out joints to brickwork or hack stone or concrete surfaces to form key. Clean off and remove all dust, oil and chemical stains and ensure that the surface is clean and dry before applying the finish.*

b. *Apply two coats 13 mm work cement:lime:sand 1:2:9 with waterproof additive to surface, wooden float finish.*

c. *Protect from hot sun and frost until rendering is fully cured.*

for brickwork	*F10*
for concrete	*E10*

M40 Tiling

GENERAL GUIDANCE

This section includes ceramic and clay tiles used for external work at the edges of outdoor swimming pools and so forth.

Whatever type of tile the landscape designer specifies, it is important to check with the manufacturer that it is frost and slip resistant, and if used in or adjacent to a swimming pool that it is resistant to swimming pool water and cleaners. Swimming pools with seawater are not included in the data given below; where this is to be used, contact the technical department of the chosen manufacturer.

Laying tiles

Tiles are laid on a cement:sand bedding or with special adhesives designed for the purpose; they must be fully bedded without air pockets, and for cement:sand bedding both the tiles and the base should be lightly wetted before tiling. The joints are usually pointed afterwards with a contrasting or matching cement:sand grout, and therefore they are at least 3 mm wide. The surplus grout must be cleaned off immediately after grouting without the use of chemicals or abrasives. Special sealants are often used for swimming pool tiling, or where there is likely to be extra stress on the tiles such as very hot water or frost, or where chemical spills may occur.

Tiled floors should be laid to good falls to ensure that rain, splashes or cleaning water can drain away quickly, as wet tiles are considerably less slip resistant than dry ones. Drain grids or gullies should be provided at frequent intervals to avoid ponding. Ribbed or chequered tiles are usually used at the edges of pools or steps to improve the slip resistance.

Because tiles are a very rigid form of paving, movement joints must be provided at not more than 3.0 m intervals, unless both the bedding and jointing consist of a flexible material. Joints should be made of flexible waterproof sealing strips or compound, which will take up any movement without impairing the water-resistance properties of the surface; the sealant manufacturer should be consulted. Movement joints should follow the joints in the concrete base below.

Any acceptable sealants used must be water resistant, and applied to a dry surface. Once the tiles are fixed and grouted they must be matured for two weeks before filling with water or other use.

Note that ceramic tiles are measured in centimetres, not millimetres.

BRITISH STANDARDS

5385 Pt 4 : 1992   Wall and floor tiling: Code of practice for tiling and mosaics in specific conditions

6431 Pt 1 : 1983 (EN 87)   Ceramic floor and wall tiles: definitions, classifications, characteristics, marking

6431 Pt 2 : 1984 (EN 121 : 1991)   Ceramic floor and wall tiles: extruded ceramic tiles with a low water content (Group A1)

6431 Pt 6 : 1984 (EN 176 : 1991)   Ceramic floor and wall tiles: extruded ceramic tiles with a low water content (Group B1)

DATA

Ceramic tiles to BS 6431 Pt 1 : 1983 (EN 87)

Group   l: less than 3%   low water absorption
     lla: between 3% and 6%   medium water absorption
     llb: between 6% and 10%   medium water absorption
     lll: more than 10%   high water absorption

Type   A: quarry (cut from a single extruded column, then either pressed or not pressed);
     B: dust-pressed tiles (body material reduced to powder or small grains and moulded at high pressure;
     C: cast (unlikely to be used by the landscape designer).

Identification marks

To include manufacturer's name or symbol, BS/EN reference, type/group reference (this may be on the package or tile).

Sizes of tiles commonly available

Modular sizes

Nominal joint width 3–11 mm. Coordinated sizes in centimetres:

30 × 15	25 × 25	20 × 20	15 × 15	10 × 10
	25 × 12.5	20 × 10		
		20 × 15		

Non-modular sizes

Nominal joint width 3 mm. Coordinated sizes in cm:

28 × 14	15.2 × 15.2	14 × 14
26 × 13	15.2 × 7.6	13 × 13
22.9 × 22.9	15 × 15	
20.3 × 20.3		

Conditions for use of ceramic tiles

Tiles to BS 5385 Pt 4 : 1992. Tiles suitable for landscape work

Use/condition	Acceptable	Check with manufacturer
Walls		
Infrequent contact with water	AI AIIa BI BIII	
Frequent or continuous immersion	AI BI	AIIa BIII
Thermal change (climatic)	AI BI	AIIa BIII
Floors		
Light to moderate traffic*	AI AIIa BI	
Infrequent contact with water	AI AIIa BI BIII	
Frequent or continuous immersion	AI BI	AIIa BIII
Thermal change (climatic)	AI BI	AIIa BIII

*Low-density pedestrian and soft-wheeled light-weight vehicles.

Bedding of tiles

For light to moderate traffic,* infrequent contact with water, thermal change (climatic):

- cement and sand;
- cement and sand (semi-dry mix);
- cement and sand (semi-dry mix) over separating layer;
- cementitious adhesives;
- organic adhesives;
- chemically resistant adhesives.

*low-density pedestrian and soft-wheeled light-weight vehicles.
For frequent or continuous immersion:

- cement and sand;
- cement and sand (semi-dry mix) when bonded to the base;
- cementitious adhesives;
- chemically resistant adhesives.

OUTLINE SPECIFICATION

Tiling to wet area

a. *Lay external-quality clay quarry tiles 20 cm × 20 cm to BS 6431 Pt 1 : 1983 (EN 87), Type A pressed, Group I with less than 3% water absorption, colour to match approved sample, on semi-dry cement:sand 1:3 bedding 20 mm thick minimum, bonded to base on concrete slab specified in E10.*

b. *Base to be smooth, clean and free from injurious substances, concrete to be fully cured before laying tiles. Tiles and base to be wetted before laying and surplus water removed, joints to be 3 mm wide grouted with cement:sand 1:2 grout coloured to match tiles and cleaned off immediately after laying.*

for concrete   *E10*

Tiling to dry area

a. *Lay external-quality clay quarry tiles 20 cm × 20 cm to BS 6431 Pt 1 : 1983 (EN 87), Type B pressed, Group I with less than 3% water absorption, colour to match approved sample, on cement:sand 1:3 bedding 20 mm thick minimum, on concrete slab specified in E10.*

b. *Base to be smooth, clean and free from injurious substances, concrete to be fully cured before laying tiles. Tiles and base to be wetted before laying and surplus water removed, joints to be 5 mm wide grouted with cement:sand 1:2 grout coloured to match tiles and cleaned off immediately after laying.*

for concrete   *E10*

M60 Painting and clear finishing

GENERAL GUIDANCE

This section covers work carried out on site, and includes finishes to metal, wood and masonry; it covers paint, clear finishes and masonry paint, together with site-applied timber preservatives.

A prerequisite of all painting and decorative finishes is that they should be compatible not only with each of the constituent layers used on site, but also with the 'delivery' or 'factory' finish. As items in different batches have been known to have different 'factory' finishes, the landscape designer is advised to get all the information available from the supplier's technical department before specifying a paint or decorative regime. Although protective paint should be applied as soon as the surface is ready for it, the final coat of paint should be left until the end of the contract to avoid damage and patchy making good. All paint of one type and colour must come from the same batch, because colour can vary noticeably from one batch to another.

Timber

The traditional painting system is four-coat work: the primer (now often the 'factory' finish on made-up items), two undercoats, and a top coat. The standard system omits one of the undercoats, but both systems can be applied only to wrought or finished timber that has been sanded to a smooth finish. (The landscape designer will find it difficult to decide the number of undercoats that have been applied, and will have to rely on the contractor's carrying out the work as specified.) Where knots occur in softwood they must be treated before priming to prevent them bleeding though the paint finish. If this does occur, it is particularly noticeable in white or light-coloured finishes; when the client asks for these colours, the landscape designer is advised to specify timber with a low resin content and few knots. Wood stains can be applied to either wrought or unwrought timber, and clear protective finishes can be used to maintain the grain of the timber in its natural colour. CCA and OS (organic solvent) preservatives are usually suitable for painting, but this should be checked with the paint supplier. Most external timber is now supplied pressure impregnated with a preservative, but many of these are inimical to plants, and it is advisable to check the type of preservative used with the supplier if the timber is to used as a support for climbers or as a planter. Timber preservative must be applied to any timber exposed on site by cutting or jointing.

Metal

All bare metalwork should be cleaned on site immediately before painting; this can be done by hand or by sandblasting, which needs to be carried out with care. Steel railings and other metalwork can be supplied factory finished or primed ready for painting. Farm gates with ironmongery, and heavy security railings, are usually supplied hot-dipped galvanized, which gives a thicker coating than electrolytic methods. The galvanizing can be left as supplied, or it can be treated with a mordant and a calcium plumbate primer so that an ordinary paint system can be used on site. Ornamental metalwork can be factory powder-coated, and although this finish is durable, any damage will have to be made good with paint. The traditional red lead and white lead paints are no longer legal (with a few exceptions), and should not be specified. Bitumen-based paints to BS 3416 are suitable for concealed ironwork where it is not readily accessible to the public.

Brickwork, stonework and concrete

Brick and stone walls are seldom painted in landscape contracts, but there may be a need to provide a reflective finish to walls to improve lighting conditions. The traditional whitening was lime wash – a compound of slaked lime and tallow, which had a preservative effect – but the modern treatment is to apply one of the proprietary cement-based paints or masonry paints in one or two coats. These are made in white and pale shades of buff, yellow or pink, and they may have sand or grit incorporated to give a textured surface.

Water-repellent clear finishes are manufactured for preserving stonework from acid rain and chemical run-off, but these should be specified only on the advice of a stonework expert, as damage can be done by incorrect application. Plain concrete walls may be painted to enhance their appearance, or to conceal formwork marks and so forth; here a textured paint should be used.

BRITISH STANDARDS

1070 : 1993	Black paint (tar based)
3416 : 1991	Bitumen-based coatings for cold application, suitable for use in contact with potable water
4072 Pt 2 : 1987	Wood preservation: copper/chromium, arsenic compositions
4072 Pt 1 : 1987 & AMD	Wood preservation: copper/chromium, arsenic compositions
4652 : 1995	Metallic zinc-rich priming paint (organic media)
4756 : 1971	Ready mixed aluminium priming paints for woodwork
4800 : 1989	Paint colours for building purposes
5082 : 1993	Water-borne priming paints for woodwork
5358 : 1993	Solvent-borne priming paints for woodwork: low lead content
5493 : 1977	Code of practice for protective coating of iron and steel against corrosion
5589 : 1989	Code of practice for preservation of timber
6150 : 1991	Code of practice for painting of buildings
6477 : 1992	Water repellents for masonry surfaces
6949 : 1991	Bitumen based coatings for cold application, excluding use in contact with potable water
6952 Pt 1 : 1988	Exterior wood coating systems: guide to classification and selection
7773 : 1995	Code of practice for cleaning and preparation of metal surfaces (replaces CP 3012)

DATA

Durability of timber

Species	Natural durability	Treatment Heartwood	Sapwood
Ash – European	P	MR	MR
Beech – European	P	P	P
Birch – European	P	P	P
– Paper	ND	MR	MR
Cedar, Western Redwood imported	D	R	R
UK	MD	R	R
Chestnut – European horse	P	P	P
– sweet	D	ER	n/a
Elm – Dutch	ND	MR	P
– English	ND	MR	n/a

Durability of timber *continued*

Species	Natural durability	Treatment Heartwood	Sapwood
Fir – Douglas (UK)	MD	R	MR
– Douglas (N. American coast)	MD	R	n/a
– Douglas (N. American mountains)	MD	ER	n/a
Hem-fir (N. American)	ND	R	n/a
Hemlock – western	ND	R	MR
Iroko	VD	ER	n/a
Jarrah	VD	ER	n/a
Karri	D	ER	n/a
Larch – Dunkeld (UK)	MD	R	MR
– European	MD	R	MR
Lime	P	P	P
Oak – European	D	ER	P
– Turkey	MD	ER	P
Padauk – Burma	VD	ER	n/a
Pine – American pitch	MD	R	P
– Canadian red	ND	MR	P
– Corsican	ND	MR	P
– Scots	ND	MR	P
– yellow	ND	MR	P
Redwood – European			
Sapele	MD	R	MR
Spruce – European (whitewood)	ND	R	n/a
– Sitka	ND	R	MR
Sycamore	P	P	P
Utile	D	ER	n/a
Walnut – African	ND	R	n/a

Treatment key

Durability		Ability to accept treatment	
VD	very durable	ER	extremely resistant
D	durable	R	resistant
MD	moderately durable	MR	moderately resistant
		P	permeable
ND	non-durable	n/a	data not available
P	perishable		

Note: The above information is based on the timber preservatives listed below.

Types of preservative to BS 5589:1989

Creosote (tar oil) can be 'factory' applied
    by pressure to BS 144 Pts 1 and 2
    by immersion to BS 144 Pt 1
    by hot and cold open tank to BS 144 Pts 1 and 2
Copper/chromium/arsenic (CCA)
    by full cell process to BS 4072 Pts 1 and 2
Organic solvent (OS)
    by double vacuum (vacvac) to BS 5707 Pts 1 and 3
    by immersion to BS 5057 Pts 1 and 3
Pentachlorophenol (PCP)
    by heavy oil double vacuum to BS 5705 Pts 2 and 3
    Boron diffusion process (treated with disodium octaborate to BWPA Manual 1986)
Note: Boron is used on green timber at source, and the timber is supplied dry.

Durable timbers for fencing

Preservation of timbers for fencing based on BS 5589 : 1989.
    Timbers durable enough to be used without treatment when:

	Category A 40 year life	Category B 20 year life
In the ground	VD	D VD
Not in contact with the ground	D VD	MD D VD

Durable timbers for agricultural and horticultural use

Preservation of timbers for agricultural and horticultural use based on BS 5589 : 1989.
    Timbers durable enough to be used without treatment when:

	Category A 50 year life	Category B 20 year life
In the ground, or in/near manure	VD	D VD
Subject to intermittent wetting	D VD	MD D VD

VD = very durable, D = durable, MD = moderately durable

Suitable preservatives:
- CCA water borne, leave 7 days then hose down thoroughly
- Coal tar creosote, check for bleeding before commissioning

- PCP in heavy oil
- OS to BS 5707 Pt 1 class F or F/N, allow to evaporate before use

Note: Some fruits are very susceptible to tainting; check with the manufacturer before using preservatives.

OUTLINE SPECIFICATION

Painting steel and ironwork

On factory-primed steel and ironwork, clean off all surfaces, touch up primer, paint two undercoats and one finishing coat of gloss oil paint to specified colours.
or
On bare steel and ironwork, clean off all surfaces, paint one coat primer, two undercoats and one finishing coat of gloss oil paint to specified colours.

for railings	*Q40*
for security fencing	*Q40*
for pedestrian guard rails	*Q40*
for street furniture	*Q50*

Painting galvanized steelwork

On galvanized steelwork, apply one coat mordant solution, one coat special primer, one coat enamel finishing coat to specified colours.
or
On galvanized steelwork, apply two coats bituminous paint to BS 6949.

for railings	*Q40*
for security fencing	*Q40*
for pedestrian guard rails	*Q40*
for trip rails	*Q40*
for street furniture	*Q50*

Painting timber

On factory-primed timber, touch up primer, paint two undercoats and one finishing coat of oil gloss paint to specified colours.
or
Rub down and remove dust, knot, paint one coat primer, stop, paint two undercoats and one finishing coat of oil gloss paint to specified colours.
or
Rub down and remove dust, apply three coats proprietary clear synthetic varnish.
or
Rub down and remove dust, apply two coats proprietary wood stain to specified colours.

for timber fencing	Q40
for trip rails	Q40
for street furniture	Q50

Timber preservation

To dry timber, apply two coats proprietary timber preservative well brushed in.
or
To dry timber, apply two coats light golden creosote.

| *for timber fencing* | Q40 |

Painting concrete

Brush off all loose material, apply two coats bituminous paint, first coat blinded with fine sand.
or
Brush off all loose material, apply two coats proprietary textured masonry paint to specified colours, all in accordance with manufacturer's instructions.

| *for street furniture* | Q50 |

Painting brick and stonework

To dry surface, clean down, apply proprietary masonry paint to specified colours all in accordance with manufacturer's instructions.

for brickwork	F10
for cast stone walling	F22
for precast concrete fencing	Q40
for trip rails	Q40

P30 Trenches/pipeways/pits for buried engineering services

GENERAL GUIDANCE

This section gives guidance on trenches for buried services.

Drains and sewers inside the site boundary are normally the responsibility of the owner of the land, and the pipes are laid by a contractor.

Water from the boundary stopcock and thence over the site are normally the responsibility of the owner/occupier, and the pipes are laid by a contractor.

Electricity, and if required gas, are the responsibility of their respective utilities, and all work to the meter is carried out by the utility, or subcontracted out by them.

Underground services should be in place, and all trenches backfilled, before any road or paving work is started; unless instructed otherwise, the utility contractors and general contractor will lay these under any path, foot-way or road if one can be found, and this is understandable as it provides good protection. However, if the landscape designer has the choice where underground services are to run, he or she should ensure that any future maintenance requiring a trench to be dug should be positioned where it causes the least possible damage; a relaid pavement or other hard surface can never be as good as the original work. If the pipes are to run in a joint trench, water pipes must be specified to be at 750 mm below ground level and all others at 450 mm below ground level. Once the water pipe has been inspected, the trench should be backfilled to 450 mm below ground level ready to receive the electricity, gas or other service tubes if required.

Most utilities prefer a central position on the footway for services, as this does not affect the kerb, or suffer from the weight of vehicle overrun; if there are boundary walls on the inner side of the footway they will not be affected either. It is also recommended that street furniture is kept 450 mm from the kerb to allow connection to lamps, signs and so forth to be made without disturbing the kerb. Where possible, all services in this position should be laid in ducts to enable maintenance to be carried out without dis-turbing the traffic. If this is done, the landscape designer must specify that the position be marked and the service indicated on site, as well as being shown on the drawings.

Safety precautions

Where water and electricity run either in parallel or in the same trench, the horizontal distance between water and electricity tubes should be at least 200 mm. As an extra safety measure, specify identification tapes to be positioned at least 150 mm above all actual tubes and bedded in as the work proceeds.

Both water and gas mains have been known to be fractured by tree roots, and the landscape designer should take care when specifying trees in these areas.

Note: Landscape design in ignorance of the relationship with existing and proposed external services is a recipe for potential conflict – contractually, professionally and aesthetically. When tree-planting location is vital to a par-ticular design purpose, e.g. where geometry is important or where space for tree screening is at a premium or needs to relate to a particular viewpoint, external services design should defer to the tree-planting layout. In such cir-cumstances the landscape designer will need to specify the minimum under-ground clearances (vertically and horizontally) necessary for the successful tree planting and growth to maturity, taking into account the likelihood of a requirement for re-excavation or access to the service line within the life-span of the trees concerned.

There are occasions when clearances can be achieved that would satisfy the soil space criteria for tree planting and growth (i.e. the protection of trees from service construction) but do not meet the requirements for the protection of ordinarily constructed underground services from direct damage resulting from tree growth. In such circumstances, the responsibility for ensuring that whatever additional protective measures are necessary to safeguard the undergrowth services must be placed squarely with the relevant services designer.

BRITISH STANDARDS

65 : 1991 & AMD Vitrified clay drain and sewer pipes
1710 : 1994 Identification of pipelines and services
4660 : 1989 PVC-U underground drain pipes and fittings 110 and 160 mm
5837 : 1991 Trees in relation to construction
6437 : 1984 Polythene pipes (type 50) for general purposes
6700 : 1987 Design, installation, testing and maintenance of services supplying water for domestic use within buildings and their curtilages

DATA

Service tubes and identifying tapes from utilities

Utility	Colour	Size	Identity tape
Electricity	Black	38 mm o.d.	Yellow with black lettering
Gas	Yellow	42 mm o.d. rigid	Yellow with black lettering
Water	Blue	(normally untubed)	Blue

Depth of cover required for services under carriageways:
   Water   900 mm
   Gas   750 mm
   Electricity 600 mm minimum
Depth of cover required for services under footways:
   Water   900 mm
   Gas   600 mm
   Electricity 600 mm minimum
Stopcocks for water normally 750 mm deep

Bedding flexible pipes: PVC-U or ductile iron

Type 1 = 100 mm fill below pipe, 300 mm above pipe: single-size material
Type 2 = 100 mm fill below pipe, 300 mm above pipe: single-size or graded material
Type 3 = 100 mm fill below pipe, 75 mm above pipe with concrete protective slab over

Type 4 =   100 mm fill below pipe, fill laid level with top of pipe
Type 5 =   200 mm fill below pipe, fill laid level with top of pipe
Concrete =   25 mm sand blinding to bottom of trench, pipe supported on chocks, 100 mm concrete under the pipe, 150 mm concrete over the pipe

This method should be used where pipes pass close to buildings or walls. If the side of the pipe trench is less than 1 m from the foundation of a wall, Building Regulations require that the pipe be bedded in concrete to the level of the underside of the wall foundation.

Sizes of granular fill material

10 mm granular material for 100 mm pipes
10 or 14 mm granular material for 150 mm pipes
10, 14, 20 mm granular material for 200–300 mm pipes
14 or 20 mm granular material for 375–500 mm pipes

Protecting pipes

Rigid pipes:
less than   1000 mm below road foundation   must be protected
            900 mm below fields and gardens   must be protected
Ductile iron pipes:
less than   1000 mm below road foundation   no protection
            300 mm below road foundation   concrete protection
            1000–300 mm below road foundation   granular surround
Flexible pipes:
less than   600 mm below soft landscape, footways must be protected
            900 mm below road foundation   reinforced concrete

---

OUTLINE SPECIFICATION

Trenches for pipework

a. *Excavate trench for pipework 900 mm deep × width to suit pipework including earth support, remove spoil to specified dump on site, dig out weak spots and fill with consolidated granular DoT Type 1 material, grade bottom of trench by hand to given levels.*

b. *After pipelaying, backfill trench with granular fill specified by the utility contractor, lay marker tape appropriate to utility, backfill with excavated material screened to remove all stones over 40 mm.*

for water pipes   *S10*
for electric cables   *V41*

Trenches for pipe ducts

a. *Excavate trench for pipe duct 900 mm deep × width to suit pipe including earth support, remove spoil to specified dump on site, dig out weak spots and fill with consolidated granular DoT Type 1 material, grade bottom of trench by hand to given levels.*

b. *Lay 160 mm PVC-U pipe to BS 4660 : 1989, Type 1 bedding of DoT Type 1 granular material, install drawline specified by utility contractor as the work proceeds, backfill with 300 mm DoT Type 1 material, backfill with excavated material screened to remove all stones over 40 mm, lay marker tape appropriate to utility 300 mm below finished ground surface, complete backfill to underside of paving base.*

    for electric cables   *V41*

Stop valve, drawpit or meter pit

a. *Excavate hole for pit and spread spoil round pit to 10 m distance.*

b. *Lay FND2 concrete slab 450 mm × 450 mm × 75 mm thick, construct half-brick-thick wall to pit 235 mm × 235 mm internal × 750 mm high in engineering brick Class B, laid in Group 1 sand:cement mortar 1:3, supply and fix cast iron hinged box cover set in Class 1 mortar as above.*

    for water services   *S10*
    for electric cables   *V41*

Q Paving generally

Unit pavings of stone have been used for paving footways for many hundreds of years. They varied in size from the vast, door-sized slabs used in Caithness to the hand-sized setts used in many town and cities throughout Great Britain. Edinburgh's Old and New Towns still have both footways and carriageways, as well as pedestrian precincts of setts, and in the latter many of the original patterns that were designed into them have been retained. End-grain elm blocks were also a common feature on the carriageways of many towns and cities. Both setts and blocks were introduced during the era of horse-drawn transport, and are not suited to the motor vehicle – or for that matter the bicycle. Setts are noisy at all times, and treacherous in wet or frosty weather; elm blocks become saturated with oil and again are treacherous in wet or frosty weather. In order to overcome these problems both setts and blocks were coated with macadam, and most have now been taken up and replaced by a safer, quieter monolithic finish.

As local quarries were worked out and cement was manufactured in vast quantities, precast concrete flags became the economic option, and were used to pave the footways of the new residential areas that appeared between the two World Wars; the broken pieces were bought by householders and used to form crazy paving in their gardens.

PAVING MATERIALS

Paving in the landscape contract may be divided into three main types: monolithic, unit, and granular. Heavy-duty road construction is outside the scope of this book, as it is the concern of the highway engineer.

Monolithic paving

Monolithic paving is laid as a continuous sheet of material, and includes tar macadam (now little used), asphalt, bitumen macadam, and *in situ* concrete.

Monolithic paving can be laid quickly, so that the final coat can be laid at the end of the contract without delaying the hand-over. Monolithic paving is also usually cheaper to construct than unit paving, but patterns are limited to surface texturing. Ribbing, studding, or other patterns can be formed on the surface when the material is newly laid.

Unit paving

Unit paving covers ordinary bricks, special paving bricks, pavers, rectangular paving blocks, interlocking blocks, cobbles set in concrete, concrete or stone setts, stone flags, precast concrete flags, and grass concrete blocks. Unit paving offers the widest range of prices, sizes, textures and colours, from the economical plain concrete flag used for city pavements to the very expensive natural slate or Yorkshire stone flags. The wide range available makes unit paving the best choice where the paving is an important element in the design. Unit paving can be laid in small and complicated areas such as internal courtyards and shopping malls, and it is useful for paving that adjoins buildings, as the small scale and bonded patterns are in harmony with brick or stone walls. One disadvantage is that unless the paving base is soundly constructed, even slight settlement will cause some units to sink or rise, while some unit paving, in particular ordinary bricks and natural stone flags, are not resistant to oil or other stains, and should not be specified where chemical handling and vehicle servicing are expected.

Granular paving

Granular paving includes loose gravel, bound gravel, hoggin, shale (or blaes), cobbles laid loose, shingle, ballast, bark and wood chips, and stone chippings. These paving materials are suitable for rural or domestic pedestrian paths and bridle-ways, but they should not be used where vehicle overrun is likely to occur regularly. Granular materials are adaptable for slopes and changes of angle; they absorb settlement and frost heave without permanent damage, and the appearance is not spoilt by minor scarring and scraping, but they must be provided with edge restraints both during construction and permanently to prevent the material from spreading over the surrounding ground. Loose material can be washed away on steep slopes, and must be controlled by stepping or reinforcement. Loose cobbles, shingle or heavy ballast must never be used in public areas where vandalism is likely to occur.

SAFETY

The Construction, Design and Management Regulations came into force on 31 March 1995, and it is now more important than ever that the landscape designer specifies materials fit for their purpose, and provides the client with a commissioning document describing the use for which the project was designed and the maintenance requirements of all surfaces. The slip resistance of any paved area is dependent on the finish, the gradient, the drainage system and the efficient maintenance of the area throughout its economic life. At the time of going to press the British Standard tests for slip resistance were under review, and there will be an ISO standard published that gives coefficients of friction. Testing stations use two main systems: the 'Tortoise system', which can test samples both in the laboratory and on site, but is unable to test materials with blistered, dimpled or ridged finishes; and the German ramp test, which can only carried out in a laboratory but will give readings for any type of finish.

Lifting any kind of paving for repairs almost invariably leaves dangerous raised edges or hollows, which can trip pedestrians and cause accidents to prams and wheelchairs. It is essential that all underground services should be in place and all trenches backfilled before any road or paving work is started.

For paving subsections refer to:

Kerbs, edgings, channels	Q10
Bases for all paving	Q20
In situ concrete roads	Q21
Macadam and asphalt roads	Q22
Gravel, hoggin and bark paving	Q23
Interlocking blocks	Q24
Slabs, bricks, setts and cobbles	Q25

Q10 Kerbs, channels and edgings

GENERAL GUIDANCE

This section covers all types of kerb and edging used to restrain light roadways and paved areas, as well as the units used as drainage channels in paving, including prefabricated surface-water drains. It does not include the heavy kerbs used in major highway construction.

Kerb is the term used for the heavy, dense, precast concrete sections associated with the junction between footways and carriageways, while **edging** is the term used for the lighter smaller sections associated with park and garden work, but both need to be firmly supported to prevent them from overturning and subsiding under pressure. **Edging strips** is the term used for longer lengths of material including timber; their main use is to retain grass, bark chips and gravel from the adjacent surface. Brick, stone and many other materials can be used as edgings.

The primary function of both kerbs and edgings is to retain the surface being laid, poured, vibrated and so forth.

The secondary function is to:

- separate the footway from the carriageway, thus segregating pedestrians from vehicles;
- provide a safe demarcation at the edge of a roadway by a slip-resistant and/or reflective surface;
- accentuate a change of surface for disabled people;
- deter vehicles from mounting the footway while accepting the safe over-run (two wheels on and two off) of the occasional emergency or heavy vehicle;
- separate different materials from each other to prevent spillage and damage, e.g. grass from paving or gravel from grass.

Channels

The function of channels is to remove water from a surface and lead it to drainage outlets. Channels may be formed in the paving material itself by sinking the channel below the surface, but it is more satisfactory to lay purpose-made channels first to accurate falls and then lay the paving to the channels. Precast concrete channel units are covered in BS 7263 Part 1. While these are used for most plain road work, the landscape designer may wish to make the channel a part of his paving design, and the channel can then be constructed of bricks, cobbles, setts, natural stone, or concrete blocks set at a lower level than the paving. The paving should grade into the channel without sharp edges, which might trip pedestrians and are difficult to cross for prams, trolleys and wheelchairs. Where a completely flat surface is required, proprietary channels can be used. These consist of precast concrete or polyester concrete units that incorporate a fall, so that while the invert falls the surface remains level, although the fall is limited by the design and is not likely to exceed 1 in 50. These channels are fitted with grids of various patterns, silt pits, inspection points, and gullies.

Mowing stones

Mowing stones are laid at the edge of grassed areas where they meet a wall or step, and allow the grass to be cut without damage to the machine or vertical surface. Stones are set 25 mm below the level of the grass.

Kerbs

Natural stone kerbs are still used in conservation and repair work. British Standard 435 for edge and flat kerbs is still in current use, and specifies sizes from 125 mm × 250 mm to 150 mm × 300 mm deep for edge kerbs, and 250

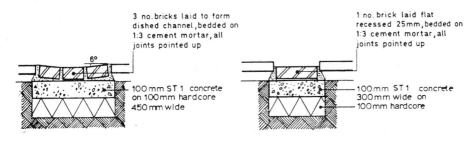

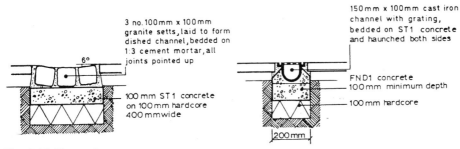

Fig. 4.13 Channels.

mm × 125 mm to 300 mm × 150 mm deep for flat kerbs. Both have a minimum length of 600 mm. Those used today are normally granite or whinstone, and they are often old kerbs removed from town streets and re-dressed. They are laid and bedded in the same way as precast concrete kerbs, but they are very much heavier.

Most kerbs specified are natural grey concrete to British Standard 7263 or to British Standard sizes but with a selection of textures, colours and finishes. Straight or curved sections may be battered, splayed or bull-nosed on one side and in lengths between 450 mm and 915 mm, the latter being the most commonly used. Quadrants have radii of 305 mm or 455 mm. Radial kerbs to match most sections are available, allowing neat circles, curves and serpentines to be made without cutting; small units have radii of 1000 mm and 2000 mm, and larger units have radii from 3 m to 12 m in 1.5 m stages.

Dropper kerbs are made right or left handed and in standard lengths. They should be specified for house, factory, service, shop entrances and so forth to provide a safe transition between the footway and road surface. They are designed to match a range of kerb profiles at one end, and to be flush with the roadway at the other.

If great resistance is to be put on the kerb, for example where a substantial amount of material is to be retained and/or vehicle overrun is expected, holed kerbs should be specified, with 12 mm steel reinforcing rods set vertically into the foundation concrete and inserted into the precast holes in the kerb and grouted; there must be a 50 mm cover to the steel in the foundation. Alternatively, where unreinforced sections are specified, and there is then some doubt as to their ability to cope with the maintenance vehicles and machinery that could back into them, it is advisable to reinforce the haunching with 8–12 mm steel reinforcing rods positioned vertically behind the kerb and set into the foundation, covered by at least 50 mm of concrete on all faces.

In pedestrian areas and where a few light slowly driven vehicles move through the area, much smaller lighter sections should be specified.

Where movement joints are required they should follow the line of the joints in the main paving.

All kerbs to British Standard 7263 have reference letters that are linked together to describe the type of kerb; thus BN = bull-nose, HB = half battered, SP = 45° splay, Q = quadrant, IA = internal angle and XA = external angle. For example, the reference BNXA tells the contractor that a bull-nosed external angle quadrant has been specified.

Edgings

Edgings are incorporated into the British Standard for kerbs (BS 7263). They are all specified to be 50 mm thick, and are either bull-nosed, or round or flat topped. Although the standard contains tests for transverse strength, none is reinforced. Edgings should not be used as an economic option for any other purpose, such as drainage channels or mowing stones, except where no vehicle traffic is expected, as they are not designed to be laid flat; neither are they intended for use where vehicles are likely to pass over them. They should only be used as an edging to grassed areas, shrubberies, unit paving or gravel paths intended for foot traffic only or at most light maintenance vehicles and machinery.

Proprietary kerbs and edgings

Paving manufacturers make special kerbs to match their own proprietary paving systems; though not to British Standard, they are often of the same dimensions and profiles. Specialist manufacturers make Victorian replica rope-edged and other profiles; these are thin sections made of hard brick clay or terracotta.

BRITISH STANDARDS

435 : 1975	Dressed natural stone kerbs, channels, quadrants and setts
5328: Pt 1 : 1991 & AMD	Guide to specifying concrete
5328: Pt 2 : 1991 & AMD	Methods for specifying concrete mixes
7263 Pt 1 : 1994	Precast concrete flags, kerbs, channels, edgings and quadrants
7263 Pt 2 : 1990	Code of practice for laying precast concrete flags, kerbs, channels, edgings and quadrants

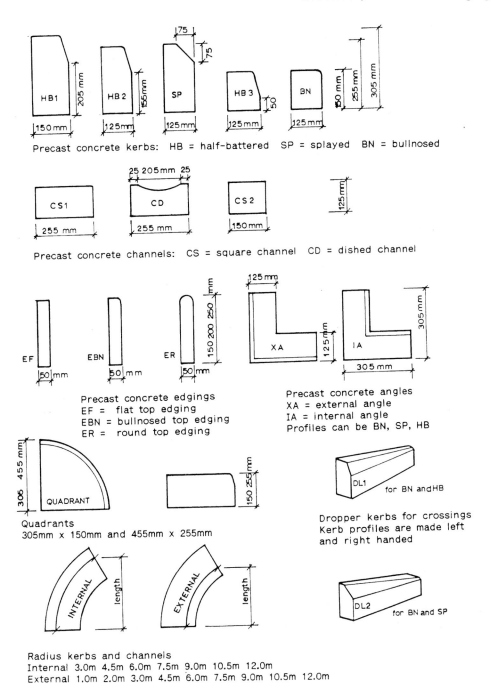

Precast concrete kerbs: HB = half-battered   SP = splayed   BN = bullnosed

Precast concrete channels: CS = square channel   CD = dished channel

Precast concrete edgings
EF = flat top edging
EBN = bullnosed top edging
ER = round top edging

Precast concrete angles
XA = external angle
IA = internal angle
Profiles can be BN, SP, HB

Dropper kerbs for crossings
Kerb profiles are made left
and right handed

Quadrants
305mm x 150mm and 455mm x 255mm

Radius kerbs and channels
Internal 3.0m 4.5m 6.0m 7.5m 9.0m 10.5m 12.0m
External 1.0m 2.0m 3.0m 4.5m 6.0m 7.5m 9.0m 10.5m 12.0m

Fig. 4.14 Kerbs, channels and edgings to BS 7263 Pt 2 : 1994.

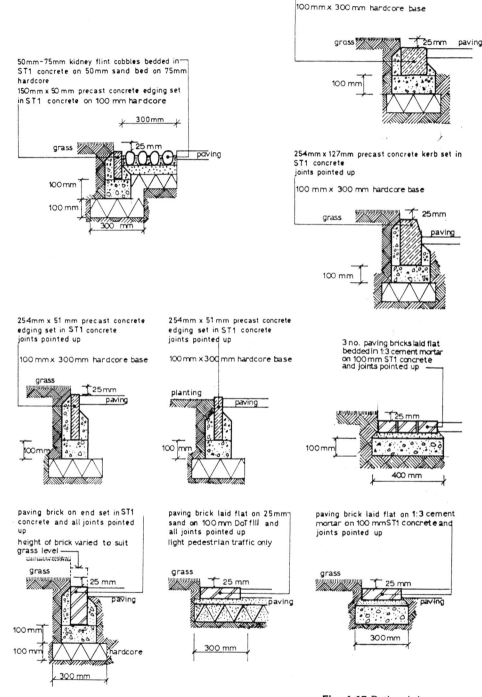

Fig. 4.15 Path edgings.

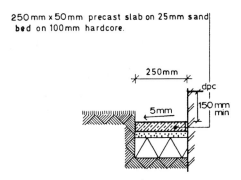

250 mm x 50mm precast slab on 25mm sand
bed on 100mm hardcore.

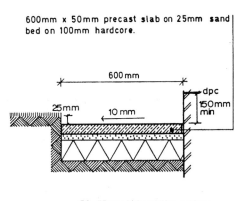

600mm x 50mm precast slab on 25mm sand
bed on 100mm hardcore.

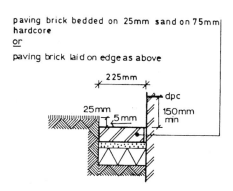

paving brick bedded on 25mm sand on 75mm
hardcore
or

paving brick laid on edge as above

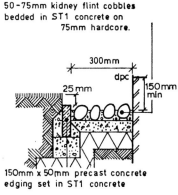

50-75mm kidney flint cobbles
bedded in ST1 concrete on
75mm hardcore.

150mm x 50mm precast concrete
edging set in ST1 concrete

Fig. 4.16 Mowing margins.

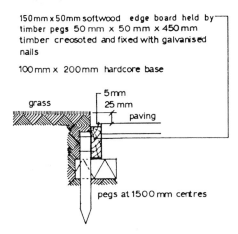

150mm x 50mm softwood edge board held by
timber pegs 50 mm x 50 mm x 450 mm
timber creosoted and fixed with galvanised
nails

100mm x 200mm hardcore base

grass

5mm
25 mm
paving

pegs at 1500 mm centres

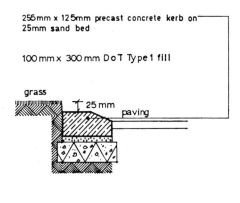

255mm x 125mm precast concrete kerb on
25mm sand bed

100 mm x 300 mm D o T Type 1 fill

grass

25 mm
paving

Fig. 4.17 Edgings to paths.

DATA

Precast concrete kerbs to BS 7263

Straight kerb units: length from 450 to 915 mm
150 mm high × 125 mm thick
    bull-nosed   type BN
    half battered   type HB3
255 mm high × 125 mm thick
    45° splayed   type SP
    half battered   type HB2
305 mm high × 150 mm thick
    half battered   type HB1

Quadrant kerb units
150 mm high × 305 and 455 mm radius to match type BN   type QBN
150 mm high × 305 and 455 mm radius to match type HB2, HB3
                                   type QHB
150 mm high × 305 and 455 mm radius to match type SP   type QSP
255 mm high × 305 and 455 mm radius to match type BN   type QBN
255 mm high × 305 and 455 mm radius to match type HB2, HB3
                                   type QHB
225 mm high × 305 and 455 mm radius to match type SP   type QSP

Angle kerb units
305 mm × 305 mm × 225 mm high × 125 mm thick
    bull-nosed external angle   type XA
    splayed external angle to match type SP   type XA
    bull-nosed internal angle   type IA
    splayed internal angle to match type SP   type IA

Channels
255 mm wide × 125 mm high flat   type CS1
150 mm wide × 125 mm high flat   type CS2
255 mm wide × 125 mm high dished   type CD

Transition kerb units
From kerb type SP to HB   left handed   type TL
                           right handed   type TR
From kerb type BN to HB   left handed   type DL1
                           right handed   type DR1
From kerb type BN to SP   left handed   type DL2
                           right handed   type DR2

Radial kerbs and channels

All profiles of kerbs and channels

External radius (mm)	Internal radius (mm)
1000	3000
2000	4500
3000	6000
4500	7500
6000	9000
7500	1050
9000	1200
1050	
1200	

A quarter circle can be laid without cutting units.

Precast concrete edgings to BS 7263

All dimensions in mm.

Round top, type ER	Flat top, type EF	Bull-nosed top, type EBN
150 × 50	150 × 50	150 × 50
200 × 50	200 × 50	200 × 50
250 × 50	250 × 50	250 × 50

Natural stone kerbs

Straight kerbs (all dimension in mm):

Laid on edge Width Depth	Laid flat Width Depth	Channels Width Depth
200 × 300	300 × 200	300 × 150
150 × 300	300 × 150	250 × 150
150 × 250	250 × 150	
150 × 200	250 × 125	
125 × 250		

Minimum length 600 mm

Quadrants, radial kerbs and radial channels are manufactured to order.

Stone dressing

Type A Fine picked   This is the smoothest
Type B Fair picked
Type B Single axed
Type C Rough punched   This is the coarsest

Because of the cost of dressing very hard natural stone, only the exposed faces are dressed.

Concrete mixes to BS 5328: Pt 1 & AMD

Application	Standard mix	Designated mix	Slump (mm)
Mass concrete fill and blinding	ST 2 (C10P)	GEN 1	75
Mass concrete foundations in non-aggressive soils	ST 2 (C10P)	GEN 1	75
Trench fill foundations in non-aggressive soils	ST 2 (C10P)	GEN 1	125
Foundations			
in Class 2 sulphate soils	–	FND 2	75
in Class 3 sulphate soils	–	FND 3	75
in Class 4A sulphate soils	–	FND 4A	75
in Class 4B sulphate soils	–	FND 4B	75

GEN concretes with low cement content may not be suitable for good cast and direct finished surfaces.

Traditional concrete mixes

Application	Name of mix	Cement:aggregate:sand (parts by volume)
Setting posts, filling weak spots in ground, backfilling pipe trenches:	one, two, three	1:2:3
	or	or
	one to five all-in	1:5
Paving bases, bedding manholes, setting kerbs, edgings, precast channels:	one, two, four	1:2:4
	or	or
	one to six all in	1:6
In situ concrete paving:	one, one & half, two & half	1:1.5:2.5

'All-in' is mixed aggregate as dug from the pit.

OUTLINE SPECIFICATION

Precast concrete kerbs

a. *Construct foundation of FND2 concrete 300 mm wide × 150 mm deep.*

b. *Lay precast concrete kerb units to BS 7263 Part 1, 255 mm high × 125 mm thick bull-nosed type BN, bedded and haunched both sides in semi-dry GEN1 concrete, slump 35 mm maximum, haunching to be minimum two-thirds of height of kerb and full width of foundation, kerb units dry jointed.*

c. *Variation in level of kerb not to exceed 3 mm.*

for excavation for foundations	*D20*
for concrete mixes	*E10*
for paving bases	*Q20*

Alternative specification

a. *Construct foundation of FND2 concrete 250 mm wide × 150 mm deep.*

b. *Lay precast concrete kerbs to BS 7263 Part 1, 150 mm high × 125 mm thick splayed type SP, bedded in 1:3 lime:sand mortar 10 mm thick, haunched with ST1 concrete one side, haunching to be minimum two-thirds of height of kerb and full width of foundation, 5 mm joints filled with compacted 1:4.5 cement:sand mortar.*

c. *Variation in level of kerb not to exceed 3 mm.*

Edging

a. *Construct foundation of FND2 concrete 200 mm wide × 100 mm deep.*

b. *Lay precast concrete edging units to BS 7263 Part 1, 200 mm high × 50 mm thick round top type ER bedded in 1:3 lime:sand mortar, haunched with ST1 concrete one side, units dry jointed, haunching to be two-thirds of height of kerb and full width of foundation.*

c. *Variation in level of edging not to exceed 3 mm.*

for excavation for foundations	*D20*
for paving bases	*Q20*

Channels

a. *On main paving base previously laid, supply and lay precast concrete channel units to BS 7263 Part 1, 255 mm wide × 125 mm high dished type CD, laid to falls shown on drawing, bedded in 1:3 lime:sand mortar, jointed in Group 2 cement:sand mortar 1:3.*

b. *Variation in level along line of channel not to exceed 3 mm.*

for excavation for foundations   *D20*
for concrete mixes   *E10*
for paving bases   *Q20*

Brick channel

a. *Construct FND3 concrete foundation 600 mm wide × 200 mm deep.*
b. *Lay channel of five courses of Class B engineering bricks to depths and falls shown on drawing, bricks to be laid as headers along the channel, bedded in 1:3 lime:sand mortar bricks close jointed in Group 2 cement:sand mortar 1:3, joints compacted as laid.*
c. *Variation in level between bricks not to exceed 3 mm.*

for excavation for foundations   *D20*
for concrete mixes   *E10*
for paving bases   *Q20*

Precast concrete proprietary channel

a. *In trench previously excavated, lay 100 mm FND3 concrete bed laid level.*
b. *Supply and lay proprietary precast concrete channel units internally tapered to fall, haunched with 100 mm FND2 concrete both sides, complete with galvanized slotted steel grating Type XXX laid level with paving.*

for excavation for foundations   *D20*
for concrete mixes   *E10*
for paving bases   *Q20*

Q20 Bases and sub-bases to roads and paving

GENERAL GUIDANCE

This section covers the construction of bases to light roadways, paved areas and pedestrian areas, but not bases to major highways.

The layers of material that make up a full-scale road are:

- **Substrate** or **subgrade**: natural or engineered ground level
- **Sub-base**: main structural and levelling layer
- **Road base**: secondary structural and levelling layer (not always necessary in light road work)
- **Surfacing**: the finished surface that carries the traffic

Substrate

When dealing with sub-bases the soil can be divided into six types and two categories depending on the level of the water-table. Each type and category is given a CBR percentage number (CBR is short for **California Bearing Ratio**): the higher the CBR figure, the stronger the substrate. The substrate (or subgrade) may be strong enough as excavated, but if not, the weak areas must be dug out and backfilled with compacted DoT Type 1 or 2 granular fill. If the substrate is weak, and also where vehicles may overrun paving, a 'capping layer' of well-rammed hardcore is placed on the substrate. With a CBR between 2% and 5% a 300 mm capping layer should be laid; between 1% and 2% a 600 mm capping layer should be laid; if the CBR is below 1%, a highway engineer should be consulted.

The substrate may be a naturally free-draining soil such as gravel or coarse sand, but if it is water-retentive, such as clay, land drainage should be installed, because continual water percolation loosens the soil structure and allows it to deform. The sub-base must also drain freely if the paving is not to become waterlogged with the risk of freezing and therefore spalling or lifting the paving units. The area must then be drained with perforated plastic or clay agricultural drains, and connected to a soakaway, a lagoon, or the site land-drainage system.

Where there is a high water-table, or a very fine soft substrate, a geotextile filter membrane should be laid between the substrate and the sub-base. Geotextile filter material is made in various thicknesses and densities, which allow the passage of varying amounts of water.

Sub-base

The sub-base is the main layer of material that forms the structural foundation for the roadway or paving, and establishes the paving or road levels. For roads the sub-base is usually made of cement-bound material (CBM) such as hardcore, gravel, hoggin, brick-earth, or broken stone, hard brick or concrete bound together with cement. For pedestrian areas or light vehicles the alternatives are: a solid concrete base, a cement bound sub-base, or a granular sub-base.

Solid concrete

This is required where the paving must be smooth, even, and stable under all foreseeable conditions. The areas where the landscape designer should consider a concrete sub-base include the entrances and paths around hospital and other health care buildings, pedestrian areas in shopping malls, old people's homes, schools, or other areas where wheelchairs and disabled people are regular visitors.

The concrete should be ST4 or GEN4 for normal soils, or FND2, 3 or 4 for sulphate soils, with 20 mm aggregate, laid on well-rammed hardcore of broken brick, concrete or stone. Hardcore should be 100 mm thick as a minimum, and 150 mm is preferable. Open-textured hardcore should be blinded with sand or ash before the concrete is laid. The concrete sub-base should be

100 mm for areas restricted to pedestrians only, and 150 mm for vehicle over-run. If the substrate is filled ground, or if there are many changes of level, light steel mesh reinforcement may be needed in the sub-base. For reinforced sub-bases, 100 mm is the minimum thickness, and the mesh must be placed halfway between the top and bottom of the concrete. Concrete 100 mm thick with reinforcing mesh 200 mm × 200 mm ref. A/142 to BS 4483, with head laps 300 mm and side laps 150 mm, will be adequate for light traffic, while concrete 150 mm thick with ref. A/193 mesh is usual for heavier loading.

Cement-bound macadam

This is clean, hard aggregate material, at least 15% of which passes a 75 mm sieve, bound with cement and laid *in situ*; the material is then compacted by vibrating rollers. Cement-bound material sub-bases are more suitable for large areas of paving not subject to heavy loading, such as urban pedestrian areas or lightly trafficked estate roads, especially as expansion joints are not necessary. The proportion of cement to aggregate for the commoner materials is:

cement per cubic metre: 80–120 kg for evenly graded hard aggregate
80–140 kg pulverized fuel ash
130–190 kg unevenly graded hard aggregate
170–200 kg brick earth
180–225 kg chalk

The thicknesses given below for granular sub-bases will be appropriate, but the material must be compacted in layers not exceeding 150 mm, though better results are obtained with 100 mm layers.

Granular material

A granular sub-base consists of aggregate to the specification of Department of Transport Type 1 or Type 2 to Clause 803. Type 1 consists of crushed rock, crushed concrete or well-burnt non-plastic shale, and graded to the standard given in Clause 803 of the Department of Transport Specification for Highway Works. Type 2 may be used for lightly loaded sub-bases; this type covers the materials for Type 1 but also includes natural sand, gravel, and crushed sound blast-furnace slag. All these materials must pass a 75 mm sieve, and they must be free from dust or any pollutant such as oil or other chemicals.

Wet lean concrete

Another specification for sub-bases or road bases is 'wet lean concrete', which has less cement than ordinary concrete, and it is classified in four grades equivalent in strength to the cement-bound material grades CBM 1, CBM 2, CBM 3, and CBM 4. These are specified in DoT Specification for Highway Works, clauses 1001–1006.

Calculation of sub-base

The thickness of bases for pedestrian paving may be taken from published tables, but when trafficked roads form part of the landscape contract proper, calculations for the bases must be made. Full-scale highways are the responsibility of the highway engineer. The procedure is set out below.

1. Determine the California Bearing Ratio percentage value from the table below:

CBR percentages for soil types and construction conditions

Soil type	Plasticity index	Water-table more than 600 mm down			Water-table 600 mm or less down		
		Construction conditions			*Construction conditions*		
		Poor	Average	Good	Poor	Average	Good
Heavy clay	70	1.5–2.0	2.0	2.0	1.5–2.0	2.0	2.0–2.5
	60	1.5–2.0	2.0	2.0–2.5	1.5–2.0	2.0	2.0–2.5
	50	1.5–2.0	2.0–2.5	2.0–2.5	2.0	2.0–2.5	2.0–2.5
	40	2.0–2.5	2.5–3.0	2.5–3.0	2.5	3.0	3.0–3.5
Silty clay	30	2.5–3.5	3.0–4.0	3.5–5.0	3.0–3.5	4.0	4.0–6.0
Sandy clay	20	2.5–4.0	4.0–5.0	4.5–7.0	3.0–4.0	5.0–6.0	6.0–8.0
	10	2.5–3.5	3.0–6.0	3.5–7.0	2.5–4.0	4.5–7.0	7.0–8.0
Silt	–	1.0	1.0	2.0	1.0	2.0	2.0
Sand:	–						
unevenly graded		20.0	20.0	20.0	20.0	20.0	20.0
evenly graded		40.0	40.0	40.0	40.0	40.0	40.0
Sandy gravel:	–						
evenly graded		60.0	60.0	60.0	60.0	60.0	60.0

The **plasticity index** indicates the difference in the moisture content of a soil when it is neither too liquid nor too dry to be plastic. The figures are calculated for evenly graded soils on level ground, though very wet or dry conditions may affect the CBR findings.

Good site conditions means good subsoil drainage, good weather during construction, and efficient and rapid completion of the paving.

Average site conditions means adequate drainage and normal construction.

Poor site conditions means little or no drainage, poor weather, or standard of work less than average.

2. Determine the expected volume of traffic. This is rated as a number of standard axles, and is calculated by taking the actual number of commercial vehicles expected to use the paving, and then multiplying this by a given factor to get a notional number expressing the number of 'standard' axles.

The factors are:

- Roads with less than 250 commercial vehicles per day: multiply by 0.45.
- Roads with 250–1000 commercial vehicles per day: multiply by 0.72.

In narrow roadways where vehicles must follow the same wheel tracks every time, the number of standard axles should be multiplied by 2 or even 3.

3. Determine the sub-base thickness required according to the CBR:

| CBR | Standard axles | | | | | | |
	20 000	40 000	60 000	80 000	100 000	200 000	400 000
under 2%	510 mm	520 mm	530 mm	530 mm	540 mm	560 mm	570 mm
2%	360	370	380	380	390	410	420
3%	270	280	290	290	290	310	320
4%	200	210	220	220	230	240	260
5%	160	170	180	180	190	200	210
6%	120	130	140	140	140	150	160
7%	100	100	100	110	110	120	130

The minimum sub-base thickness is 150 mm where a capping layer is laid on the substrate, or where the substrate material is as strong as Type 1 granular material. In all other cases 200 mm is the minimum if no capping layer is used.

An example of the design procedure is shown for a commercial vehicle access roadway:

Number of lorries expected per day	= 50
Multiply by working days per year	= 50 mm × 312 days = 15 600 vehicles
Multiply by assumed life of roadway	= 15 600 mm × 20 years = 312 000 vehicles
Convert to standard axles	= 312 000 mm × 0.45 = 140 400 s.a.

The substrate is sandy clay with a low water-table, plasticity index of 20, and the site conditions are good. The CBR is therefore between 4.5% and 7%. The paving system chosen is 80 mm blocks on 50 mm sand bed, which would be a 'thin' paving system, with a CBR of 4.5%. By interpolation, the table gives a sub-base of 200 mm.

BRITISH STANDARDS

63 Pt 1 : 1987	Single sized aggregate for general purposes
882 : 1992 & AMD	Aggregates from natural sources for concrete
1377 Pt 1 : 1990 & AMD	Soil tests for civil engineering. Sample preparation
1377 Pt 2 : 1990	Soil test for civil engineering. Classification requirements
1377 Pt 4 : 1990 & AMD	Soil tests for civil engineering. Compaction related tests

1377 Pt 5 : 1990 & AMD	Soil tests for civil engineering. Compressibility, permeability, durability
1377 Pt 9 : 1990 & AMD	Soil tests for civil engineering. *In situ* tests
3892 Pt 2 : 1993	PFA for use in grouts and for miscellaneous uses in concrete
5075 Pt 1 : 1982 & AMD	Concrete admixtures: accelerating, retarding, and water reducing
5075 Pt 2 : 1982 & AMD	Concrete admixtures: air-entraining admixtures
5328 Pt 1 : 1991 & AMD	Guide to specifying concrete
5328 Pt 2 : 1991 & AMD	Methods for specifying concrete mixes
5328 Pt 3 : 1990 & AMD	Producing and transporting concrete
5328 Pt 4 : 1990 & AMD	Sampling and testing
5837 : 1991	Code of practice for trees in relation to construction
5930 : 1981	Code of practice for site investigations for civil engineering operations
6543 : 1985	Industrial by-products and waste materials: road making

DATA

Permeability of soils

Soil type	Drainage	Frost risk	Use as road foundation
Boulders and cobbles	Good	Very slight	Good–excellent
Hardcore			
broken stone, brick	Excellent	Very slight	Very good–excellent
chalk, soft rock	Fair–poor	Medium–high	Good–excellent
Gravel			
well graded (no fines)	Excellent	None–very slight	Excellent
with fines	Fair–poor	Slight–medium	Good–excellent
Sand			
well graded (no fines)	Excellent	None–very slight	Good–excellent
with fines	Fair–poor	Slight–high	Fair–good
Silt	Poor	Medium–high	Poor–fair
Clay			
low plasticity	Very poor	Medium–high	Poor–fair
moderate plasticity	Fair–bad	Slight	Poor
high plasticity	Bad	Very slight	Bad–poor
Peat	Fair–poor	Slight	Very bad

Estimated sub-base thickness

For pedestrian paving and light traffic the thickness of the sub-base can be taken from the table without reference to the number of standard axles:

CBR	Footways, patios garden paths, house parking (mm)	Pedestrian areas, occasional vehicle overrun (mm)	Car parks, minor roadways with no through traffic (mm)
Under 2%	380	500	550
2%	230	350	400
3%	180	260	300
4%	160	200	230
5%	140	160	190
6%	120	120	140
7% and over	100	100	100

Concrete road layers

Roadway layer	Concrete mix to BS 5328
Sub-base or road base:	
wet lean concrete 1	ST 1 (C7.5, C7.5P)
wet lean concrete 2	ST 2 (C10, C10P)
wet lean concrete 3	ST 3 (C15, C15P)
wet lean concrete 4	ST 4 (C20, C20P)
Road base:	
continuously reinforced:	C40
Surface slab (wearing course):	
unreinforced	C40
jointed reinforced	C40
continuously reinforced	C40

Concrete mixes to BS 5328: Pt 1 & AMD

Application	Standard mix (mm)	Designated mix	Slump
Mass concrete fill and blinding	ST 2 (C10P)	GEN 1	75
Mass concrete foundations in non-aggressive soils	ST 2 (C10P)	GEN 1	75
Trench fill foundations in non-aggressive soils	ST 2 (C10P)	GEN 1	125

Concrete mixes to BS 5328: Pt 1 & AMD *continued*

Application	Standard mix (mm)	Designated mix	Slump
Foundations			
in Class 2 sulphate soils	–	FND 2	75
in Class 3 sulphate soils	–	FND 3	75
in Class 4A sulphate soils	–	FND 4A	75
in Class 4B sulphate soils	–	FND 4B	75
Kerb bedding and haunching	ST 1 (C7.5P)	GEN 0	10

GEN concretes with low cement content may not be suitable for good cast and direct finished surfaces.

Traditional concrete mixes

Application	Name of mix	Cement:aggregate:sand (parts by volume)
Setting posts, filling weak spots in ground, backfilling pipe trenches:	one, two, three	1:2:3
	or	or
	one to five all-in	1:5
Paving bases, bedding manholes, setting kerbs, edgings, precast channels:	one, two, four	1:2:4
	or	or
	one to six all in	1:6
In situ concrete paving:	one, one & half, two & half	1:1.5:2.5

'All-in' is mixed aggregate as dug from the pit.

Cement-bound material for bases and sub-bases

- CBM1: very carefully graded aggregate from 37.5 mm to 75 μm, with a 7-day strength of 4.5 N/mm^2
- CBM2: same range of aggregate as CBM1 but with more tolerance in each size of aggregate with a 7-day strength of 7.0 N/mm^2
- CBM3: crushed natural aggregate or blast-furnace slag, graded from 37.5 mm to 150 μm for 40 mm aggregate, and from 20 to 75 μm for 20 mm aggregate, with a 7-day strength of 10 N/mm^2
- CBM4: crushed natural aggregate or blast-furnace slag, graded from 37.5 mm to 150 μm for 40 mm aggregate, and from 20 to 75 μm for 20 mm aggregate, with a 7-day strength of 15 N/mm^2

Concrete pavement sealants to BS 5212

Cold-applied joint sealants can be either one-part or two-part materials.

- Type N: normal sealant
- Type F: resistant to engine fuel
- Type FB: resistant to flame and fuel

Joint fillers are preformed compressible strips of material.

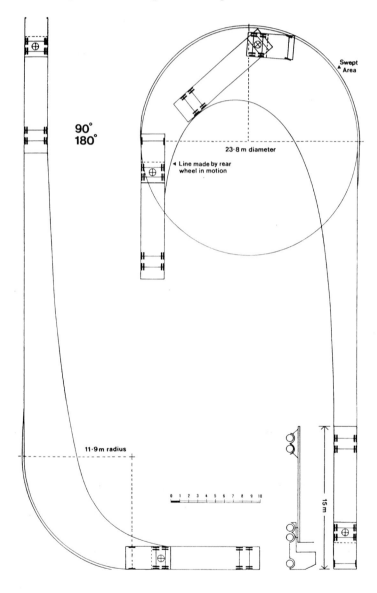

Fig. 4.18 Turning circles: 15 m articulated vehicle.

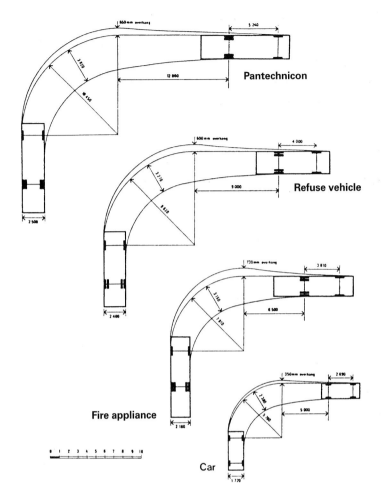

Fig. 4.19 Turning circles: vehicles turning through 90°.

OUTLINE SPECIFICATION

Subgrade

a. *Cut back tree roots in subsoil well clear of the paved area, and treat subsoil with total herbicide.*

b. *Identify loose soil, extra damp areas, recent fill, old drains, or other break in the continuity of the subgrade, dig out and replace with well-rammed hardcore laid in layers not more than 150 mm thick.*

c. *Blind substrate with sand, roll entire substrate with a 2.5 tonne roller, or hand compact with mechanical rammer.*

for excavation   *D20*

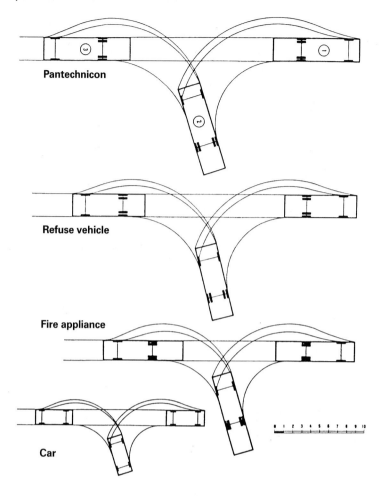

Pantechnicon

Refuse vehicle

Fire appliance

Car

Fig. 4.20 Turning circles: reverse side turn.

Concrete sub-base

a. *Supply and lay reinforced concrete sub-base on subgrade previously specified 100 mm thick of ST4 concrete with 20 mm aggregate, including all necessary formwork, laid to given levels on 150 mm well-rammed hardcore of broken brick, concrete or stone blinded with sand or ash.*

b. *Reinforcement to be steel mesh fabric 200 mm × 200 mm Type A 193 to BS 4483, head lap 300 mm, side lap 150 mm, laid 50 mm from bottom of concrete with concrete tamped and vibrated round reinforcement.*

c. *Wet-form movement joints of cold-applied sealant Type N to BS 5212 at 10 m intervals.*

d. *Maximum permitted deviation from given levels 15 mm. Concrete to be cured for seven days, and no site traffic allowed on sub-base before*

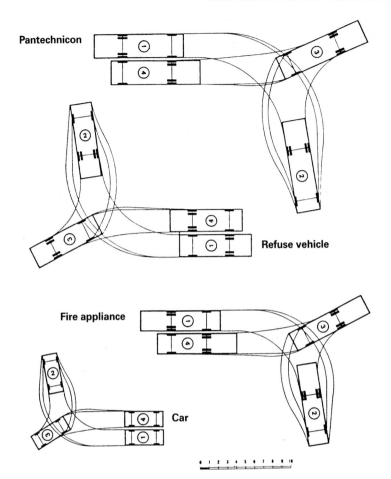

Pantechnicon

Refuse vehicle

Fire appliance

Car

Fig. 4.21 Turning circles: forward side turn.

road base or paving is laid. Protect surface of concrete from run-off or other contamination. Concrete not to be poured from a height greater than 1.0 m.

for kerbs	*Q10*
for in situ *concrete paving*	*Q21*
for macadam and asphalt paving	*Q22*
for granular material paving	*Q23*
for interlocking brick/block paving	*Q24*
for flag, brick, sett, cobble paving	*Q25*

Cement-bound macadam

Supply and lay sub-base 300 mm thick of CBM 3 on subgrade previously specified, compacted with vibrating rollers in layers not exceeding 150 mm to given levels. Maximum permitted deviation from given levels 15 mm.

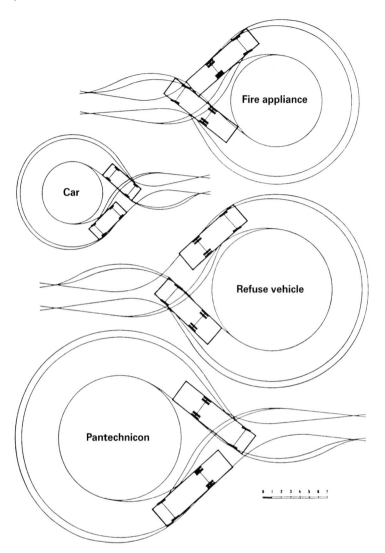

Fig. 4.22 Turning circles: full lock forward.

for kerbs	*Q10*
for in situ *concrete paving*	*Q21*
for macadam and asphalt paving	*Q22*
for granular material paving	*Q23*
for interlocking brick/block paving	*Q24*
for flag, brick, sett, cobble paving	*Q25*

Granular material

a. *Supply and lay sub-base 250 mm thick of DoT Type 1 granular material on subgrade previously specified, compacted with vibrating rollers in layers not exceeding 150 mm to given levels.*

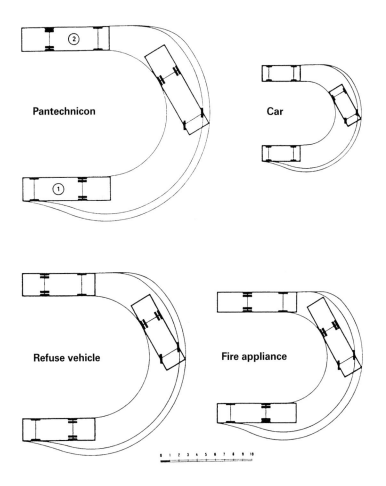

Pantechnicon

Car

Refuse vehicle

Fire appliance

Fig. 4.23 Turning circles: full lock reverse.

b. *Maximum permitted deviation from given levels 15 mm. No site traffic allowed on sub-base before road base or paving is laid.*

for kerbs	*Q10*
for in situ *concrete paving*	*Q21*
for macadam and asphalt paving	*Q22*
for granular material paving	*Q23*
for interlocking brick/block paving	*Q24*
for flag, brick, sett, cobble paving	*Q25*
for land drainage	*R13*

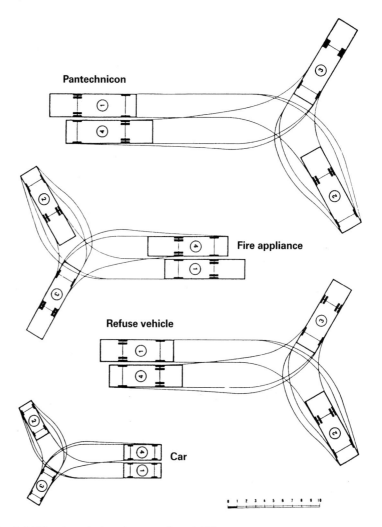

Pantechnicon

Fire appliance

Refuse vehicle

Car

Fig. 4.24 Turning circles: hammerhead, Y-form.

Q21 Concrete roads and paved areas

GENERAL GUIDANCE

This section deals with plain and reinforced concrete light roadways and concrete pedestrian and car park areas. It excludes concrete highways and heavily loaded concrete areas, which require civil engineering expertise.

Plain self-finished concrete paving has largely been superseded by interlocking concrete block paving where macadam paving is unsuitable, but simple specifications are given here for areas where plain concrete is required.

The design and construction of concrete roads is specified in great detail in the Department of Transport Specification for Highway Works, clauses

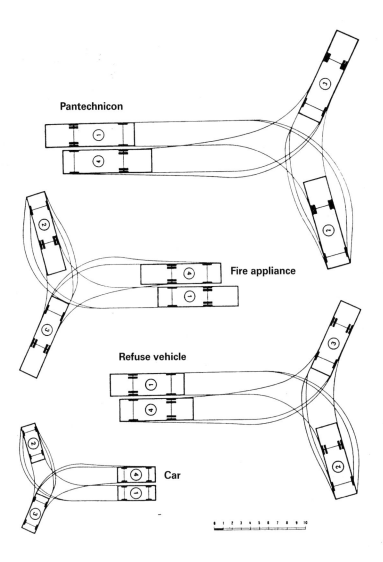

Fig. 4.25 Turning circles: hammerhead, T-form.

1000–1042. Although the minor roadways usual in landscape contracts do not have to take the stresses of major highways, the principles of construction are similar. Sub-bases and road bases can be made of 'wet lean concrete', which has less cement, and can be specified in four grades equivalent in strength to the cement-bound macadam grades CBM 1, CBM 2, CBM 3 and CBM 4. The grades of concrete for these layers and the wearing course are specified in DoT Specification for Highway Works, clauses 1001–1006. The maximum size of aggregate that can be used is 40 mm.

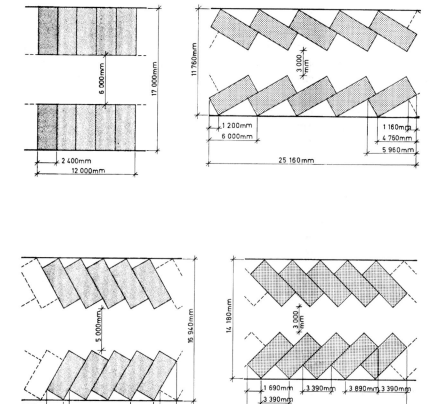

Fig. 4.26 Typical parking layouts: (a) 90° parking; (b) 30° parking; (c) 60° parking; (d) 45° parking. All car parking spaces shown are 5500 mm × 2400 mm.

Road bases

The bases of concrete roads are made of normal building concrete, but concrete made with a blend of Portland cement and PFA (pulverized fuel ash) gives greater density, less permeability, and greater resistance to chemical attack. Plasticizers and retarders can be used to extend the periods when road work can be carried out.

Wearing surface

The wearing surface is less permeable if it is made with air-entrained concrete, though this is not as strong as ordinary concrete. The strength of the wearing course should be not less than 48.2 N/mm², which allows for variations in mixing and laying.

Water must not drain out of the road base concrete before it can set, so a waterproof membrane must be laid on the sub-base; reinforced road bases should be laid on a bitumen emulsion spray instead, to bond the upper layers.

Fresh concrete road bases must be protected from frost, heat and drying winds by covering them with plastic sheet, spraying them with water, or laying wet hessian sheeting topped with plastic sheet.

The surface of the wearing course must be textured to a depth of 1 mm to give a grip to tyres. The concrete is brushed immediately after it has set, and the assessment of the exact time and brush setting for this operation requires experience and judgement on the part of the operator.

Reinforced concrete

Reinforced concrete is used for roads where there is a poor substrate, or where large service pipes pass under the road. Reinforcement may be steel bars laid in both directions and tied; light roads are usually reinforced with prefabricated steel mesh laid halfway through the thickness of the slab. The steel must be properly supported and kept rigid while the concrete is poured and set.

Joints

Transverse joints 12.5 mm or 25 mm wide are filled with softboard and sawn out after the concrete has just set, as preforming joints in the wet concrete may give less accurate joints, and the exact shape and size of the joints is very important to prevent the sealant from either rising above or sinking below the road surface. There must be no opportunity for water to penetrate the road surface and freeze. The joints are filled after sawing with special sealants to BS 2499 *Hot applied sealants*, or BS 5212 *Cold applied sealants*.

BRITISH STANDARDS

882 : 1992 & AMD	Aggregates from natural sources for concrete
2499 Pt 1 : 1993	Hot-applied joint sealants for concrete pavements
3892 Pt 2 : 1993	PFA for use in grouts and for miscellaneous uses in concrete
4449 : 1988	Carbon steel bars for the reinforcement of concrete
4483 : 1985	Steel fabric for the reinforcement of concrete
5212 Pt 1 : 1990	Joint sealants for concrete pavements
5212 Pt 2 : 1990	Code of practice for application and use of joint sealants for concrete pavements
5328 Pt 1 : 1991 & AMD	Guide to specifying concrete
5328 Pt 2 : 1991 & AMD	Methods for specifying concrete mixes
5328 Pt 3 : 1990 & AMD	Producing and transporting concrete
5328 Pt 4 : 1990 & AMD	Sampling and testing

DATA

Concrete road layers

Roadway layer	Concrete mix to BS 5328
Sub-base or road base:	
Wet lean concrete	
wet lean concrete 1	ST 1 (C7.5, C7.5P)
wet lean concrete 2	ST 2 (C10, C10P)
wet lean concrete 3	ST 3 (C15, C15P)
wet lean concrete 4	ST 4 (C20, C20P)
Road base:	
continuously reinforced:	C40
Surface slab (wearing course):	
unreinforced	C40
jointed reinforced	C40
continuously reinforced	C40

Movement joints

Longitudinal joints:
Unreinforced concrete: maximum distance apart 4.2 m
Reinforced concrete: maximum distance apart 6.0 m
Offset between layers: 300 mm
Transverse joints in unreinforced slabs at spacings giving a longitudinal:transverse ratio no more than 2:1.

Reinforcement BS 4483 steel sheet fabric

Designated fabric has standard reference numbers and letters that specify the fabric.
Square mesh:

A393	200 mm × 200 mm mesh	10 mm wire
A252	200 mm × 200 mm mesh	8 mm wire
A193	200 mm × 200 mm mesh	7 mm wire
A142	200 mm × 200 mm mesh	6 mm wire
A98	200 mm × 200 mm mesh	5 mm wire

Rectangular mesh is available, but square mesh is most commonly used for road work.

OUTLINE SPECIFICATION

Concrete paving

a. *On sub-base previously specified lay 100 mm ST3 concrete road base including all formwork.*

b. *Lay 100 mm PAV1 concrete wearing course to given falls and levels with movement joints at 4.2 m, textured to provide slip-resistant finish.*

c. *Kerbs to be set and cured before road construction is carried out.*

for kerbs	*Q10*
for concrete mixes	*E10*
for formwork	*E20*
for reinforcement	*E30*

Reinforced concrete paving

a. *On sub-base previously specified lay 150 mm ST3 concrete road base including all formwork.*

b. *Lay 150 mm C40 reinforced concrete wearing course with steel mesh reinforcement to BS 4483 Type A252, side laps 150 mm, head laps 300 mm, minimum cover to reinforcement 50 mm, to given falls and levels, textured to provide slip-resistant finish.*

c. *Construct movement joints at 6.0 m in both directions.*

d. *Kerbs to be set and cured before road construction is carried out.*

for kerbs	*Q10*
for concrete mixes	*E10*
for formwork	*E20*
for reinforcement	*E30*

Q22 Macadam and asphalt roads and paved areas

General guidance

This section covers light roadways and paved areas constructed with bitumen-based road materials. It excludes major highway construction requiring civil engineering expertise.

Road materials

Tarmac, asphalt, and macadam roads are all mixtures of various types of bitumen and aggregate in different proportions. Four types of bituminous mixture are normally used for road construction:

- coated macadam;
- hot-rolled asphalt;
- mastic asphalt;
- dense road tar (DTS – dense tar surfacing).

Coated macadam
This is the most commonly used mixture. It is an evenly graded aggregate mixed with bitumen of various grades; it is fairly open textured, and when

chippings are rolled into the top surface it provides a good grip for tyres. Open-graded and fine-graded bitumen macadams are suitable only for footways and pedestrian areas, as they are less skid-resistant than the coarser surfaces.

Coated macadam is laid hot, and is composed of:

- a binder of petroleum bitumen, thinned or 'cut-back' bitumen, road tar, or a mixture of tar and bitumen (treated as bitumen);
- fine aggregate of crushed hard rock or flint, sand, or sand and fines mixed, all to be under 3.35 mm;
- filler of any of the above passing 75% of 3.35 mm sieve, Portland cement, or hydrated lime;
- coarse aggregate of crushed hard rock such as granite, basalt, limestone, gritstone, or gabbro; gravel of hard rock; blast-furnace or steel slag. The material is accurately graded in standard proportions of each size from the maximum size specified in each type down to 3.35 mm.

Textured macadam is made by applying a fine coat of bitumen emulsion or tar spray to the wearing course and rolling in decorative chippings, but it is not as durable as plain bitumen macadam, and should be used where only pedestrian traffic is expected.

Coloured macadam can be made either by adding a pigment to the hot mix and using a coloured aggregate, or by using a coloured aggregate only. There are clear resin binders manufactured by specialist firms, which can be used instead of black bitumen; these allow the true colour of the aggregate to be seen. Light-coloured aggregates can therefore be used with the resins to produce a more light-reflecting surface, but these surfaces are not intended for heavily trafficked roads, as the tyre wear would soon discolour them.

Hot-rolled asphalt

For pavements, and where a smoother surface is required, hot-rolled asphalt can be used. This is a close-graded mixture using hard bitumen and aggregate with small and large sizes but not many intermediate sizes. Asphalt produces a smoother surface than macadam, and is therefore very useful for wheelchair paths, trolley areas, and cycle paths. Natural asphalt mined in Trinidad used to be used, but although superior to man-made asphalt, it is too expensive for most projects. Asphalt is composed of:

- a binder of bitumen or lake-asphalt bitumen;
- fine aggregate of fines under 2.36 mm of hard rock or flint, sand, or sand and fines mixed;
- filler of very finely ground limestone or Portland cement;
- coarse aggregate of material graded in standard proportions of each size from the maximum specified in each type down to 2.36 mm, composed of crushed hard rock such as granite, basalt, limestone, gritstone or gabbro; gravel of hard rock; blast-furnace slag or steel slag.

Asphalt is very sensitive to temperature variations while it is being laid, so it is important to arrange for the asphalting work to proceed without hindrance on the site.

Mastic asphalt

This is a softer and finer material, which is used mostly for balconies, roof gardens, enclosed courtyards and similar small areas. This material can be hand-laid, unlike hot-rolled asphalt and bitumen macadam, which both need heavy road equipment to lay and compact the layers. The asphalt may be lightly reinforced by steel mesh or grids to prevent point wheel loads deforming the asphalt.

Mastic asphalt is made of fine and coarse aggregate and 'asphaltic cement', which forms a cohesive impermeable material:

- asphaltic cement: bitumen, lake asphalt, asphaltite, or mixtures of these materials;
- fine aggregate: natural rock asphalt (limestone naturally impregnated with bitumen) or fine crushed limestone;
- coarse aggregate: crushed granite or limestone, or grit.

Dense road tar (DTS)

The traditional tar and gravel surfacing is rarely used now, as tar was a by-product of town gas manufacture, but a modern version of tar surfacing is preferable where there is likely to be oil spillage, as it is more resistant to oil damage than bitumen. Tar-based wearing courses are most suitable for loading bays and garage forecourts where regular oil dripping can be expected. The construction of the sub-base, road base and wearing courses is the same as for coated macadam, and the same requirements for laying and detailing are applicable.

Road construction

A roadway is constructed of bituminous material comprising a sub-base, road base, base course and wearing course, though for pedestrian areas the road base can be omitted. The **sub-base** is the main structural layer forming the shape of the road, which is laid on the substrate or subgrade (the natural or made-up ground). The **road base** is the main structural layer for levelling and forming an even bed, while the **base course** provides a foundation for the **wearing course**, which has to take the wear and tear of traffic. The maximum size of aggregate permitted for road work is 40 mm, while the strength of the wearing course should be not less than 48.2 N/mm^2.

Sub-base

Sub-bases can be made of 'wet lean concrete' specified in four grades equivalent in strength to the cement-bound macadam grades CBM 1, CBM 2, CBM 3 and CBM 4. The grades of concrete for these layers and the wearing course are specified in detail in DoT Specification for Highway Works, clauses 1001–1006. The maximum size of aggregate permitted for road work is 40 mm, while the strength of the wearing course should be not less than 48.2 N/mm², which allows for deviations in mixing and laying.

Road base

The thickness is determined by the expected traffic and the CBR (California Bearing Ratio) of the substrate. This layer is usually made of dry coated macadam or cement bound macadam, but it can be dry hardcore for pedestrian-only areas.

Base course and wearing course

The final surface of the road can be laid in two courses: a base course under a top wearing course. The base course can be made of cheaper materials than the wearing course.

The wearing course or 'carpet' is usually made of close-graded macadam or hot-rolled asphalt, with coated chippings added to give better grip and abrasion resistance for trafficked roads and parking bays.

BRITISH STANDARDS

63 Pt 1 : 1987	Single sized aggregate for general purposes
63 Pt 2 : 1987	Single sized aggregate for surface dressing
76 : 1994	Tars for road purposes
434 Pt 1 : 1984	Bitumen road emulsions
434 Pt 2 : 1984	Code of practice for use of bitumen road emulsions
594 Pt 1 : 1985	Hot-rolled asphalt for roads and other paved areas
594 Pt 2 : 1992	Transport, laying, and compaction of hot-rolled asphalt
873 Pt 4 : 1987	Road studs
1446 : 1973	Mastic asphalt (natural rock) for roads and footways
1447 : 1988	Mastic asphalt (limestone) for roads, footways and paving in buildings
3262 Pt 1 : 1989	Hot applied thermoplastic road-marking materials
3262 Pt 2 : 1989	Hot applied thermoplastic road-marking materials: road performance
3262 Pt 3 : 1989	Hot applied thermoplastic road-marking materials: road surfaces
3690 Pt 1 : 1989	Bitumens for roads and other paved areas
4987 Pt 1 : 1993	Coated macadam for roads and other paved areas
4987 Pt 2 : 1993	Coated macadam for roads and other paved areas: transport, laying and compaction
5273 : 1975	Dense tar surfacing for roads and other paved areas
6044 : 1987	Pavement marking paints
6543 : 1985	Industrial by-products and waste materials: road making

DATA

Coated macadam

Open-textured mixture using soft bitumen and evenly graded aggregate.

Hot-rolled asphalt

Close-graded mixture using hard bitumen and aggregate with small and large sizes but not many intermediate sizes.

Dry-coated macadam

40 mm aggregate with fine material vibrated into it with an 8–10 tonne vibrating roller to give finished thickness not less than 100 mm.

Wet-mix macadam

Aggregate bound with bituminous material.

Aggregates for macadam

The aggregates recommended for road bases and wearing courses are:
Dense road base:
    40 mm aggregate graded
    28 mm aggregate graded thickness 70–100 mm
Open-graded base course:
    20 mm aggregate graded thickness 45–75 mm (this is for light traffic only)
Single combined wearing course and base course:
    40 mm aggregate graded thickness 75–105 mm with a minimum thickness
    of 80 mm if it is used without a separate wearing course
Dense base course:
    40 mm aggregate graded thickness 95–140 mm
    20 mm aggregate graded thickness 50–80 mm
Open-graded wearing course:
    14 mm aggregate graded thickness 40–55 mm
    10 mm aggregate graded thickness 30–35 mm
Close-graded wearing course:
    14 mm aggregate graded thickness 40–55 mm
    10 mm aggregate graded thickness 30–35 mm
    (better than open-graded for heavier traffic)
    6 mm aggregate graded thickness 20–25 mm
    (suitable surfacing for estate roads, playgrounds, and footways)
Medium-graded wearing course:
    6 mm aggregate graded thickness 20–25 mm
    (only for light traffic, footways, playgrounds and cycle paths)
Fine-graded wearing course:

3 mm aggregate thickness 15–25 mm

(urban footways, courtyards, light car parking, footways; use coated chippings on roadways)

Pervious wearing course:

20 mm aggregate graded thickness 45–60 mm

10 mm aggregate graded thickness 30–35 mm

(fast traffic where water must drain rapidly to the side of the road and therefore a steeper camber is necessary. An impermeable substrate is necessary to direct the water away from the surface)

Standard asphalt mixtures to BS 594 Pt 1 : 1985

Road base, base course, regulating course (a regulating course is used to regulate the levels):

Mixture 50/14; (preferred) 14 mm aggregate thickness 35–65 mm

Mixture 60/28; (preferred) 28 mm aggregate thickness 60–120 mm

Wearing course, design type F:

Mixture 30/14; (preferred) 14 mm aggregate thickness 40 mm

Mixture 40/14; (preferred) 14 mm aggregate thickness 50 mm

Wearing course, design type C:

Mixture 0/3; 2.36 mm aggregate thickness 25 mm

Mixture 30/14; 14 mm aggregate thickness 40 mm

Mixture 40/14; 14 mm aggregate thickness 50 mm

Mixture 55/10; 10 mm aggregate thickness 40 mm

Mixture 55/14; 14 mm aggregate thickness 45 mm

(Although type F and type C look similar, the proportions of different-sized aggregates vary between the two types. Mixtures 55/10 and 55/14 are not coated with chippings and are used only for secondary roads, estate roads, play areas and other lightly trafficked roadways where high-speed skidding is not likely.)

Design type R:

Mixture R30/14F; 14 mm aggregate thickness 40 mm

Mixture R40/14F; 14 mm aggregate thickness 50 mm

Mixture R30/14C; 14 mm aggregate thickness 40 mm

Mixture R40/14C; 14 mm aggregate thickness 50 mm

R types are 'enriched mixtures', which are designed to prevent breakup caused by fatigue and for cold wet conditions, though they are more liable to deformation than types F and C.

Chippings for asphalt surfaces

Coated chippings rolled in hot after the first layer of asphalt has been laid and compacted.

Chippings for low skid resistance: 60% of total coverage.

Chippings for high skid resistance: 80% of total coverage.

Dense tar surfacing

Depending on the traffic loading, DTS wearing course thicknesses can be:

- 40 mm thick with 10 mm aggregate
- 40 mm thick with 14 mm aggregate
- 30 mm thick with 10 mm aggregate

Overall thickness of road construction

Lorry parking bays:
    650–425 mm for clay subgrades
    425–350 mm for loam subgrades
    350–300 mm for gravel subgrades
    Where lorries with point-load jacks or heavy steel ramps are working, steel reinforcement should be included in the specification.

Concrete roads

Roadway layer	Concrete mix to BS 5328
Sub-base or road base:	
wet lean concrete 1	ST 1 (C7.5, C7.5P)
wet lean concrete 2	ST 2 (C10, C10P)
wet lean concrete 3	ST 3 (C15, C15P)
wet lean concrete 4	ST 4 (C20, C20P)
Road base:	
continuously reinforced:	C40
Wearing course (surface slab):	
unreinforced	C40
jointed reinforced	C40
continuously reinforced	C40

Levels

Variations from sub-base to road base:	+10 mm or -30 mm
Variations from road base to base course:	±15 mm
Variations from base course to wearing course, playgrounds, private car parks, cycle paths:	±10 mm
Variations from base course to wearing course, roads:	±6 mm
Straight cross-fall:	not less than 2% or more than 3%
Camber from crown to channel:	not less than 2% or more than 3%
Fall in the channel:	not less than 0.8%
Mastic asphalt:	not less than 1:80

Joints

Joints offset not less than 300 mm from joints in layers below

---

OUTLINE SPECIFICATION

Pedestrian-only roadways

(Footways, cycle paths and pedestrian areas)
a. *On sub-base previously specified in section Q20 lay road base of wet-mix bitumen macadam 75 mm thick 28 mm aggregate to given falls and levels, roll with 2.5 tonne roller.*
b. *Lay base course of hot-rolled asphalt 50 mm thick 20 mm aggregate.*
c. *Lay wearing course of hot-rolled asphalt 20 mm thick 6 mm aggregate. Kerbs to be set and cured before road construction is carried out.*

for kerbs	*Q10*
for bases	*Q20*

Occasional-traffic roadways

(Small private car parks, driveways, pedestrian areas and cycle paths)
a. *On sub-base previously specified in section Q20 lay road base of wet-mix bitumen macadam 100 mm thick 40 mm aggregate to given falls and levels.*
b. *Lay base course of open-graded bitumen macadam 60 mm thick 20 mm aggregate.*
c. *Lay wearing course of close-graded bitumen macadam 25 mm thick 10 mm aggregate. Kerbs to be set and cured before road construction is carried out.*

for kerbs	*Q10*
for bases	*Q20*

Category 1 roadways

(Light traffic: non-through residential roads, public car parks, and delivery areas for light vans)
a. *On sub-base previously specified in section Q20 lay road base of wet-mix bitumen macadam 150 mm thick 40 mm aggregate to given falls and levels.*
b. *Lay base course of hot-rolled asphalt 60 mm thick 20 mm aggregate.*
c. *Lay wearing course of hot-rolled asphalt 35 mm thick 10 mm aggregate. Kerbs to be set and cured before road construction is carried out.*

for kerbs	*Q10*
for bases	*Q20*

Category 2 roadways

(Regular traffic with up to 75 commercial vehicles per day in each direction)

a. *On sub-base previously specified in section Q20 lay road base of close-graded bitumen macadam 135 mm thick 40 mm aggregate to given falls and levels.*

b. *Lay base course of dense-coated macadam 60 mm thick 20 mm aggregate.*

c. *Lay wearing course of close-graded bitumen macadam 35 mm thick 10 mm aggregate. Kerbs to be set and cured before road construction is carried out.*

Heavy-duty roadways

(Goods vehicle delivery access roadways and loading bays where slewing and braking may occur)

a. *On sub-base previously specified in section Q20 lay road base of close-graded bitumen macadam 150 mm thick 40 mm aggregate to given falls and levels.*

b. *Lay base course of close-graded bitumen macadam 60 mm thick 20 mm aggregate.*

c. *Lay wearing course of close-graded tar surfacing 40 mm thick 14 mm aggregate. Kerbs to be set and cured before road construction is carried out.*

for kerbs	Q10
for bases	Q20

Mastic asphalt for lightly loaded areas

(Play areas, courtyards, roof gardens)

a. *Lay base layer of one coat roofing-grade asphalt.*

b. *Lay glass-fibre separating membrane.*

c. *Lay top layer of one coat roofing-grade asphalt with 5–10% grit rolled in, overall thickness 25 mm.*

for kerbs	Q10
for bases	Q20

Mastic asphalt for heavily loaded areas

(Lorry loading and standing)

a. *Lay base layer 20 mm thick of two coats roofing-grade asphalt to given falls and levels.*

b. *Lay glass-fibre separating membrane.*

c. *Lay top layer 40 mm thick of paving-grade asphalt.*

for kerbs	Q10
for bases	Q20

Q23 Granular surfaces

GENERAL GUIDANCE

This section covers loose or compacted granular materials used for footpaths and informal paving in pedestrian-only areas.

All the materials listed below make attractive informal pathways, and all require a base course and edge restraints. Most will drain naturally when laid to adequate falls and cross-falls, but should not be used in areas where the water-table is close to the surface, as this will inevitably cause ponding. Unless they are heavily compacted, these materials make an excellent surround for trees and shrubs, because they allow the roots to take in air as well as water. The finer materials – particularly those with natural clay binders – require a geotextile layer to prevent them from leaching into the base course, while granular materials laid on a subgrade with a high fine-particle content may need a geotextile layer to prevent the particles from squeezing up into the granular paving. It is worth remembering that some gravels and aggregates are aggressively acidic or alkaline and liable to damage planting inimical to these conditions unless the run-off is diverted from planted areas.

All granular material is difficult to clean, and regular maintenance is essential if it is to be kept free of weeds, potholes and ruts. Where there is likely to be any difficulty in replenishing the material for repair and maintenance, the landscape designer should advise the user to keep a suitable stockpile to hand.

Loose granular materials, especially gravels, make suitable alternatives to grass where it does not grow well under trees, or in small walled gardens and courtyards where mowing grass is not a practical proposition; here they make an excellent base for large pots and tubs. They should not be used on slopes greater than 1 in 20 or in areas subject to wind scour or fast surface-water run-offs. They are not suitable for disabled people, prams or any narrow-rimmed wheeled appliance. Therefore only wide-wheeled slow-moving vehicles should be allowed on them for maintenance and so forth; mountain scramble bikes must be actively discouraged, because they will churn up the surface very quickly by fast riding and hard braking. Where loose material is placed next to grassed areas it is advisable to specify an upstanding edge restraint or to fix the level of the path at least 50 mm below the grass; this should prevent damage to the mower. Large loose pebbles and ballast are not recommended in areas where children or vandals will be tempted to use them as missiles.

Compacted granular material is suitable for private drives, small country car parks and hard play/kick-about areas for older children.

Bark in coarse grades is the material of choice for bridleways, and finer bark is suitable for running tracks and supervised play areas. Edgings of any sort should not be used for riding or running tracks, because there is a danger of tripping.

Definition of surfacings

There is no official standard for defining gravel, ballast, shingle, or hoggin for non-structural uses, and each producer has his own classification, which may vary from pit to pit. Neither geological definitions nor definitions of fine aggregate used for concrete given in British Standard 882 are relevant to the aggregates used in landscape surfacing.

Terms in general use include:

- gravel;
- bound gravel;
- hoggin;
- chippings;
- ballast;
- shingle;
- shale or blaes;
- bark and wood chips.

Gravel
Most gravels are neutral in colour, ranging from grey to brown, and available from crushed hard rock or natural pebbles ranging in size from 5 mm to 15 mm. They are normally specified as fine, medium or coarse, but as there is no agreed standard for these grades, they should be to sample from a named supplier. Gravels may be pit, river or beach, and should be specified as washed prior to delivery. Pea gravel is, as the name implies, pea sized, and is dug from naturally occurring beds or from the coast areas; it too should be washed prior to delivery and be to agreed sample. 'Inert' gravels should be specified for use adjacent to lime-hating plants or where root structures need to breathe.

Bound gravel
Ordinary loose gravel is liable to work into ruts or pits where heavy foot traffic is normal, but specially selected and blended gravels are now available that are bound either with special sands or with chemical binders. These are much more hard-wearing than loose gravel, but they must be very carefully laid, and they still need regular maintenance to keep them in good condition.

Hoggin
Good hoggin has a long life if properly drained and weeded regularly. It is a naturally occurring well-graded gravel in a clay matrix, and forms a hard compact surface when laid and rolled. Any attempt to emulate the mix of clay, sand and gravel on site is doomed to failure and should not be allowed.

Chippings
These are sharp-edged stone particles produced in a range of sizes from 6 mm to 20 mm and of even size within the range. They are often prepared

from a strongly coloured natural stone, though artificially coloured material is available. They should be specified to agreed sample.

Ballast

The colour of ballast ranges from pale yellow to mid-brown. It is a mixture of sand and gravel supplied as dug from the pit. The proportion of sand to gravel may vary, and it should be specified to comply with the agreed sample.

Shingle

Shingle consists of large rounded beach or raised beach pebbles, from 20 mm to 60 mm; beach or marine shingle should be supplied washed. Because shingle is derived from a number of different rocks, the colour and shape will vary from pink-red, through grey or black, to white/grey. As shingle is not permitted to be extracted indiscriminately, ensure that the shingle comes from an approved named source and to sample.

Shale or blaes

This is no longer produced, but is still available in quantity from mine dumps, mainly in southern Scotland. It provides a useful, economical surface for hard playgrounds and small parking areas for light vehicles, though like all loose surfacing it can be rutted and pitted by vehicles driving, accelerating, and braking too fast.

Bark and wood chips

Surfacing bark or chips are usually obtained from forestry conifer industries and then shredded, broken, or ground to various sizes and grades, but there are firms that supply hardwood or wood fibre surfacing material from other sources. Bark and wood chip should be specified by type and grade and checked for weedkillers and insecticides; all wood products should be checked with the supplier to ensure that they have not been treated with prohibited chemicals, particularly methyl bromide. Playground material must conform to BS 5696 Pt 3 : 1979 *Play equipment: Code of practice for installation and maintenance*, and BS 7188 : 1989 *Testing impact absorbing playground surfaces*.

Falls

To be reasonably sure that loose surfacing will not migrate, it should not be laid to falls greater than 1:20 cross-fall, and 1:40 longitudinal fall, but steeper gradients are acceptable in sheltered areas where these falls cannot be fulfilled.

BRITISH STANDARDS

As gravel, hoggin and bark are natural materials used 'as dug' or specially graded, there is no specific British Standard for them.

882 : 1992	Aggregates from natural sources for concrete
4962 : 1989	Plastics pipes and fittings for subsoil drains: land drainage for light wheel loads
5696 Pt 3 : 1979 & AMD	Play equipment: Code of practice for installation and maintenance
7188 : 1989 & AMD	Testing impact absorbing playground surfaces

DATA

Typical manufacturers' ranges

Gravel
10 mm Golden gravel
2–5 mm Alpine gravel
8–16 mm Golden quartzite gravel
4–8 mm Golden quartzite gravel
5–10 mm Granite gravel
Graded limestone gravel
Thames Valley pea gravel

Shingle and ballast
10–20 mm Rounded flint stones
20–30 mm Rounded flint stones
30–50 mm Scottish beach stones
25–45 mm Quartz

Chippings
10 mm Derbyshire Spar
10 mm Pink granite
10 mm Grey granite
10 mm Blue Gritstone

10 mm White limestone
13-20 mm Cotswold stone
20 mm Flint nuggets
10 mm Diorite chippings

Bark and wood chips
5–30 mm natural hardwood chips
5–30 mm conifer wood chips
10–40 mm natural hardwood chips
10–50 mm dust-free playground conifer bark
15–60 mm playground conifer bark
1–30 mm conifer wood fibre
1–100 mm conifer wood peelings

OUTLINE SPECIFICATION

Gravel pedestrian-only areas

a. *Excavate for gravel area, remove all topsoil to dump on site, treat area with agreed herbicide, and backfill to given levels with approved excavated material or approved hardcore.*

b. *Lay precast concrete edging round top 150 mm × 50 mm to BS 7263 as specified in section Q10 to both sides of path.*

c. *Lay base of 150 mm compacted DoT Type 1 granular fill.*

d. *Supply and lay 60 mm washed 10 mm Thames Valley pea gravel in moist condition, roll once with 500 kg roller, water once, roll again with 500 kg roller. Roll and water twice more, not earlier than 2 days and not later than 6 days after initial laying.*

e. *All levels to be adjusted in base layer. No material to be stored on the finished surface, and no site traffic permitted after completion. No work to be done in frosty weather.*

for excavation	*D20*
for edgings	*Q10*

Alternative specification

As above but with geotextile filter fabric 1.0 mm thick, water flow 33 l/m²/s, laid on subgrade with 200 mm overlaps.

Initial maintenance

The gravel area must be watered and rolled for three months after completion of the work in order to ensure proper compaction. Rolling and watering should take place every 2 weeks.

Shingle or ballast

a. *Excavate for shingle area, remove all topsoil to dump on site, treat area with agreed herbicide, and backfill to given levels with approved excavated material or approved hardcore.*

b. *Lay 100 mm clean hardcore, blind with sand to BS 882 Grade C, lay 0.7 mm geotextile filter fabric water flow 50 l/m²/s.*

c. *Supply and lay 20–40 mm washed beach shingle 100 mm thick by hand and bring to level surface. Protect finished surface from dust and contamination till completion of works.*

for excavation	*D20*
for edgings	*Q10*

Bark or woodchip

a. *Excavate for bark area, remove all topsoil to dump on site, treat area with agreed herbicide, and backfill to given levels with approved excavated material or approved hardcore.*

b. *Lay 150 mm clean hardcore, blind with sand to BS 882 Grade C, lay 0.7 mm geotextile filter fabric water flow 50 l/m²/s.*

c. *Supply and lay 100 mm hardwood chips 10–40 mm size. No chemicals to be used on chips or bark.*

d. *Supply and fix treated softwood edging boards 50 mm × 150 mm to bark area on hardcore base extended 150 mm beyond the bark area, boards fixed with galvanized nails to treated softwood posts 750 mm × 50 mm × 75 mm driven into firm ground at 1.0 m centres. Edging boards to finish 25 mm above finished bark surface. Tops of posts to be flush with edging boards and once weathered.*

| *for excavation* | *D20* |
| *for edgings* | *Q10* |

Bark or woodchip riding track

a. *Excavate drain trench in centre of track 300 mm wide × 600 mm deep to specified falls, excavate for riding track 3.0 m wide × 300 mm deep, spread topsoil on adjoining area, remove subsoil to dump on site.*

b. *Lay 0.7 mm geotextile filter fabric water flow 50 l/m²/s in drain trench, lay 100 mm dia. flexible perforated plastic drain pipe to BS 4962 : 1989.*

c. *Backfill drain trench with gravel waste or similar, backfill track trench 100 mm depth 40 mm clean washed aggregate, lay filter membrane 3.0 m wide on fill, backfill to ground level with 150 mm conifer bark 5–30 mm size.*

for excavation	*D20*
for edgings	*Q10*
for land drainage	*R13*

Q24 Interlocking flexible brick and block paving

GENERAL GUIDANCE

This section covers both clay bricks and concrete blocks that are laid in an interlocking pattern to form a vehicular paving. Slabs, paving bricks, brick pavers, setts and cobbles laid on rigid bases are dealt with in section Q25.

Interlocking paving describes not so much the shape of the blocks or bricks as the method of locking them very firmly together on a bed of sand by means of a vibrating plate. This makes the total area so treated into a homogeneous (monolithic) flexible unit; the most substantial specification produces one of the few surfaces that can withstand the continuous turning and tracking of buses and heavy goods vehicles in confined spaces. In this respect they have much in common with the nineteenth century roads made of setts, and like them once the surface is uplifted to renew underground services it is difficult to replace them to a perfect surface. Before working on these projects, the landscape designer should assess the likely damage to the surface by the vehicles, machinery or spillages. It is advisable to plan the paving schedule so that the contractor has access to all parts of the site without travelling on the finished block paving; the final areas can be laid just before completion of the contract.

The size and strength of precast concrete paving blocks is covered by British Standard 6717 Pt 1 and brick pavers by BS 6677 Pt 1. Both come in grades for light and industrial work, and both must have an adequate edge restraint to prevent lateral spread. Blocks and bricks should have good frost, abrasion and skid resistance. The recommended fall is not less than 1.25%

(1:80) lengthways and 2.5% (1:40) crossways, and these falls must be established in the sub-base; expansion joints are not required in the block or brick surface as movement is taken up by the individual units. The strength of the paving depends almost entirely on the quality of the workmanship, and therefore the laying of interlocking paving should be well supervised.

Laying blocks

For precast concrete blocks, the most suitable patterns for heavy use are either the plain rectangular block laid in a herring-bone pattern or the fully interlocking 'jigsaw'-shaped block. Either should be 80–100 mm thick; 100 mm is better for areas where wheel tracking is expected. All block paving should be laid on a bed of sand and with a substantial road base and sub-base. It is usual to use a DoT Type 1 granular sub-base for block paving, though blocks can be bedded in mortar on a concrete base, in which case they will lose the flexibility given by the sand-bedding method.

Sand for bedding and jointing should comply with Table 2 of BS 6717 Part 2, which gives a mix of sand particles producing a suitable sand. On the base a layer of loose sand is laid and screeded to the given levels, and then compacted with a vibrating plate compactor to 50 mm thickness. The blocks are then bedded in 15 mm of loose sand screeded to the same profile, and laid with open joints 2–5 mm wide. The edge restraint must be laid before the blocks are laid in order to prevent the blocks from moving during vibration. After laying, the blocks are vibrated again, and the joints are filled with dry sand and given a last vibrating. Special limestone grit can be used for jointing, which tends to consolidate, or special bonding liquid can be used.

Fig. 4.27 Veta block paving, Russell Street, Leeds. (Marshalls Mono Ltd)

Fig. 4.28 Bianca and Mistral block paving, Swansea.
(Marshalls Mono Ltd)

Fig. 4.29 Arcadian block paving, Hamilton.
(Marshalls Mono Ltd)

Fig. 4.30 Keyblock, Paradise Circus, Birmingham. (Marshalls Mono Ltd)

For pedestrian areas and private drives where only light traffic is expected, blocks can be rectangular or shaped units 60 mm thick laid on sand on a much less substantial road and sub-base. The bond for light use can vary and be, for example, stack, half, quarter or basket weave bond. The introduction of colour, logos, directional lines and so forth helps to make an attractive identifiable space.

Blocks in small areas are hand laid, but where a large area of paving is easily accessible to machinery, blocks can be laid by machine more rapidly, though edging units are laid by hand.

Brick pavers for heavy industrial areas are laid in a herring-bone pattern on a base similar to the blocks, and if possible the V of the herring bone should point in the direction of travel. It may be advisable to lay a plain strip next to the edge restraint where this will save cutting, or the manufacturer's special edging blocks may be used. For light domestic use, any of the bonds or patterns mentioned above may be used.

BRITISH STANDARDS

5328 Pt 1 : 1991 & AMD	Guide to specifying concrete
5328 Pt 2 : 1991 & AMD	Methods for specifying concrete mixes
6677 Pt 1 : 1986	Clay and calcium silicate pavers for flexible paving
6677 Pt 2 : 1986	Code of practice for the design of lightly trafficked pavements
6677 Pt 3 : 1986	Method of construction of pavements
6717 Pt 1 : 1993	Precast concrete paving blocks

6717 Pt 3 : 1989	Code of practice for laying precast concrete paving blocks
7370 Pt 2 : 1994	Grounds maintenance. Maintenance of hard areas (except sports surfaces)
7533 : 1992	Guide for structural design of pavements constructed with clay or concrete blocks

DATA

Sizes of clay brick pavers to BS 6677 Part 1

200 mm × 100 mm × 50 mm thick
200 mm × 100 mm × 65 mm
210 mm × 105 mm × 50 mm
210 mm × 105 mm × 65 mm
215 mm × 102.5 mm × 50 mm
215 mm × 102.5 mm × 65 mm

Type PA: 3 kN
Footpaths and pedestrian areas, private driveways, car parks, light vehicle traffic and overrun.

Type PB: 7 kN
Residential roads, lorry parks, factory yards, docks, petrol station forecourts, hardstandings, bus stations.

Sizes of precast concrete paving blocks to BS 6717 Part 1

Type R blocks
200 mm × 100 mm × 60 mm
200 mm × 100 mm × 65 mm
200 mm × 100 mm × 80 mm
200 mm × 100 mm × 100 mm

Type S
Any shape within a 295 mm space

Block paving bonds

- Herring-bone – laid in alternate V-shaped patterns;
- stack bond – straight joints in both directions;
- half-bond – straight joints one way, staggered joints the other way;
- quarter-bond – as half-bond but smaller stagger;
- basket weave – two bricks laid one way, then two laid at right angles.

Concrete mixes to BS 5328 Pt 1 & AMD.

Application	Standard mix	Designated mix	Slump (mm)
Mass concrete fill and blinding	ST 2 (C10P)	GEN 1	75
Mass concrete foundations in non-aggressive soils	ST 2 (C10P)	GEN 1	75
Trench fill foundations in non-aggressive soils	ST 2 (C10P)	GEN 1	125

Concrete mixes to BS 5328 Pt 1 & AMD *continued*

Application	Standard mix	Designated mix	Slump (mm)
Foundations			
in Class 2 sulphate soils	–	FND 2	75
in Class 3 sulphate soils	–	FND 3	75
in Class 4A sulphate soils	–	FND 4A	75
in Class 4B sulphate soils	–	FND 4B	75
Kerb bedding and haunching	ST 1 (C7.5P)	GEN 0	10
Drain pipe support in non-aggressive soils	ST 2 (C10P)	GEN 1	10
Manholes etc. in non-aggressive soils	ST 2 (C10P)	GEN 1	50
In-situ concrete paving:			
pedestrian or light vehicle	–	PAV 1	75
heavy vehicle	–	PAV 2	50

GEN concretes with low cement content may not be suitable for good cast and direct finished surfaces.

Traditional concrete mixes

Application	Name of mix	Cement:aggregate:sand (parts by volume)
Setting posts, filling weak spots in ground, backfilling pipe trenches:	one, two, three	1:2:3
	or	or
	one to five all-in	1:5
Paving bases, bedding manholes, setting kerbs, edgings, precast channels:	one, two, four	1:2:4
	or	or
	one to six all in	1:6
In situ concrete paving:	one, one & half, two & half	1:1.5:2.5

'All-in' is mixed aggregate as dug from the pit.

Mortar mixes

Mortar group	Cement:lime:sand	Masonry cement:sand	Cement:sand with plasticizer
1	1:0–0.25:3		
2	1:0.5:4–4.5	1:2.5–3.5	1:3–4
3	1:1:5–6	1:4–5	1:5–6
4	1:2:8–9	1:5.5–6.5	1:7–8
5	1:3:10–12	1:6.5–7	1:8

Group 1: strong inflexible mortar
Group 5: weak but flexible

All mixes within a group are of approximately similar strength. Frost resistance increases with the use of plasticizers. Cement:lime:sand mixes give the strongest bond and greatest resistance to rain penetration.

OUTLINE SPECIFICATION

Precast concrete blocks

a. *On previously laid subgrade lay sub-base of 150 mm DoT Type 1 compacted granular material as specified in section Q20, lay sand to BS 6717 Part 2 and screed to given levels, compact with vibrating plate compactor to 50 mm thickness.*

b. *Lay 15 mm of loose sand as specified above screeded to given levels, supply and lay specified manufacturer's rectangular blocks complying with BS 6717 Pt 1 Type R. Blocks to be laid with open joints 5 mm wide, and edge restraints must be in position before the blocks are laid.*

c. *Vibrate blocks to bring to finished levels, fill joints with dry sand well brushed in, vibrate again and brush off surplus sand.*

d. *Variation of 20 mm maximum permitted in sub-base, 15 mm maximum in bedding course, 10 mm maximum in finished surface. No material to be stored on the finished surface, and no site traffic permitted after completion. Bases containing cement left exposed for more than 2 hours after pouring to be protected.*

for sub-bases   *Q20*

Q25   Rigid clay brick paving

GENERAL GUIDANCE

This section covers the lighter of the two brick pavers (type PA) specified in BS 6677 and standard or special building bricks covered by British Standards 3921 and 4729 and which comply with the durability, frost and slip resistance for type PA pavers. Both are intended for pedestrian areas, private drives and courtyards. The stronger clay pavers, which are suitable for use in flexible interlocking roadways capable of taking the loads of heavily laden lorries, buses and cargo, are dealt with in section Q24.

Areas of rigid paving are a boon to the landscape designer, as he or she can use many different bonds and colours to indicate direction and so forth, and standard special bricks can be used for symbols and logos if the project is not large enough to have 'specials' made to order. Bonds used for brick paving and pavers include herring-bone, basket weave, half, quarter and stack bonds.

Four types of brick are suitable for paving:

- hard frost-resistant and sulphate-resistant clay bricks;
- special clay brick 'pavers';

- clay engineering bricks;
- some classes of calcium silicate brick.

Clay paving bricks

These are standard building bricks, which are very hard, frost and water resistant, and slip resistant. Clay engineering bricks with chequered or ribbed faces are also suitable, though rather heavy in appearance. They must be to the standards of Type FL of BS 3921 without perforations or frogs, and Type PA of BS 4729, and they are laid in the same way as brick pavers. The colours are those of ordinary facing bricks or engineering bricks. BS 4729 lists the special brick shapes normally available that can be used for edging, corners and radial work.

Calcium silicate bricks

These should be not less than Class 4, with a strength of 27.5 N/mm² as specified in BS 187: 1978. The colours are mainly yellow, light buff or grey, and although these bricks give a lighter-coloured paving, the overall effect is more monotonous than that of natural clay bricks.

Fig. 4.31 Flame-textured light and dark grey granite setts and slabs. Sawn Yorkstone setts – good colour and textural contrast.
(Derek Lovejoy Partnership)

Fig. 4.32 Cropped Yorkstone setts and dark grey granite setts – careful attention to detail and workmanship around recessed manhole cover. (Derek Lovejoy Partnership)

Clay brick pavers

Besides the standard rectangular pavers specified in BS 6677, some manufacturers also make standard specials to the same requirements; this saves cutting pavers on site, which is not always a satisfactory exercise. The landscape designer is advised to use uncut pavers when at all possible. As with all small paving units, replacement after service runs have had to be renewed is rarely satisfactory, as the pavers either get broken and the colour and/or texture replacements differ, or the units are uneven and a hazard to users. The landscape designer may be able to arrange for the statutory undertakers to run their services in trenches or ducts, which can be covered with unit paving or another more easily replaceable surface.

Laying clay paving bricks and pavers

Accurate landscape drawings and setting out are essential to a successful job, as edge restraints, channels, ducts, gullies and so forth should be built in before the paving surface is laid. Where possible, standard specials should be used to avoid cutting, but if bricks must be cut, diamond-disc power saws should be used. The allowed tolerance between pavers is ±2 mm. Movement joints for brick paving bedded in mortar should be 10 mm wide at 4500 mm centres and at restrictions such as walls, piers and planters; they are constructed of solid polyethylene foam, a flexible two-part polysulphide sealant to BS 4254, or a silicone sealant to BS 5889.

There are two standard methods of laying rigid clay brick paving. The first is to lay a concrete sub-base and bed the pavers in mortar, and the second is to lay a granular sub-base and bed the pavers in sand. It is not advisable to mix the two systems, because the granular base is too flexible to support the mortar bedded units successfully. Mortar bedding is used for slopes steeper than 1 in 10, edging, drainage channels, ramps, or where the paving is continuously wet. If vehicle overrun is expected, it is advisable to incorporate a steel reinforcing mesh in the concrete sub-base.

Where site traffic is to use the area to be paved during the contract period, it is better to defer laying the pavers until the end, as they are damaged by diesel or oil spillages or by blobs of paint, cement and sealant, and cleaning them off is not always easy.

BRITISH STANDARDS

12 : 1996	Portland cement
187 : 1978	Calcium silicate (sandlime and flintlime) bricks
890 : 1996	Building limes
1200 : 1976 & AMD	Building sands from natural sources. Sands for mortar for brickwork
3921 : 1985	Clay bricks and blocks
4027 : 1996	Sulphate-resisting Portland cement
4721 : 1981 (1986) & AMD	Ready mixed building mortars
4729 : 1990	Shapes and dimensions of special bricks
4887 Pt 1 : 1986	Mortar admixtures. Air-entraining (plasticizing) admixtures
4887 Pt 2 : 1987	Mortar admixtures. Set retarding admixtures
5224 : 1995	Masonry cement
6677 Pt 1 : 1986	Clay and calcium silicate pavers for flexible paving
6677 Pt 2 : 1986	Code of practice for the design of lightly trafficked pavements
6677 Pt 3 : 1986	Method of construction of pavements
7583 : 1996	Portland limestone cement (A type of cement with up to 20% of specified limestone. Limestone cement can be used as ordinary Portland cement in designed mixes)

BRICK PAVING

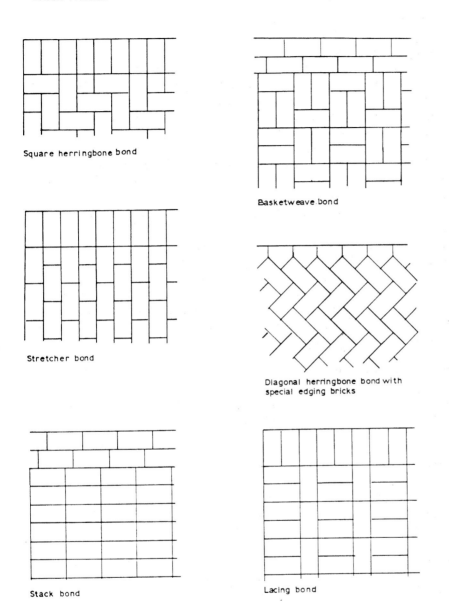

Square herringbone bond

Basketweave bond

Stretcher bond

Diagonal herringbone bond with
special edging bricks

Stack bond

Lacing bond

Brick paving bedded in 25mm Group 3 cement:lime:sand mortar on 100mm PAV1
concrete on 100mm hardcore, jointed in Group 1 cement:lime:sand mortar
or
Brick paving bedded in 25mm sand, on 50mm compacted sand, on 150mm DoT
Type 1 compacted fill, 5mm joints filled with limestone grit and vibrated

Fig. 4.33 Brick paving.

DATA

Sizes of pavers to BS 6677 Part 1

200 mm × 100 mm × 50 mm thick
200 mm × 100 mm × 65 mm
210 mm × 105 mm × 50 mm
210 mm × 105 mm × 65 mm
215 mm × 102.5 mm × 50 mm
215 mm × 102.5 mm × 65 mm

Type PA: 3 kN
footpaths and pedestrian areas, private driveways, car parks, light vehicle traffic and overrun.
Type PB: 7 kN
residential roads, lorry parks, factory yards, docks, petrol station fore-courts, hardstandings, bus stations.

Durability of bricks

FL   Frost resistant with low salt content
FN   Frost resistant with normal salt content
ML   Moderately frost resistant with low salt content
MN   Moderately frost resistant with normal salt content

Typical sizes of proprietary clay brick pavers

215 mm × 102 mm × 38 mm
215 mm × 102 mm × 47 mm
215 mm × 102 mm × 50 mm
215 mm × 102 mm × 65 mm

Engineering brick ('stable') pavers

Stable paver 215 mm × 115 mm × 75 mm thick, and thicker to order

Plain surface	Patterned
Single panel	Diamond chequer
Two panel	Dimpled non-slip
Four panel	Maltese cross
Six panel	
Eight panel	

Brick paving bonds

- Herring-bone: laid in alternate V-shaped patterns;
- stack bond: straight joints in both directions;
- half-bond: straight joints one way; staggered joints the other way;
- quarter-bond: as half-bond but smaller stagger;
- basket weave: two bricks laid one way, then two laid the other way.

Mortar mixes

Mortar group	Cement:lime:sand	Masonry cement:sand	Cement:sand with plasticizer
1	1:0–0.25:3		
2	1:0.5:4–4.5	1:2.5–3.5	1:3–4
3	1:1:5–6	1:4–5	1:5–6
4	1:2:8–9	1:5.5–6.5	1:7–8
5	1:3:10–12	1:6.5–7	1:8

Group 1: strong inflexible mortar
Group 5: weak but flexible.

All mixes within a group are of approximately similar strength. Frost resistance increases with the use of plasticizers. Cement:lime:sand mixes give the strongest bond and greatest resistance to rain penetration. Masonry cement to BS 5224: 1995 covers a wider range of cement types and defines three strength classes. Non-air-entrained cements are included.

Class	28-day strength N/mm^2	Air entrainment (X = non-air-entraining)
MC 5	5-15	yes
MC 12.5	12.5-32.5	yes
MC 12.5X	12.5-32.5	no
MC 22.5X	22.5-42.5	no

Masonry cement equals ordinary Portland cement plus a fine neutral minerla filler; it is easier to work and is more frost resistant than a cement:lime mixture. Masonry cement should never have lime added to it, and should not be used on walls subject to heavy loading.

Concrete mixes to BS 5328 Pt 1 & AMD

Application	Standard mix	Designated mix	Slump (mm)
Mass concrete fill and blinding	ST 2 (C10P)	GEN 1	75
Mass concrete foundations in non-aggressive soils	ST 2 (C10P)	Gen 1	75
Trench fill foundations in non-aggressive soils	ST 2 (C10P)	GEN 1	125
Foundations			
in Class 2 sulphate soils	–	FND 2	75
in Class 3 sulphate soils	–	FND 3	75

Concrete mixes to BS 5328 Pt 1 & AMD *continued*

Application	Standard mix	Designated mix	Slump (mm)
in Class 4A sulphate soils	–	FND 4A	75
in Class 4B sulphate soils	–	FND 4B	75
Kerb bedding and haunching	ST 1 (C7.5P)	GEN 0	10
In-situ concrete paving:			
pedestrian or light vehicle	–	PAV 1	75
heavy vehicle	–	PAV 2	50

GEN concretes with low cement content may not be suitable for good cast and direct finished surfaces.

Traditional concrete mixes

Application	Name of mix	Cement:aggregate:sand (parts by volume)
Setting posts, filling weak spots in ground, backfilling pipe trenches:	one, two, three	1:2:3
	or	or
	one to five all-in	1:5
Paving bases, bedding manholes, setting kerbs, edgings, precast channels:	one, two, four	1:2:4
	or	or
	one to six all in	1:6
In situ concrete paving:	one, one & half, two & half	1:1.5:2.5

'All-in' is mixed aggregate as dug from the pit.

OUTLINE SPECIFICATION

Brick pavers

a. *Construct 100 mm FND2 concrete base on 150 mm well-rammed hardcore.*

b. *Supply and lay specified manufacturer's clay brick pavers Type PA to BS 6677 in stack bond bedded and jointed in 1:¼:4 cement:lime:sand mortar, sand to BS 882 Grade M, with 10 mm joints struck flush and compacted as the work proceeds.*

c. *No subsequent pointing permitted. Perimeter bats and cut pavers adjoining structures to be cut with diamond-disc power saws, or manufacturer's specials to be used.*

d. *Finished work to be protected by impervious sheeting for three days after completion. No material to be stored on the finished surface and no site traffic permitted after completion.*

for excavation	*D20*
for concrete	*E10*
for kerbs and edgings	*Q10*
for bases	*Q20*

Paving bricks

a. *Construct base of 150 mm DoT Type 1 granular material compacted.*

b. *Supply and lay specified manufacturer's clay paving bricks, quality FN, in herring-bone bond, bedded in 25 mm sand to BS 882 Grade F on 50 mm Grade F compacted sand, vibrated, 5 mm joints filled with lime-stone grit and vibrated.*

c. *Perimeter bats and cut pavers adjoining structures to be cut with dia-mond-disc power saws, or manufacturer's specials to be used.*

d. *Finished work to be protected by impervious sheeting for three days after completion. No material to be stored on the finished surface and no site traffic permitted after completion.*

for excavation	*D20*
for concrete	*E10*
for kerbs and edgings	*Q10*
for bases	*Q20*

Q25 Slab pavings

GENERAL GUIDANCE

The term **flag** is used in British Standards, while the term **slab** is used in the Common Arrangement system of classification. This section deals with small unit paving laid on a previously constructed base; it includes all types of pre-cast concrete and natural stone flags. Brick paving, cobbles and setts are dealt with in separate sections under Q25. Interlocking blocks for pedestrian and vehicular traffic are covered in section Q24. Artificial cobble and sett pat-terning in wet concrete are not considered. Bases for all type of paving are in section Q20.

Unit pavings are, as the name implies, individual small pieces used in pat-terns to make a more interesting finish than can be obtained from a mono-lithic surface, though they may be an expensive option, as laying them is a labour-intensive operation.

The original precast concrete paving slabs (which are still used for local authority pavements) were large, thick, extremely heavy, and hard wearing, with a natural grey finish. Since then, the interest in unit paving for pedestrian precincts, footways, private and public gardens and so forth has produced a wealth of different types of unit paving.

In respect of paving of all types, the landscape designer should check with the manufacturer the slip resistance of any paving likely to be used. It is obvious from personal experience that the slip resistance of a wet paving surface is considerably lower than that of a dry surface, but it is something that can easily be overlooked. The proposed use of the paving should be confirmed by the client, and the manufacturer should be asked to provide slip-resistance figures for both wet and dry conditions.

British Standard precast concrete flags

The current British Standard for flags provides for hydraulically pressed plain precast concrete units in varying-sized rectangles, but a range of colours and finishes can be obtained from manufacturers. Both British Standard and non-British Standard flags vary in strength and quality, but all are intended for pedestrian areas, and are liable to crack and subside when vehicles drive over or park on them. Neither they nor the sub-base on which they are laid are designed to take these weights, with the result that the uneven and damaged surface can cause a hazard to wheel-chair users, prams and pedestrians alike. Where vehicles will overrun the paving, stronger bases and thicker flags must be specified. The small flags E, F and G used to be recommended for vehicle overrun, but a heavy vehicle can punch the 300 mm × 300 mm flags through into the base layer; the 400 mm × 400 mm or 450 mm × 450 mm flag is preferred.

For heavy use, hydraulically pressed flags should be specified. The lighter flags are suitable only for light use in public areas or in domestic gardens. Now that flags are available in more sizes, and if the landscape designer takes care over the design, the labour-intensive exercise of cutting them can easily be dispensed with. Where small pieces do need to be used to complete an awkward-shaped pattern, the small pieces should be set at the back away from traffic, thus ensuring a stronger construction. Manhole covers are available with recesses to take flags so that the pattern of the paving is not interrupted.

Brick pavers, cobbles or setts can usefully be used in conjunction with flags to form patterns or to infill awkward corners. Special paving flags can be commissioned from manufacturers that carry raised or sunk logos, symbols, or direction arrows. Precast concrete flags can be moulded from casts of natural stone slabs, and faced with stone dust; the surface looks like the original stone, but the face can be worn down by heavy use. Exposed aggregate flags are made with a surface of natural aggregate such as gravel or stone chippings of spar or granite.

Natural stone slabs

If the landscape designer has the opportunity to use natural stone paving, it is vital that either the quarry should be visited, or samples of the paving are sent from the quarry selected. The compressive strength and slip resistance of natural stone pavings can vary enormously from one quarry to another.

British Standard tactile flags

Tactile flags are used to aid blind people on road crossings. They are made in three sizes with the references TA/E 450 mm × 450 mm, TA/F 400 mm × 400 mm, and TA/G 300 mm × 300 mm. They have a pattern of raised dots, which can be felt by the sole of the foot, and they should be specified for laying at road crossings, edges of steps or platforms, and other hazardous places.

Laying and bedding flags

The design of sub-bases for flags is described in section Q20. Flags may be laid on a concrete sub-base or a granular sub-base. Concrete sub-bases should be a minimum of 75 mm thick of ST 5 (C25P) concrete with 20 mm aggregate, laid on 100 mm minimum of compacted hardcore. CBM 1, 2 or 3 bases should be a minimum of 100 mm thick with 20 mm aggregate. Granular sub-bases should be estimated or calculated in accordance with the information in section Q20. As for all paving, the drainage falls must be formed in the substrate and sub-base, not in the paving surface.

Flags for pedestrian-only areas may be laid on bases of concrete or granular material and bedded in mortar. Semi-dry mortar bedding makes it easier to adjust the position of the flags, but wet mortar bedding gives the firmest support. In either case, the mortar should be 25 mm thick, with a maximum of 3 mm difference of level between adjoining flags. The bedding mortar should be a lime:sand mix of 1:3 or 4, usually supplied ready mixed, and the joints are made 5–10 mm wide, pointed with 1:3 or 4 cement:sand mortar.

Flags for paving subject to light traffic or occasional overrun may be laid on a base of granular material, CBM or structural concrete. They are laid on 25 mm moist compacted sand to BS 882 Grade C or M with 2–4 mm joints pointed with dry concreting sand (grade F to BS 882) brushed into the joints. Small unit paving should be bedded on sand and vibrated as for concrete block paving.

Natural stone slabs are usually bedded in mortar because of their uneven thickness; the minimum bed is 25 mm.

Flags for paving work are preferably laid with a broken bond, and the use of different-sized flags together with narrower or wider staggering between joints can produce a more interesting pattern.

BRITISH STANDARDS

12 : 1996	Portland cement
882 : 1992	Aggregates from natural sources for concretes
890 : 1996	Building limes
1200 : 1976 & AMD	Building sands from natural sources: sands for mortar for brickwork
4027 : 1996	Sulphate-resisting Portland cement
4721 : 1981 (1986) & AMD	Ready mixed building mortars
4887 Pt 1 : 1986	Mortar admixtures. Air-entraining (plasticizing) admixtures
4887 Pt 2 : 1987	Mortar admixtures. Set retarding admixtures
5224 : 1995	Masonry cement
7263 Pt 1 : 1994	Precast concrete flags, kerbs, channels, edgings and quadrants
7263 Pt 2 : 1990	Code of practice for laying precast concrete flags, kerbs, channels, edgings and quadrants

DATA

Sizes of flags to BS 7263

British Standard flags are manufactured to BS 7263 Part 1. The Standard includes requirements for transverse strength and moisture absorption, and the following sizes are available, specified either by letter reference or by size and thickness:

Reference	Nominal size (mm)	Thickness (mm)
A	600 × 450	50 and 63
B	600 × 600	50 and 63
C	600 × 750	50 and 63
D	600 × 900	50 and 63
E	450 × 450	50 and 70 chamfered top surface
F	400 × 400	50 and 65 chamfered top surface
G	300 × 300	50 and 60 chamfered top surface

Nominal size means that there is a working clearance so that the laid flags meet the dimensions given.

Location of flags

Pedestrian-only areas:	flags of any size or thickness
Vehicular crossings – light traffic:	B: 63 mm, E: 70 mm, F: 65 mm, G: 60 mm
Occasional vehicular overrun:	E: 70 mm, F: 65 mm, G: 60 mm

Types E, F and G are slightly chamfered.

Tactile flags for blind guidance

Tactile flags: TA/E 450 mm × 450 mm, TA/F 400 mm × 400 mm, and TA/G 300 mm × 300 mm.

Rubber tiles for laying on existing paving: 400 mm × 400 mm in three types to Department of Transport standards:

- wide-spaced raised dots for pedestrian crossings to roads (red for controlled crossings, buff for uncontrolled crossings);
- close-spaced raised dots for platform edges (black);
- ribs for directional guidance through wide open pedestrian areas; the ribs follow the direction of travel.

Non-standard flags

These are available in a wider range of sizes and finishes than the BS flags.

Typical non-standard range: 400 mm × 400 mm, 450 mm × 450 mm, 500 mm × 500 mm, 600 mm × 400 mm, 600 mm × 600, thicknesses from 50 to 70 mm.

- Exposed aggregate made with crushed natural rock of 6 mm and 10 mm sizes.
- Riven to resemble split natural stone in various colours.
- Natural cobble surfaced.
- Tactile ribbed or dots.
- Reproduction setts, brick pavers and slate slabs.
- Square edged or chamfered edged.

Most proprietary paving systems include edgings, kerbs, edge restraints, tree grilles and channels to match or complement the paving pattern.

Skid resistance of paving

BS 8204 Part 2 : 1987 states that: 'Categories of slip resistance cannot be defined with any degree of accuracy'.

The figures in this table relate to values devised by the former Greater London Council Technical Department, and are still used by some firms:
Skid resistance:
75 or above excellent
40–74 satisfactory
20–39 marginal
0–19 dangerous

Slip resistance for York stone paving and plain concrete paving

Diamond sawn	dry 84–82
	wet 74–76
Shot sawn	dry 87
	wet 77–78

Shot sawn against the grain	dry 94–96	
	wet 81–82	
Rustic	dry 93–95	
	wet 81–82	
Concrete paving – plain	dry 63	
	wet 62	

The figures in this table relate to the values given in the GLC table above.

Natural stone paving

York stone: commonly available sizes:
    Lengths: random up to 1200 mm

Use	Thickness (mm)	Widths (mm)							
Pedestrian (light use)	40	450	500	550	600	610	650		
Pedestrian (heavy usage)	50	450	500	550	600	661	650	700	
Vehicles (light)	65	450	500	550	600	661	650	700	750
Vehicles (occasional heavy vehicles)	75	450	500	550	600	661	650	700	750

Note: Paving can be diamond sawn. This produces a fine finish for special contracts; slabs are available up to 3000 mm × 1750 mm.

As with all natural pavings the stone from different quarries varies enormously, and the landscape designer should check the slip resistance and compressive strength of actual samples. The difference between the wet and dry compressive strength can be about 60%.

Concrete mixes to BS 5328 Pt 1: & AMD

Application	Standard mix	Designated mix	Slump (mm)
Mass concrete fill and blinding	ST 2 (C10P)	GEN 1	75
Mass concrete foundations in non-aggressive soils	ST 2 (C10P)	GEN 1	75
Trench fill foundations in non-aggressive soils	ST 2 (C10P)	GEN 1	125
Foundations			
in Class 2 sulphate soils	—	FND 2	75
in Class 3 sulphate soils	—	FND 3	75
in Class 4A sulphate soils	—	FND 4A	75
in Class 4B sulphate soils		FND 4B	75
Kerb bedding and haunching	ST 1 (C7.5P)	GEN 0	10
In-situ concrete paving:			
pedestrian or light vehicle	—	PAV 1	75
heavy vehicle	—	PAV 2	50

GEN concretes with low cement content may not be suitable for good cast and direct finished surfaces.

Traditional concrete mixes

Application	Name of mix	Cement:aggregate:sand (parts by volume)
Setting posts, filling weak spots in ground, backfilling pipe trenches:	one, two, three	1:2:3
	or	or
	one to five all-in	1:5
Paving bases, bedding manholes, setting kerbs, edgings, precast channels:	one, two, four	1:2:4
	or	or
	one to six all in	1:6
In situ concrete paving:	one, one & half, two & half	1:1.5:2.5

'All-in' is mixed aggregate as dug from the pit.

Mortar mixes

Mortar group	Cement:lime:sand	Masonry cement:sand	Cement:sand with plasticizer
1	1:0–0.25:3		
2	1:0.5:4–4.5	1:2.5–3.5	1:3–4
3	1:1:5–6	1:4–5	1:5–6
4	1:2:8–9	1:5.5–6.5	1:7–8
5	1:3:10–12	1:6.5–7	1:8

Group 1: strong inflexible mortar
Group 5: weak but flexible

All mixes within a group are of approximately similar strength. Frost resistance increases with the use of plasticizers. Cement:lime:sand mixes give the strongest bond and greatest resistance to rain penetration.

Masonry cement to BS 5224: 1995 covers a wider range of cement types and defines three strength classes. Non-air-entrained cements are included.

Class	28-day strength N/mm^2	Air entrainment (X = non-air-entraining)
MC 5	5-15	yes
MC 12.5	12.5-32.5	yes
MC 12.5X	12.5-32.5	no
MC 22.5X	22.5-42.5	no

Masonry cement equals ordinary Portland cement plus a fine neutral mineral filler; it is easier to work and is more frost resistant than a cement:lime mixture. Masonry cement should never have lime added to it, and should not be used on walls subject to heavy loading.

OUTLINE SPECIFICATION

Paving flags to BS 7263

a. *Supply and lay precast concrete paving flags to BS 7263 Part 1, 600 mm × 600 mm × 50 mm and 450 mm × 450 mm × 50 mm as shown on drawing, on 150 mm DoT Type 1 compacted granular fill laid to falls shown, flags bedded on 25 mm compacted sand to BS 882 grade C or M, jointed in sharp sand to BS 882 Grade F.*

b. *Joints to be 4 mm wide and completely filled, and surplus sand cleaned off. No projections to exceed 4 mm in any direction. No material to be stored on finished paving, and no site traffic allowed thereon.*

for concrete mixes	*E10*
for kerbs and edgings	*Q10*
for sub-bases	*Q20*

Small unit flags

a. *Supply and lay precast concrete small unit paving flags to BS 7263 Part 1, Type E70 as shown on drawing, on 150 mm DoT Type 1 compacted granular fill laid to falls shown, units bedded on 30 mm compacted sand and vibrated.*

b. *Joints to be 4 mm wide filled with limestone grit, and the paving vibrated. No projections to exceed 4 mm in any direction. No material to be stored on finished paving, and no site traffic allowed thereon.*

for concrete mixes	*E10*
for kerbs and edgings	*Q10*
for sub-bases	*Q20*

Non-standard flag paving

a. *Supply and lay specified manufacturer's precast concrete paving flags 500 mm × 500 mm × 50 mm as shown on drawing, laid to falls shown and bedded in 25 mm 1:3 lime:sand mortar, on 100 mm PAV1 concrete.*

b. *Joints to be 5 mm wide filled with sharp sand to BS 882 Grade F. Joints to be completely filled and surplus sand cleaned off. No material to be stored on finished paving, and no site traffic allowed thereon.*

for concrete mixes	*E10*
for kerbs and edgings	*Q10*
for sub-bases	*Q20*

Natural stone slab paving

a. *Supply and lay Portland Whitbed flags in random sizes up to 800 mm × 800 mm × 75 mm thick, flags to be free of quarry sap and not sur-bedded, laid on 125 mm GEN4 concrete to falls shown on drawing, bedded and jointed in 25 mm 1:3 lime:sand mortar, sand to BS 882 Grade M.*

b. *Joints to be tooled as the work proceeds and left clean. No subsequent pointing permitted. Finished work to be protected by impervious sheeting for three days after completion. No material to be stored on finished paving, and no site traffic allowed thereon.*

for concrete mixes	*E10*
for kerbs and edgings	*Q10*
for sub-bases	*Q20*

Q25 Deterrent paving

GENERAL GUIDANCE

Deterrent paving can be used to prevent people from walking across it or vehicles parking on it. Whether mildly or severely deterrent, it can never be 100% successful against those ready to take up the challenge to beat the system. Although it may be tempting to design some particularly aggressive paving in difficult areas, the landscape designer must not allow the design to be potentially capable of inflicting injury to children or adults. None of these pavings is particularly easy to keep clean, as small pieces of rubbish, leaves and so forth lodge in the crevices. In large areas of paving where good maintenance is particularly important a mild form of deterrent paving should be used so that it can be cleaned regularly by machine.

Besides the precast concrete deterrent units (many of these fit into the manufacturers paving system) ranging from mild raised flat-topped studs to the more aggressive small projecting pyramids and triangles, bricks, setts, cobbles and pebbles set in concrete can all be used as deterrents, and though this is not an economical option, it gives the landscape designer a free hand in the choice of material to suit the surroundings as well as the degree of deterrence. All deterrent paving should be laid on a granular base.

Deterrent paving can be used in circles, squares or any other shape to protect trees, tubs, sculpture and other features, as well as in narrow (1 m) strips to keep people off grassed areas without the need for a visual barrier. It is particularly effective when laid to a slight slope, and this should go a long way to prevent illegal parking of vehicles or, for that matter, ice cream or fish and chip vans, whose customers will not be prepared to queue on an uncomfortable surface.

BRITISH STANDARDS

12 : 1996	Portland cement
435 : 1975	Dressed natural stone setts
890 : 1996	Building limes
1200 : 1976 & AMD	Building sands from natural sources. Sands for mortar for brickwork
3921 : 1985	Clay bricks and blocks
4027 : 1996	Sulphate-resisting Portland cement
4721 : 1981 (1986) & AMD	Ready mixed building mortars
4729 : 1990	Shapes and dimensions of special bricks
4887 Pt 1 : 1986	Mortar admixtures. Air-entraining (plasticizing) admixtures
4887 Pt 2 : 1987	Mortar admixtures. Set retarding admixtures
7583 : 1996	Portland limestone cement (A type of cement with up to 20% of specified limestone. Limestone cement can be used as ordinary Portland cement in designed mixes)

DATA

Sizes and shapes of standard bricks are given in the section on bricks.

Sizes and shapes of interlocking concrete blocks are given in the section on interlocking blocks.

Sizes of cobbles and setts are given in the section on cobbles and setts.
Proprietary deterrent paving:
precast concrete, either plain or aggregate faced, made in flag-sized units with projecting pyramids, waves, or other forms, which provide a very irregular surface.

Concrete mixes to BS 5328 Pt 1 & AMD

Application	Standard mix	Designated mix	Slump (mm)
Mass concrete fill and blinding	ST 2 (C10P)	GEN 1	75
Mass concrete foundations in non-aggressive soils	ST 2 (C10P)	GEN 1	75
Trench fill foundations in non-aggressive soils	ST 2 (C10P)	GEN 1	125
Foundations			
in Class 2 sulphate soils	–	FND 2	75
in Class 3 sulphate soils	–	FND 3	75
in Class 4A sulphate soils	–	FND 4A	75
in Class 4B sulphate soils	–	FND 4B	75
Kerb bedding and haunching	ST 1 (C7.5P)	GEN 0	10

Traditional concrete mixes

Application	Name of mix	Cement:aggregate:sand (parts by volume)
Setting posts, filling weak spots in ground, backfilling pipe trenches:	one, two, three	1:2:3
	or	or
	one to five all-in	1:5
Paving bases, bedding manholes, setting kerbs, edgings, precast channels:	one, two, four	1:2:4
	or	or
	one to six all in	1:6
In situ concrete paving:	one, one & half, two & half	1:1.5:2.5

'All-in' is mixed aggregate as dug from the pit.

OUTLINE SPECIFICATION

Cobbles

a. *Supply and lay South Coast flint origin cobbles, sandy buff colour, 100–75 mm sizes and 50–75 mm sizes laid on 100 mm FND2 concrete and bedded by hand in ST1 semi-dry concrete.*

b. *Paving to be laid with random spacing and sizes of cobbles to form an irregular surface with maximum variation in height, no space between cobbles to exceed 150 mm. No materials to be stored on finished surface.*

   for concrete   *E10*

Natural granite setts

a. *Supply and lay natural granite setts 100 mm × 100 mm × 250 mm and 100 mm × 100 mm × 150 mm, laid on 100 mm FND3 concrete on 150 mm well-rammed hardcore, set on end in random pattern to form an irregular surface with maximum variation in height, bedded in 25 mm 1:1/4:4 cement:lime:sand mortar, no space between cobbles to exceed 150 mm.*

b. *No materials to be stored on finished surface.*

   for concrete   *E10*

Proprietary paving

Supply and lay specified manufacturer's precast concrete pedestrian deterrent units 600 mm × 600 mm × 250 mm overall, wave pattern, laid to falls and pattern shown on drawing, on 100 mm ST4 concrete, on 150 mm well-rammed hardcore, bedded in 25 mm 1:4 lime:sand mortar and close jointed in 1:2:9 cement:lime:sand mortar as the work proceeds.

for concrete   *E10*

Q25 Cobble and sett paving

GENERAL GUIDANCE

This section deals with natural cobble paving, natural stone setts and precast concrete setts. The difference between concrete block paving and concrete sett paving is largely one of size and function: the setts are more expensive to lay because of the larger number of units per square metre, and they are usually manufactured to imitate natural stone setts with a rougher texture than the concrete blocks, which makes them less suitable for trafficked areas.

Modern concrete interlocking blocks can be laid by machine, which makes them even more economical to lay than setts.

Cobbles

The collection of cobbles – whether from coastal beaches or inland raised beds – is controlled, and many sites are legally protected geological features; it is an offence to use cobbles taken from these areas. The landscape designer would therefore be advised to specify cobbles from acceptable areas.

The charm of cobbles is their deep texture. Cobbles, particularly the small ones, are not the friend of high heels and walking aids; they do, however, act as an attractive means of discouraging people from walking on them, or if the band is wide enough from crossing over them. Unfortunately they do harbour small pieces of rubbish and natural detritus.

Cobbles make an attractive surface in small areas, where they can demarcate an area, provide formal patterns, accentuate the existence of a tree, planter or sculpture, or form the surround to gullies. A further use for the larger cobbles is the shallow dished areas associated with bubble fountains. Cobbles must be set firmly into concrete to prevent them from being used as ammunition for rioters and vandals.

The colour, shape and size of cobbles vary depending on the base rock from which they were originally formed. Colours vary from shades of grey through brown to buff and pink; some cobbles also have stripes or speckles. Commonly used sizes vary from 450 mm down to 20 mm.

As it is all too easy for an inexperienced pavior to convert a beautifully executed landscape drawing into a mass of cobbles smothered in cement, the landscape designer should therefore make sure that either an experienced layer is available, or ask for a sample panel to be laid before work commences.

When laying cobbles on a slope, the first five or six rows are laid first in stiff mortar, and where they abut soft landscape an edge restraint should be used.

An alternative fine-weather laying method for small areas that can be closely supervised involves packing the cobbles tightly together in dry concrete or mortar with their tops protruding. When the design is completed the whole area is gently watered through a fine rose to prevent the tops of the cobbles from being marked by the cement; the area is then left to set.

Setts

The traditional grey and occasionally pink granite and natural grey stone setts are made to British Standard sizes, which range from 100 mm × 100 mm × 100 mm to 100 mm × 150 mm × 150–250 mm. Reclaimed setts are often split to present a clean face; they are normally 100 mm × 100 mm and of varying thickness. Imported dark grey setts from Portugal are also available in sizes from 100 mm × 100 mm × 100 mm to 100 mm × 150 mm × 200 mm. Because natural granite and stone setts are expensive, their use for other than conservation work is mainly reserved for small areas where patterns or borders to other unit paving, or accents to statues and other features, show them to advantage.

The need for more economical setts has resulted in manufacturers' producing them in coloured precast concrete. Most are modular in order to form patterns and circles; examples of sizes for 70 mm thick setts are 120 mm × 120 mm, 120 mm × 180 mm, with radial setts 135–60 mm or 115–90 mm on opposite sides giving splays of 22.5° and 8° respectively. Because these are cheaper, they can be used in larger areas such as drives and forecourts.

All setts in heavily trafficked areas must be laid close together on a reinforced sub-base and bedded in 25 mm of cement and sand 1:3; all joints must be filled with fine sand. In pedestrian areas with the occasional light traffic, or in private drives, they can be set on a granular base with the setts bedded in 50 mm of loose sand and all joints filled with fine dry sand; the whole should then be compacted with a vibrating plate compactor. Where joints are filled with fine sand, the client should be advised to avoid the use of powerful suction-cleaning equipment.

BRITISH STANDARDS

435: 1975   Dressed natural stone setts
882: 1992   Aggregates from natural sources for concrete

There is no separate British Standard for precast concrete setts; they should be specified to comply with BS 6717 Part 1 : 1986 for quality and strength.

There is no specific British Standard for cobbles.

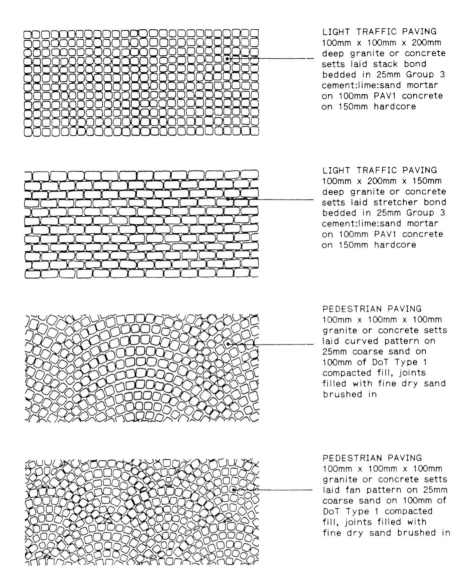

LIGHT TRAFFIC PAVING
100mm x 100mm x 200mm
deep granite or concrete
setts laid stack bond
bedded in 25mm Group 3
cement:lime:sand mortar
on 100mm PAV1 concrete
on 150mm hardcore

LIGHT TRAFFIC PAVING
100mm x 200mm x 150mm
deep granite or concrete
setts laid stretcher bond
bedded in 25mm Group 3
cement:lime:sand mortar
on 100mm PAV1 concrete
on 150mm hardcore

PEDESTRIAN PAVING
100mm x 100mm x 100mm
granite or concrete setts
laid curved pattern on
25mm coarse sand on
100mm of DoT Type 1
compacted fill, joints
filled with fine dry sand
brushed in

PEDESTRIAN PAVING
100mm x 100mm x 100mm
granite or concrete setts
laid fan pattern on 25mm
coarse sand on 100mm of
DoT Type 1 compacted
fill, joints filled with
fine dry sand brushed in

Fig. 4.34 Paving: granite setts.

DATA

Sizes of natural stone setts to BS 435

Width (mm)		Length (mm)		Depth (mm)
100	×	100	×	100
75	×	150–250	×	125
75	×	150–250	×	150
100	×	150–250	×	100
100	×	150–250	×	125
100	×	150–250	×	150

Typical sizes of precast concrete setts

Width (mm)		Length (mm)		Depth (mm)
140	×	140	×	80
210	×	140	×	80
70	×	70	×	70
172	×	115	×	60
115	×	115	×	60
112	×	115	×	60

Radial, circular segment, and round setts are available.

Engineering brick setts

100 mm × 100 mm × 100 mm square edged
100 mm × 100 mm × 100 mm double bull-nosed
100 mm × 100 mm × 100 mm battered edge

Typical sizes of cobbles

South Coast flint origin cobbles, grey or sandy, 100–75 mm, 75–50 mm, and 50–40 mm
Norfolk cobbles, dark grey, 100–75 mm
Cumbrian granite cobbles, dark grey, pink, and brown mixed, over 100 mm, 100–75 mm, 75–50 mm, 50–20 mm
Scottish granite, gneiss and quartz mixed, light pink-brown speckled and banded, over 100 mm, 100–75 mm, 75–50 mm, 50–30 mm
Scottish granite, light grey and light brown, 450–300 mm, 300–150 mm
Shropshire quartzite, red-brown, 150–100 mm, 125–50 mm, 75–50 mm

Concrete mixes to BS 5328 Pt 1 & AMD

Application	Standard mix	Designated mix	Slump (mm)
Mass concrete fill and blinding	ST 2 (C10P)	GEN 1	75
Mass concrete foundations in non-aggressive soils	ST 2 (C10P)	GEN 1	75
Trench fill foundations in non-aggressive soils	ST 2 (C10P)	GEN 1	125
Foundations			
in Class 2 sulphate soils	–	FND 2	75
in Class 3 sulphate soils	–	FND 3	75
in Class 4A sulphate soils	–	FND 4A	75
in Class 4B sulphate soils	–	FND 4B	75
Kerb bedding and haunching	ST 1 (C7.5P)	GEN 0	10
Drain pipe support in non-aggressive soils	ST 2 (C10P)	GEN 1	10
Manholes etc. in non-aggressive soils	ST 2 (C10P)	GEN 1	50
In-situ concrete paving:			
pedestrian or light vehicle	–	PAV 1	75
heavy vehicle	–	PAV 2	50

GEN concretes with low cement content may not be suitable for good cast and direct finished surfaces.

Traditional concrete mixes

Application	Name of mix	Cement:aggregate:sand (parts by volume)
Setting posts, filling weak spots in ground, backfilling pipe trenches:	one, two, three	1:2:3
	or	or
	one to five all-in	1:5
Paving bases, bedding manholes, setting kerbs, edgings, precast channels:	one, two, four	1:2:4
	or	or
	one to six all in	1:6
In situ concrete paving:	one, one & half, two & half	1:1.5:2.5

'All-in' is mixed aggregate as dug from the pit.

OUTLINE SPECIFICATION

Cobble paving

a. *Supply and lay South Coast flint origin cobbles, sandy buff colour, 100–75 mm sizes laid on 100 mm FND2 concrete and bedded by hand in ST1 semi-dry concrete.*

b. *Dry grout with rapid-hardening 1:3 cement:sand mortar lightly watered by hand with spray nozzle to finish 15 mm below surface of cobbles.*

c. *Paving to be laid to falls shown on drawing with flat face of cobbles in line of fall.*

d. *No materials to be stored on finished surface, and no site traffic allowed on paving after completion.*

for concrete mixes	*E10*
for paving bases	*Q20*

Sett paving

a. *Supply and lay proprietary precast concrete setts 70 mm × 70 mm in pattern shown on drawing, laid on 100 mm FND3 concrete with bedding layer of 25 mm compacted coarse sand and set in stabilized sand of 25 kg cement:100 litre coarse sand, joints fully filled with fine dry sand to BS 882 Grade F, and setts lightly tapped to consolidate joints.*

b. *Paving to be covered with impervious sheeting for three days after laying; no watering of joints permitted. No materials to be stored on finished surface, and no site traffic allowed on paving after completion.*

As above but with granite setts 100 mm × 100 mm.

for concrete mixes	*E10*
for paving bases	*Q20*

Q25 Tree grilles

GENERAL GUIDANCE

Tree grilles are laid in expanses of non-absorbent hard landscape to allow free movement of air and water to the tree pit. This does imply that all the run-off water must be reasonably clean and free of chemicals deleterious to plant life. In car parks, washing bays and so forth, all the surrounding surfaces must fall away from the trees.

All grilles in public areas, whether preformed or *in situ*, must be firmly anchored to prevent damage or unauthorized removal. The overall size of the grille must be sufficient for the needs of the tree when fully mature; it should allow room for staking and guards during the early stages of growth.

Preformed grilles may be of bronze, cast iron, hardwood or precast concrete. Precast concrete grilles may be part of a manufacturer's paving system. As grilles are very heavily constructed, most are made in units to be linked together on site to form the complete grille; these are kept level with the surrounding paving by means of paving support frames. Some units have inserts to allow greater protection to the bole of the young tree, but these should be used only where observant ground maintenance is assured, because failure to remove an insert once the bole has filled the space will cause irreparable damage to the tree. Some cast iron grilles also incorporate an irrigation inlet for controlled root ball watering. Preformed grilles are easily fractured, and must be protected from vehicles and heavy loads; their outer edge must not overlap the adjoining surface, because this would form an easily fractured bridge to the centre.

In situ grilles are made of small units either laid directly onto the ground, or inserted into paving support frames designed to allow the free passage of water. Suitable insert material includes brick, cobbles, setts and small paving units. Grass concrete blocks or clay drain pipes can be used.

BRITISH STANDARDS

There are no British Standards specifically for tree grilles.

EN 295-5 : 1994	Perforated vitrified clay pipes and fittings
3921 : 1985 & AMD	Clay bricks and blocks
5837 : 1991	Code of practice for trees in relation to construction
5911 Pt 100 : 1988 & AMD	Unreinforced and reinforced concrete pipes and fittings with flexible joints
6717 Pt 1 : 1993	Precast concrete paving blocks
6717 Pt 3 : 1989	Code of practice for laying precast concrete paving blocks

DATA

EN 295-5 perforated clay pipes (unglazed)

These pipes can be laid close butted vertically and filled with soil or gravel to form a tree grille.

Standard sizes are: 75, 100, 125, 150 and 200 mm bore.

Larger sizes are available.

Bricks

Perforated clay bricks are available in standard brick sizes. The perforations are between 20% and 50% of the surface.

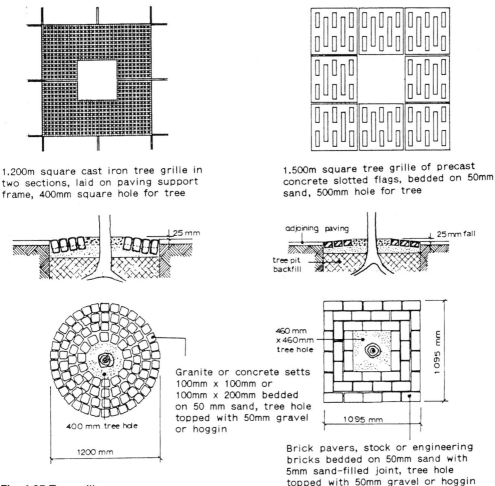

1.200m square cast iron tree grille in two sections, laid on paving support frame, 400mm square hole for tree

1.500m square tree grille of precast concrete slotted flags, bedded on 50mm sand, 500mm hole for tree

adjoining paving   25mm fall

tree pit backfill

400 mm tree hole

1200 mm

Granite or concrete setts 100mm x 100mm or 100mm x 200mm bedded on 50 mm sand, tree hole topped with 50mm gravel or hoggin

460 mm x 460mm tree hole

1095 mm

1095 mm

Brick pavers, stock or engineering bricks bedded on 50mm sand with 5mm sand-filled joint, tree hole topped with 50mm gravel or hoggin

Fig. 4.35 Tree grilles.

Typical grass concrete blocks

366 mm × 274 mm × 100 mm thick
600 mm × 400 mm × 125 mm thick with 56% grass area
600 mm × 400 mm × 125 mm thick with 75% grass area
360 mm × 415 mm × 100 mm thick hexagonal
292 mm × 338 mm × 100 mm thick hexagonal
225 mm × 112.5 mm × 60 mm thick interlocking
225 mm × 112.5 mm × 80 mm thick interlocking

Typical precast concrete tree grilles

600 mm × 400 mm × 55 mm with slot perforations to form square tree hole
600 mm × 600 mm × 50 mm thick with tree hole 460 mm diameter in 4 sections

750 mm × 750 mm × 50 mm thick with tree hole 600 mm diameter in 4 sections

900 mm × 900 mm × 50 mm thick with tree hole 800 mm diameter in 8 sections

960 mm diameter with tree hole 460 mm diameter in 4 sections

600 mm × 600 mm × 50 mm square perforated blocks laid as paving round tree; complements British Standard 600 module paving flags.

Traditional cast iron tree grilles

Ornamental pattern

800 mm × 800 mm with 300 mm tree hole in 1 section

900 mm × 900 mm octagonal with tree hole 300 mm octagonal

1200 mm × 1200 mm with 600 mm diameter tree hole in 2 sections

1200 mm × 1200 mm with 300 mm diameter tree hole in 4 sections

1217 mm diameter with 400 mm diameter tree hole in 3 sections

1700 mm × 1700 mm with tree hole 700 mm × 700 mm in 4 sections

2000 mm diameter with tree hole 500 mm diameter in 2 sections

Cast iron tree grilles are available as small units, which can be linked to form a large grille for a group of trees.

Removable two-section inner grilles are available for young trees. They must be removed when the bole grows.

Typical ductile iron tree grilles

1200 mm × 1200 mm with 460 mm diameter tree hole in 2 sections

1500 mm × 1500 mm with 600 mm diameter tree hole in 2 sections

1800 mm × 1800 mm with 800 mm diameter tree hole in 4 sections

1200 mm diameter with 460 mm diameter tree hole in 2 sections

Typical paving support frames

1217 mm diameter in 2 sections

1220 mm × 1220 mm in 2 sections

OUTLINE SPECIFICATION

Cast iron tree grille

a. *Supply and lay specified manufacturer's ductile iron tree grille 1200 mm × 1200 mm with 600 mm diameter tree hole Type XXX complete with irrigation inlet and locking points, on matching steel angle paving support frame with flanges laid on paving base.*

b. *Lay paving round tree grille, insert tree grille units and backfill to top of grille with topsoil to BS 3882 premium grade or approved equivalent.*

c. *Top up soil as necessary until completion of maintenance period. Provide protection to tree during laying of grille.*

for in situ *concrete paving*   *Q21*
for interlocking block paving   *Q24*
for flag, brick, sett or cobble paving   *Q25*
for tree planting   *Q31*

Precast concrete tree grille

a. *Supply and lay specified manufacturer's precast concrete tree grille 1500 mm diameter with 800 mm diameter tree hole Type XXX on matching steel angle paving support frame with flanges laid on paving base.*
b. *Lay paving round tree grille, insert tree grille units to finish level with adjoining paving, and backfill to top of grille with topsoil to BS 3882 premium grade or approved equivalent.*
c. *Top up soil as necessary until completion of maintenance period. Provide protection to tree during laying of grille.*

     for in situ *concrete paving*   *Q21*
     for interlocking block paving   *Q24*
     for flag, brick, sett or cobble paving   *Q25*
     for tree planting   *Q31*

Small unit tree grille

a. *Supply and lay on lightly compacted soil in tree pit precast grass-concrete blocks 600 mm × 400 mm × 125 mm thick with 75% grass area bedded in 50 mm sand to BS 882 Grade C and jointed in Grade F sand, blocks to finish level with adjoining paving.*
b. *Fill blocks with approved hoggin to top of units.*
c. *Top up fill as necessary until completion of maintenance period. Provide protection to tree during laying of grille.*

     for in situ *concrete paving*   *Q21*
     for interlocking block paving   *Q24*
     for flag, brick, sett or cobble paving   *Q25*
     for tree planting   *Q31*

Q25 Grass concrete

GENERAL GUIDANCE

This section deals with grassed concrete blocks, excluding the heavy-duty units used for revetments and embankments (see section D11). It includes grass concrete used for car parking, pedestrian areas and fire paths.

The term **grass concrete** covers precast units in either concrete or plastic, and *in situ* concrete systems using plastic formers; the latter are not recommended for use in built-up areas as they involve burning off the surplus plastic once the concrete has set, and this produces a smoke hazard.

In order to produce a successfully grassed area, all types of grass concrete require a free-draining substrate and a granular sub-base, and they all require good quality regular maintenance. Where grass is not a suitable infill material, well-rammed hoggin or bound gravel may be used. Suitable grass mixes for pedestrian and light traffic, and for fire paths and permanently fire paths are given below.

All the systems are cellular; the proportion of void to solid varies from 50% to 80%. The 50% void is suitable for permanently used car parks, service areas, fire paths, embankments and revetments subject to wave action, flooding, or other erosion; the 80% void can be used for pedestrian areas and the occasional overspill car parking of light vehicles and shallow undisturbed embankments. Most manufacturers provide guidance on grass seed mixtures and sowing rates. Sowing rates recommended for conventional seeding operations should be reduced in proportion to the void:solid ratio of the grass concrete.

Precast units

As there is no British Standard for either concrete or plastic precast units, the landscape designer has to rely on previous experience or on the manufacturer's technical information. The manufacturers all produce their own designs; some have special edging units, and some can be supplied with reinforcement to enable them to take heavier point loads. As the plastics units are very much lighter than the concrete ones, they are made in larger units with a higher percentage of void to solid area; this makes them suitable for pedestrian areas and private house drives. One of the chief snags of these units is that, unless they are well maintained, the grass portion shrinks, leaving the unit projecting above the general grassed surface waiting to catch the heels of shoes and so forth. Level surfaces can be laid with plain grass concrete units, but for slopes it is better to use interlocking blocks.

The long-term appearance of grass concrete surfaces can be very unpredictable. The surface is easily marred by over-use, uneven use, spillage and grime accumulation. Turf disorders cannot be rectified by normal cultivation measures. For these reasons, grass concrete should be restricted to situations where use and maintenance can be rigorously controlled.

BRITISH STANDARDS

There are no British Standards specifically for grass concrete.
4483: 1985   Steel fabric for the reinforcement of concrete

DATA

Typical precast grass concrete units

600 mm × 400 mm × 125 mm thick heavy-weight plain block
    56% grass area for fire paths and heavy-duty embankments

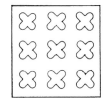

Interlocking blocks are
used on slopes
Plain edged blocks can be
used on level surfaces
Grass concrete blocks
bedded in sand on DoT
Type 1 fill
100mm minimum thickness
for light vehicular use
150mm minimum thickness
for fire paths

600 x 600 x 73 mm   406 x 406 x 103 mm

Fig. 4.36 Grass concrete.

600 mm × 400 mm × 110 mm thick medium-weight interlocking block
   40% grass area for water-washed embankments
500 mm × 300 mm × 100 mm thick light-weight plain block
   37.5% grass area for driveways and pedestrians
500 mm × 300 mm × 110 mm thick heavy-weight plain block
   75% grass area for fire paths and heavy-duty embankments
600 mm × 400 mm × 125 mm thick medium-weight plain block
   75% grass area for light car parking
500 mm × 300 mm × 95 mm thick medium-weight plain block
   75% grass area for car parking
292 mm × 292 mm × 100 mm thick light-weight hexagonal
   large grass area for light fire paths

Typical seeding mix for grass concrete

Car parks
40% Chewing's red fescue
10% Smooth-stalked meadow grass
50% Perennial ryegrass

Embankments
15% Ryegrass
25% Chewing's red fescue
35% Creeping red fescue
15% Timothy
10% Browntop

Fire paths
40% Chewing's red fescue
40% Creeping red fescue
10% Smooth-stalked meadow grass
10% Browntop

Pedestrian areas: amenity grass
70% Creeping red fescue
20% Timothy
10% Browntop bent

Residential driveways
80% Chewing's fescue
20% Browntop bent

Cartways
80% Perennial ryegrass
20% Creeping red fescue

Fire paths

Size of fire paths

1 or 2 storeys	3.8 m road, 3.8 m working point, 10 m inside turning radius
3 storeys	4.4 m road, 5.0 m working point, 13 m inside turning radius
Multi-storey	5.0 m road, 7.4 m working point, 13 m inside turning radius

---

OUTLINE SPECIFICATION

In situ grass concrete

a. *Supply and lay specified manufacturer's in situ formers with movement joints at 10 m centres on sub-base of natural subsoil or base of 150 mm DoT Type 1 granular material lightly compacted and blinded with 10 mm sand.*

b. *Lay steel mesh reinforcement to BS 4483 Type A193, pour GEN4 concrete 125 mm slump, 10 mm aggregate, and strike level.*

c. *Remove waste former material after 48 hours and fill with BS 3882 premium grade approved topsoil mixed with specified fertilizer at specified rate, and sow at recommended rate. Finished level of grass concrete to be 29 mm below adjoining grassed area.*

d. *Grass to be watered during dry weather until completion of maintenance period.*

for embankments	*D11*
for excavation and filling	*D20*
for seeding	*Q30*

Plastic interlocking grass blocks

a. *Lay base of 200 mm DoT Type 1 granular material lightly compacted, spread 50 mm 1:1 topsoil:sand mix on base.*

b. *Supply and lay specified manufacturer's interlocking grass blocks Type XXX in accordance with manufacturer's instructions, fill with BS premium grade approved topsoil mixed with specified fertilizer at specified rate.*

c. *Seed with grass mixture 15% ryegrass, 25% Chewing's red fescue, 35% creeping red fescue, 15% Timothy, 10% browntop at 50 g/m². Level of blocks to be 20 mm below adjoining grassed areas.*

d. *Grass to be watered during dry weather until completion of maintenance period.*

for embankments	*D11*
for excavation and filling	*D20*
for seeding	*Q30*

Precast grass concrete blocks for light car parking

a. *Supply and lay specified manufacturer's grass concrete blocks Type XXX on base of 150 mm DoT Type 1 granular material lightly compacted, bedded in 20 mm sand to BS 882 Grade M or C, tapped into place by hand, variation in level of adjoining blocks not to exceed 3 mm. Level of blocks to be 20 mm below adjoining grass areas.*

b. *Fill with BS 3882 premium grade approved topsoil mixed with NPK 6:9:6 fertilizer at 35 g/m².*

c. *Seed with 20% Chewing's fescue, 20% smooth-stalked meadow grass, 25% perennial ryegrass, 25% creeping red fescue, 10% browntop.*

d. *Grass to be watered during dry weather until completion of maintenance period.*

for embankments	*D11*
for excavation and filling	*D20*
for seeding	*Q30*

Q26 Special surfaces for sport

GENERAL GUIDANCE

This section describes the specialist surfaces required for club class and professional sports that must be played on surfaces with particular characteristics (non-slip or non-abrasive, for example). Most of these surfaces are laid complete with bases by specialist contractors, although in some cases the base may be laid by the general contractor to the specialist contractor's specification. Simpler surfaces for playgrounds are also included.

Special sports surfaces are designed to cope with many factors: the most important are the way the surface affects the behaviour of the ball, the risk of injury to the players from abrasion or bruising, and the grip of the players' boots on the ground. Professional players are trained to avoid accidents, but school players are more at risk through inexperience, and the sports surface must be selected with this in mind. The National Playing Fields Association publishes guidance on the best surfaces for the most common sports, and the individual sports associations can offer advice on their particular needs.

The specification for the specialist contractor should include:

- type of sports to be played;
- intensity of use and level of maintenance to be expected;
- standard of play (whether school, club or championship).

Playgrounds

Playground equipment and the surface on which it stands must comply with the relevant British Standards; not to do so would lay the owner, designer and contractor open to prosecution and damages for criminal negligence under the Health and Safety at Work etc. Act 1974. The requirements for playground surfaces are set out in BS 7188 : 1989 *Methods of test for impact absorbing playground surfaces*. BS 5696 Pt 3 : 1979 *Play equipment intended for permanent installation outdoors* requires that the impact-resisting surface must extend for a minimum of 1.75 m beyond the outside edge of static equipment, and beyond the limit of travel of moving equipment. In practice, the safest course is to allocate an area for the playground, and to specify the type of equipment to be provided, and then to obtain quotes for combined surface and equipment supply and installation. The safety surface should be designed to match the equipment selected. Both supply and installation must be certified by the contractor to comply with all safety regulations and standards.

Synthetic turf

Less expensive than the fully engineered sports surfaces, this is suitable for good-quality school or club play. It is usually composed of three layers: a solid level 'dynamic' base of compacted broken stone; a 'shockpad', which is made of various proprietary plastic/rubber compounds and bonded to the base; and a playing surface of artificial grass bonded to the shockpad. The artificial grass can be laid plain or filled with sand, which gives a better non-slip surface and is less likely to cause abrasion burns.

However, because its life is limited it may wear unevenly round goals, though real grass is even more susceptible to wear around goal-mouths. Patching artificial grass must be done very carefully to avoid snags, which will trip players.

Engineered sports surfaces

These are much more expensive than ordinary playing surfaces, and are normally used only for high-performance sports and athletics where cost is not critical. They are constructed as a complete unit by specialist contractors, though the base may sometimes be laid by the landscape contractor. There may be several base layers according to site conditions, with one or more special playing surface layers bonded to the base. The composition of the playing surface is designed for exact response to the playing conditions required for each sport, and the best authority is the governing body of the sport for which the surface is intended. Multi-sport surfaces made to a simpler and cheaper specification are also available for school sports areas.

Proprietary hard porous surfaces

These are made with a water-bound granular material carefully compacted; they are the cheapest type of surface, and they do not give such good performance as the engineered sports surfaces. They are suitable for casual kick-about areas only. They should not be specified for sports where the players are likely to be injured by contact with the ground, though sports such as tennis can be played comparatively safely. As they are water bound, they must be kept watered regularly with controlled amounts of water, and good drainage is essential to cope with heavy rainfall or over-watering. Proprietary tennis courts are made of very specialized hard porous material laid by a specialist contractor, and they can be obtained in various grades according to the standard of play required.

BRITISH STANDARDS

5696 Pt 3 : 1979 & AMD	Play equipment: Code of practice for installation and maintenance
7044 Pt 1 : 1990	Artificial sports surfaces: classification, general introduction
7044 Pt 2 : S.2.5: 1990	Artificial sports surfaces: miscellaneous
7044 Pt 4 : 1991	Surfaces for multi-sports use. Gives requirements for artificial sports surfaces
7188 : 1989 & AMD	Testing impact absorbing playground surfaces

DATA

Playground surfaces to BS 7188 : 1989

Wear index should be not less than 1.0.
Wear ratio should be not less than 1.0–3.0.
Slip resistance should be not less than 40 wet or dry.

Playground surfaces to BS 5696 Pt 3 : 1979

Impact resisting surface to be:

- minimum of 1.75 m beyond outside edge of static equipment;
- beyond the limit of travel of moving equipment.

Maximum potential free fall height (MPFFH): height of any access point or platform; heights obtained from playground equipment manufacturer.

Sizes of sports areas

Sizes given include clearances:

Association football, senior	114 m × 72 m
Association football, junior	108 m × 58 m
Rugby union pitch	156 m × 81 m
Rugby league pitch	134 m × 80 m

Hockey pitch   100.5 m × 61 m
Shinty pitch   186 m × 96 m
Men's lacrosse pitch   106 m × 61 m
Women's lacrosse pitch   110 m × 60 m
Target archery ground   150 m × 50 m
400 m running track 6 lanes 176 m × 106 m

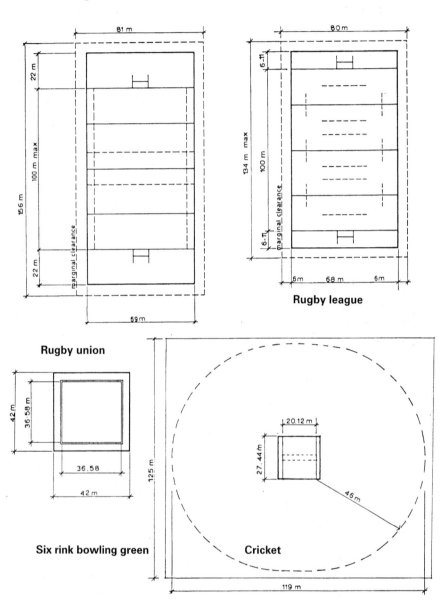

Fig. 4.37 Recreation and sports facilities.

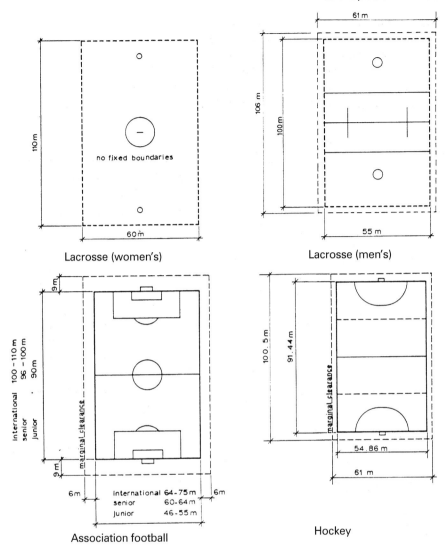

Lacrosse (women's)

Lacrosse (men's)

Association football

Hockey

Fig. 4.38 Recreation and sports facilities (continued).

OUTLINE SPECIFICATION

Playgrounds

a. *Excavate playground area to 300 mm depth and bring to grade, remove topsoil to dump on site, remove subsoil to dump on site, dig out weak areas by hand and backfill with approved excavated material.*

b. *Lay 150 mm compacted hardcore, lay filter membrane, supply and lay 10–50 mm conifer bark chips free from dust on surface 150 mm thick*

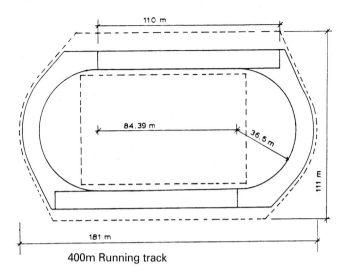

400m Running track

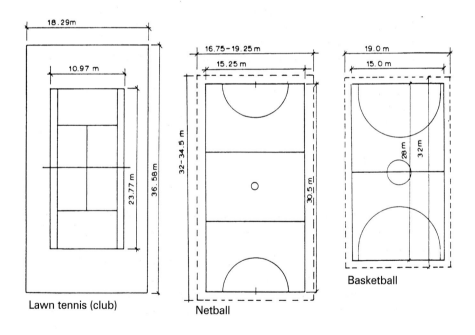

Lawn tennis (club)   Netball   Basketball

Fig. 4.39 Recreation and sports facilities (continued).

evenly spread. No chemical treatment of bark permitted. Bark must be guaranteed free from injurious chemicals.

for excavation	*D20*
for bark	*Q23*

Alternative specification

a. *Excavate playground area to 225 mm to given levels and falls, remove topsoil to dump on site, remove subsoil to dump on site, dig out weak areas by hand and backfill with approved excavated material.*

b. *Lay 150 mm DoT Type 1 granular fill compacted, lay two-coat work of bitumen macadam 75 mm thick, base course of 28 mm aggregate dense bitumen macadam 50 mm thick, wearing course of 10 mm aggregate dense bitumen macadam 25 mm thick*

for excavation	*D20*
for macadam	*Q22*

Hard porous surface

a. *Excavate sports area to 275 mm to given levels and falls, remove topsoil to dump on site, remove subsoil to dump on site, dig out weak areas by hand and backfill with approved excavated material, well compacted.*

b. *Lay 150 mm DoT Type 1 granular fill compacted and blinded with sand, supply and lay surface of well-graded crushed limestone with finished thickness 125 mm, maximum size of particles 5 mm.*

c. *Compact surface with 1.5 tonne roller, water thoroughly, roll again, repeat watering and rolling twice.*

d. *No material to be stored on the finished surface, and no site traffic permitted after completion.*

for excavation	*D20*

Artificial grass

a. *Excavate sports area to 200 mm to given levels and falls, dig out weak areas by hand and backfill with DoT Type 1 granular fill, well compacted.*

b. *Lay geotextile fabric, lay 150 mm thick base of clean broken stone and compact with 1.5 tonne roller, lay proprietary shock pad in situ 25 mm thick, lay second geotextile fabric layer, lay proprietary artificial turf surface 20 mm pile filled with approved quartz sand.*

for excavation	*D20*

Engineered sports surface

a. *Excavate sports area to 200 mm to given levels and falls, dig out weak areas by hand and backfill with DoT Type 1 granular fill, well compacted.*

b. Lay base of well-graded broken hard stone 200 mm thick to specialist contractor's requirements. Underlayers and playing surface to be laid by specialist contractor. The underlayers and playing surface to be guaranteed suitable for multi-sports use rate of play XXX games weekly.

for excavation   *D20*

Safety playground surface

a. Excavate playground area to 350 mm to given levels and falls, remove topsoil to dump on site, remove subsoil to dump on site, dig out weak areas by hand and backfill with compacted DoT Type 1 granular fill.

b. Lay 100 mm PAV1 concrete to fall 2% in all directions with raised edge to take proprietary safety tiles. Drainage holes 50 mm diameter to be formed in concrete at 1.0 m centres and filled with chippings.

c. Supply and lay on dry surface proprietary EPDM faced safety tiles 500 mm × 500 mm × 75 mm thick fixed to concrete base and adjoining tiles with manufacturer's adhesive and strictly in accordance with manufacturer's instructions. Cut tiles around positions for equipment fixings.

Q30/31 Planting generally

PESTICIDES AND THE LAW

This is of vital importance to the landscape designer, and the preliminaries to the Contract must include clauses on this subject.

Part III Pesticides etc. of the Food and Environmental Act 1985 has the power under s.16 (3) to prohibit imports, sale, offer or exposure for sale, supply or offer of supply, storage and use of pesticides.

S.16 (16) defines pesticides as: protecting plants and wood or other plant products from harmful organisms; regulating the growth of plants; giving protection against harmful creatures; rendering such creatures harmless, controlling organisms with harmful or unwanted effects on water systems, buildings and other structures, or on manufactured products; protecting animals against ectoparasites as if it was a pesticide.

This definition covers an enormous range of chemicals, including fungicides, herbicides, insecticides, rodenticides, soil sterilants and wood preservatives used in landscape work.

Schedule 2 of the Control of Pesticides Regulations 1986 (SI 1986/1510) sets out the basic conditions for the sale, supply and storage of pesticides, and Schedule 3 sets out the basic conditions for their use. This does not allow mixing two or more chemicals together unless this is permitted in the manufacturer's instructions; nor can adjuvants (substances other than water, without significant pesticidal properties, which can enhance the effectiveness of a pesticide) be added unless permitted in the manufacturer's instructions. Those who use them in the course of their business (e.g. dealing with a wasps' nest while hedge cutting) must have at least basic instruction and guidance on the safe, efficient and humane use of the pesticide used. Approved pesticides must not be used by those giving a commercial service, unless they hold a valid certificate of competence. They may be used by a person under the direct and personal supervision of a certificate holder. Those born after 31 December 1964 must hold a certificate of competence.

Specifying weedkillers and pesticides as well as their application should only be carried out by those holding a certificate of competence from the British Agrochemical Standards Inspection Scheme (BASIS).

The contractor must comply with current Control of Pesticide Regulations Code of Practice and recommendations by MAAF and HSE. Where the landscape designer leaves the choice of pesticides to the contractor, the contractor should provide a list of those intended to be used, the rate at which they will be applied, and the method of disposal, together with the manufacturer's current instructions, which should include a plant list of species that cannot tolerate or are susceptible to their use either throughout their life or when young or newly planted.

Some landscape designers may wish to construe these prohibitions as preventing them from specifying specific substances unless at least one member of staff has attended a training course and obtained a certificate of competence and by so doing complying with both MAFF aand HSE requirements. In any case the preliminaries to the contract should include a clause stating that only operators holding a current certificate of competence may handle, mix, apply and dispose of pesticides, and that these certificates will be checked prior to placing the contract.

The landscape designer must include a section in the preliminaries to the contract on the regulations and safety precautions associated with the pesticides, but for convenience they are included here.

When pesticides are applied to hard landscape areas, by law they must not be allowed to drain into gullies, channels or drains. In winter, if the application of pesticides to hard landscape areas (such as parking areas in country parks or business parks) is followed by snow and a rapid thaw, the run-off from residual chemicals can cause a great deal of environmental damage by directly or indirectly polluting watercourses.

SOIL

Whether the topsoil is *in situ* or imported, the landscape designer must make sure that the subsoil or formation material is not compacted by the contractor's plant and machinery, and that it has not been subjected to undesirable run-off from adjoining structures or contaminated water. Guidance on topsoil is given in BS 4428 : 1989 *Code of practice for general landscape operations*.

Compaction and poor drainage have been mentioned in many other sections of this book, but in respect of topsoils, the problem is exacerbated by poor handling of clayey and silty soils, both of which are liable to compaction. Clay soils are best left over winter for frost action to break them down before harrowing, but light sandy soils should be cultivated in the spring to avoid panning and leaching.

Both imported and site soil may contain undesirable weed seeds, which must be controlled. Cultivation is the preferred method of weed control, but occasionally herbicides may have to be used. The herbicide manufacturer should be fully informed of the nature of the soil and the proposed seeding or planting, and should be consulted about the most suitable chemicals for the purpose. All horticultural chemicals must be specified and applied by a certified operator, and the elapsed times between and after treatments must be strictly observed.

All topsoil should be spread in layers not exceeding 150 mm, and should be spread only when dry and in dry conditions.

EXISTING TOPSOIL

To be an efficient growing medium for vegetation, topsoil must have a good crumb structure, be free draining, and have an adequate capillary action for groundwater movement; if these properties do not exist they must be developed before planting takes place. It is essential not to consider topsoil in isolation but to take the nature of the subsoil or formation material into account. Samples of all types of soil on the site should be analysed at the beginning of the project, and the particle sizes ascertained, as these sizes determine the structure of the soil and the conditions in which it can be stripped and stockpiled. Stockpiled soil must be protected from contaminants and tracking by contractor's plant.

Topsoil that is to be stored on the site must be lifted and stacked carefully. Fine cohesive material such as clayey or silty soil (which is easily damaged by incorrect handling) should not be stacked very high, and it should not be moved while wet. Coarse material can be stacked somewhat higher, and it can be handled while wet. The size and height of stockpiles should be advised by the soil analyst.

IMPORTED TOPSOIL

BS 3882 : 1994 *Grades of topsoil* states that 'topsoil is the result of interaction between subsoil and the inherent nature of the parent material together with the prevailing environment'. It is important to remember this when importing topsoil, as it is no use expecting imported topsoil to retain its original characteristics once it has been stripped off, transported, and possibly stored, treated with weedkillers and moved again.

This new British Standard is the result of several factors. Some users wish to see greater use of waste organic materials rather than the removal of good topsoil from its original environment. The supplier has greater freedom in the use of materials, and this has resulted in the three grades of topsoil iden-tified below, together with clauses on sampling, analysis and handling. The important difference is the wider definition of topsoil, which can be either naturally occurring topsoil or a mineral-based material that will perform a similar function. The landscape designer must ensure that the correct grade for each use is incorporated in the specification, together with clauses to cover the careful treatment of imported soil and of the subsoil onto which it is to be spread. There will be times when the landscape designer has to accept imported topsoil that is not weed free, and it may be better to accept the soil as it stands and deal with emerging weeds by selective measures. Although the supplier will be expected to provide documents giving the grade and con-tent of the delivered soil, the landscape designer should be able to check that the correct grade has been delivered by following the instructions in the flow chart from the British Standard. Serious problems may arise when, for example, waste soils containing contaminants have been used to make up a soil in one of the lower grades. Such soils will need to be subjected to a full laboratory analysis.

BRITISH STANDARD 3882 : 1994 GRADES OF TOPSOIL

This Standard gives three grades of material together with recommendations for the use and handling of topsoil. It is not intended (or appropriate) for the grading, classification or standardization of *in situ* topsoil or subsoil.

- **Premium grade**: natural topsoil that has been correctly handled and spread on a compatible site. The amount available is limited. Use for plant nurseries, landscape sites and amenity areas where the topsoil is likely to be intensively (annuals) or frequently cultivated.
- **General Purpose grade**: natural topsoil or manufactured soil with suitable properties. This grades includes badly treated or stored premium-grade soil. It may require improvement by the use of lime and fertilizers. Use for silviculture, amenity and landscape areas for grass, trees, shrubs, herbaceous and other planting.
- **Economy grade**: lower quality than the General Purpose grade, or selected subsoil; also friable mineral material suitable for plant growth such as greensand, river silt or glacial moraine. Grades divided into low and high clay subgrades; the latter is less tolerant to compaction, and needs careful handling. It will probably require additional fertilizers, manure and careful cultivation for several years before it is at least as good as the General Purpose grade. Use for amenity woodland, wildlife conser-vation areas and amenity grassland not subject to intensive use.

SOIL AMELIORANTS

These are additives, not chemical fertilizers, which have the effect of improv-ing the soil's texture or nutrients to some degree. It is important that the additive is compatible with a particular soil. Additives are used to improve the following properties in soil, and possible ameliorants are listed below:

Crumb structure

By binding fine particles to form larger particles, which allow increased aeration and the introduction of oxygen, by:

- the provision of acceptable food for micro-organisms: use mature stable manure, leaf-mould, spent mushroom compost, local-authority-processed domestic waste, sewage sludge, peat;
- chemical means that cause flocculence: use calcium.

Drainage and water retention

- Balancing the percentage of sand, silt and clay in the existing soil: use mature stable manure, bark, local-authority-processed domestic waste, expanded minerals, sand and grit or clay depending on the present soil condition, and topsoil.
- Using buffers to protect the soil from an 'overdose' of chemicals, or lack of them after heavy rain.
- Increasing humus: use mature stable manure, bark, leaf-mould, local-authority-processed domestic waste, spent mushroom compost, sewage sludge, straw.
- Increasing nutrients: use mature stable manure, local-authority-processed domestic waste, spent mushroom compost, topsoil.
- Nutrient content: use composts, mature stable manure or good topsoil.

PEAT AND PEAT SUBSTITUTES

The characteristics that make peat a useful growing medium are as follows:

- It can retain and release water.
- It contains air spaces within its composition.
- It has a neutral or slightly acidic pH.
- It can accept, retain and release nutrients.
- It is dark in colour, and therefore a good heat retainer.
- It is reasonably sterile, and free of toxins and imported particles.

However, the use of peat is not considered to be environmentally acceptable by many landscape designers, and they are looking for alternatives that will meet the qualities listed above. To date, few suitable alternatives are as successful, though research is being carried out on 'composted' bark, sewage sludge, municipal compost, bracken, leaf-mould, straw bales and so forth. If these alternatives are to be used, the landscape designer should seek the advice of suppliers as to the most suitable material for a specific job.

Unless the existing soil is in good condition it may be necessary to add soil conditioners such as peat or peat substitutes. Notes on these substitutes are given here:

- Processed domestic or garden waste sold by local authorities is an excellent soil improver, though low in nitrogen.
- Spent mushroom compost can be used, though this may contain insecticides and molluscicides.
- Seaweed makes an excellent, though expensive, conditioner.
- Farmyard manure (FYM) is still specified, but unless it comes from an organic source it is no longer reliable, because it may contain weedkillers, hormones, pesticides, antibiotics and agricultural chemicals used on straw crops and stock, whose long-term effect on the grassed areas and the environment is unknown. If it can be obtained from an organic source it is excellent, but it must be fully matured before use.
- Stable manure is slightly less likely to contain a large number of chemicals, but must be mature before being applied to the soil.
- High-nitrogen soil conditioners such as poultry manure are useful for grassed areas on low-grade soils, but they may contain chemicals used to treat the poultry and their housing.

IRRIGATION AND DRAINAGE

On certain soils and for special sports areas it may be necessary to install an irrigation system to maintain the grass in good condition. Standards for the design and installation of an irrigation system are laid down by the British Turf and Landscape Irrigation Association (BTLIA). Such systems are usually the province of specialist sports turf and irrigation contractors; the landscape designer should consult them at an early stage in the design, bearing in mind that both a reliable water supply and an adequate drainage system must be included in the contract. Heavily used professional standard playing surfaces are often artificial fibre mats, but where grass is required it may be necessary to provide complete permanent underground systems of irrigation, drainage, heating, and distribution pipelines for nutrients, weedkillers and pesticides. These are outside the normal landscape contract.

WEEDS

Herbicides kill all plants; weedkillers are selective, and are designed to kill only certain types of plant.

Pesticides are controlled by legislation, and many of the chemicals likely to be used can be applied only by a certified operator, who must comply with the Control of Pesticides Regulations and relevant MAFF codes. Herbicides must also be used with great care to prevent damage to surrounding vegetation or neighbouring land by blown particles or spray. All should be specified as leniently as possible so as to reduce the risk of contaminating adjoining land. Where they are used, all empty containers and unused chemicals, including dilutions, must be safely disposed of in accordance with the manufacturer's instructions; none must be allowed to spill on to the surrounding area or watercourses.

Newly planted trees and shrubs benefit from weed-free soil, particularly during their first year; though these benefits are still noticeable in the second and subsequent years, they do decrease as the plant matures. The landscape designer should ensure that the contract maintenance includes weed control for 12 months after all planting has been completed, and that thereafter the client should be encouraged to continue this regime for at least one further year. The competition between newly planted trees and shrubs and weeds is at its greatest between April and June, when both are competing for the available water, nutrients and light required for maximum root growth.

If they are acceptable, herbicides will be required in the short term where the soil contains perennial weeds or annual seeds. For the best results, weeds should be tackled before they become a serious problem, and if the site is available during the season prior to planting, glyphosate can be applied as the weeds emerge.

The careful selection of plant material on well-cultivated soil in a compatible environment will, once the plants are established, do much to form a dense canopy that will prevent the need for continuous weed control.

Basic types of herbicide

Do not apply any of these if the client intends to run a totally organic enterprise or is hoping for a clean bill of health from the Soil Association. Before applying them on nature conservation areas discuss the chemical to be used with the client; many chemicals are inimical to the welfare of insects, fish and so forth.

The basic types of herbicides are set out below:

- **Contact**: a non-selective, non-permanent herbicide that kills off only the green growth above ground. Some, such as paraquat, come under the Poisons Act 1972 and the subsequent SI Poisons Regulations 1982, and many others should not be used, because they do not have MAFF approval.
- **Translocated**: both selective and non-selective herbicides such as glyphosate, which depend on active growth by moving into the root system. They are less effective in cold weather, but are suitable for applying to weeds prior to preparing the ground. For them to be effective, three weeks should be left between these operations.
- **Residual**: these herbicides include dichlobenil (Casaron G), and can be taken up by leaves and roots or roots alone. Their use is restricted by UK and EU regulations, because the residue can be taken into the groundwater.

Cost of weed control

The cost of weed control for soft landscaped areas varies considerably. For small areas, careful hand weeding and organic mulches are environmentally acceptable and the most expensive type of control. Hand weeding is a continual labour cost during the growing season, and organic mulches should be laid over clean ground, and require an initial depth of at least 75 mm to be effective, followed by annual topping up. The cheapest method, but arguably the least environmentally acceptable, is the use of herbicides, which clean the ground for planting and require further applications to control introduced weeds and seeds. One method that comes between these two extremes is inorganic sheet mulches, which may be too unsightly to be acceptable, and can easily be churned up by foxes, squirrels and so forth. Mechanical cutting by mower or strimmer also comes into the moderately priced bracket, but is suitable only for rough grassed areas and embankments: here care must be taken not to damage the bark of any trees and shrubs in the area. Mown or strimmed grass competes strongly for soil water, and is a most unsuitable surround for establishing plants.

CONTROL OF PESTS

These also come under the Control of Pesticides Regulations (which include herbicides) and relevant MAFF codes. Pesticides are not often specified by the landscape designer or used by the contractor unless soil pests are a problem. Whenever they are used it must be with great care, and all should be specified as leniently as possible so as to reduce the risk of contaminating adjoining land. Empty containers and unused pesticides, including dilutions, must be safely disposed of in accordance with the manufacturer's instructions; none must be allowed to spill onto the surrounding area or watercourses.

Pesticides

Do not apply any of these if the client intends to run a totally organic enterprise or is hoping for a clean bill of health from the Soil Association. Before applying them on nature conservation areas discuss the chemical to be used with the client. Most pesticides are inimical to the welfare of bees, hover-flies, ladybirds, beetles, worms and other beneficial insects, as well as fish and domestic pets. Biological controls by predators and parasites are used mainly in glasshouses.

Pesticides in this context cover fungicides and mollusicides as well as insecticides. The basic types are set out below:

- **Systemic insecticides and fungicides**: these are absorbed into the sap of plants, and kill insects that feed on the sap and fungi within the plant's structure.
- **Contact insecticides**: kill those insects that are in contact with the spray or powder, or eat the adjacent plant material that has been covered.

- **Preventive fungicides**: coat plant material to prevent spores from developing.
- **Organic pesticides**: these are derived from plant or natural chemicals.

Q30 Seeding grassed areas

GENERAL GUIDANCE

This section covers the final ground preparation, and the seeding and contract maintenance period for small and large grassed areas. Topsoil, soil amelioration, fertilizers, herbicides and pesticides are covered in *Planting generally*. Grass surfaces for serious competitive sports are usually turfed, and are laid and maintained by specialist contractors.

Turf and grass seed

Where the grassed areas are required to be in commission very quickly, it is preferable to specify turf; it is usable 6–12 months earlier than seeded areas, because the growing time has been spent in the grower's production fields. Although turf is much more expensive in initial outlay than seed, the cost of establishing the sward should be taken into consideration. The Sports Turf Research Institute produces an annual publication, *Turfgrass Seed*, which lists commercially available cultivars, to help specifiers select the correct grass cultivars for the intended use of the area.

Seeding can be done only at suitable times of the year, so the contract programme must take this factor into account. The optimum sowing time is May, though the actual periods vary with the weather from year to year. The current weather conditions should be taken into account.

Preparation of seed-bed

After previous cultivation and bringing to levels (see D20), the initially cultivated ground should preferably be left fallow and treated to prevent weed establishment, but the final seed-bed should be prepared just before seeding. The final levels can be adjusted with a blade grader. The seed-bed must be brought to a fine tilth, and raked or harrowed to the finished levels. All stones and debris over 50 mm for amenity grass areas, and 25 mm for lawns and fine playing surfaces, must be removed, the area lightly firmed, and the top 25 mm loosened.

Turf edging to grassed areas may be required where the grass abuts onto paths or planted areas. A strip 300 mm wide should be laid before seeding and the levels adjusted to bring seeded area and turf to the same level. Grass areas are usually finished about 50 mm above the adjoining paving in order to allow for settlement.

Grass seeding

Sowing rates vary from 6 g/m² to 70 g/m² according to the type of sward required and the cost of seeding. Each seed house has its own mixtures for various purposes, ranging from grassing colliery spoil heaps to fine bowling green swards, and the supplier's recommendations should be followed, though as new cultivars are developed, the mixtures and sowing rates will vary from year to year. There are now a large number of sources available for grass seed, and tailor-made mixtures can be prepared very quickly. Two important factors in seed specification are pure seed content and germination standard. Standards for these are laid down by MAFF, and reliable seed houses will provide information on their products. The landscape designer should check that the minimum MAFF standards are adequate for the type of grass sward required.

Low-maintenance areas where rapid establishment of sward is not important can be seeded at 6–15 g/m², but tennis courts and fine lawns, which need a dense sward, should be seeded at 50 g/m².

Seeding must be carried out in two passes at right angles to each other, with the rate of seeding carefully controlled. The seed should be lightly raked in by hand on small areas and harrowed in on large areas, and on light soils a gentle rolling is necessary.

Wild flora seeding

The decision to sow wild flora requires careful consideration; the main points are site characteristics, selection of appropriate mixtures, method of establishment, ground preparation, and long-term management. Wild flora may take 2–3 years to establish, and the mowing regime must be carefully controlled if seed production is to be successful. Low-fertility soils are most suitable. Wild flora seeding is carried out in much the same way as grass seeding, and at the same seasons, but with lower rates of sowing, as the flora seed is usually sown mixed with a compatible grass seed mixture, which should be short growing and non-aggressive, leaving open areas for wild flora colonization. Traditional wild flora mixtures are available for most types of soil, including chalkland, woodland, wetland, and clay soils, though subsoil is preferable to fertile topsoil as there is less competition from weeds. Wild flora seed can be supplied ready mixed with grass seed in a proportion of 80% grass to 20% flora, or pure for sowing directly, when a carrier of sand or sawdust is added to give better distribution. Wild flora seed is not subject to germination or purity standards, though a germination rate of 70% can be expected from most species. The *Department of Transport Wildflower Handbook* 1993 gives useful data and advice.

Results are relatively unpredictable, and the criteria for success can doubtfully be applied to a contractual situation in the way that homogeneous grass areas or individual trees and shrubs can be seen to be establishing well or not. Wild flora seeding can be a waste of time and money unless there is a long-term commitment to specialist monitoring and landscape management.

Hydroseeding

Hydroseeding is a method of seeding areas where machine and sometimes even workforce access is difficult or dangerous, or where normal machine work would be uneconomical. It is a useful technique for establishing vegetation on weathered rock, mine dumps, gabion retaining structures, waste disposal sites, and similar areas at a low initial cost, but the cost of after-care and possible overseeding should be borne in mind. The technique consists of spraying a mixture of grass seed, nutrients, growing medium and any necessary additives through high-pressure hoses onto the surface. Hydroseeding can be carried out to a reach of about 60 m and a height of about 25 m, but additional hoses can be used to extend the sprayed area. The spray mix will vary according to the existing surface, climate, and the type of vegetation desired, but it is only likely to succeed on wet soils if the work is carried out in the growing season. Mixes vary considerably, but in principle they contain water, a fibrous slurry, a seed mix, soil conditioner, fertilizers, selective weedkillers, pesticides, trace elements, a biodegradable adhesive (for steep or hard rock) and a soil stabilizer consisting of fibres and a bitumen emulsion. The mixes range from plain seed and fertilizer where there is adequate topsoil, to a compound designed to establish viable vegetation on bare rock, but in all cases a local supply of water is essential to keep costs down. Hydroseeding can be used for establishing ground cover, shrubs, and some species of pioneer trees as well as for grass and wild flora.

Temporary protection

The newly sown grass should be protected by chestnut paling, stock fencing, or light wire netting according to the type of protection needed (see Q40 *Fencing*). Proper access for watering and overseeding if necessary must be provided; if the contractor merely pulls down and puts back the paling the protection will soon cease to be effective.

After-care during maintenance period

Newly sown grass should be given light regular watering unless adequate regular rainfall occurs during the establishment period. The first cut for fine grass should be taken when the grass is about 50 mm high; before cutting, all stones and debris over 25 mm that have accumulated should be removed and the grass lightly rolled. A five- or seven-bladed cylinder mower should be used, and the cuttings boxed off. Amenity grass can be cut later, and only 40 mm debris need be removed; this type of grass is usually cut with a rotary mower, and the cuttings spread. During the contract maintenance period the grass should be kept cut to the specified height, and watered if necessary. In the first year, wild flora areas sown in autumn do not need cutting until spring, when vegetation over 100 mm should be cut back to 75 mm and cuttings removed; spring-sown flora

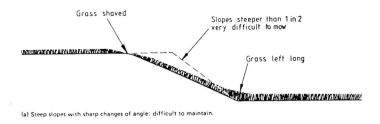

(a) Steep slopes with sharp changes of angle: difficult to maintain.

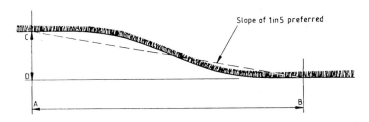

Fig. 4.40 Maintenance of grass slopes: (a) steep slopes with sharp changes of angle – difficult to maintain; (b) slack double curve – easily mown at all points. *Note*: For grades shallower than 1 in 2, AB should be less than or equal to 3.5 × CD. (BS 4428 : 1989)

should be cut after six weeks, and at intervals when growth exceeds 200 mm. If colourful annuals are included, one cut in early April and one in September or October will usually be sufficient.

Seeding for reinforced grass

Some firms manufacture a light coarse plastic mesh, which has the effect of reinforcing the grass structure, providing good soil drainage, and reducing compaction and wear. This is laid on the prepared subsoil or existing topsoil, filled and covered with fine topsoil and seeded in the usual way. Wire mesh is sometimes specified for the same purpose, but unless the surface is regularly maintained, heavy foot traffic may wear through the wire, producing dangerous jagged ends. The use of wire mesh should therefore be carefully appraised before it is selected. The technique is similar to filling erosion control mats, but the mesh has no structural strength. Reinforced turf (grass grown on reinforcement) is covered in the section on Turf and Turfing.

BRITISH STANDARDS

1831 : 1969 & AMD Recommended common names for pesticides. This
standard is continuously updated
3882 : 1994 Specification for topsoil

4156 : 1990	Peat for horticultural and landscape work
4428 : 1989 & AMD	Code of practice for general landscape operations
5551 s1.1 : 1984	Classification scheme for fertilizers and soil conditioners
7370 Pt 1 : 1991	Grounds maintenance. General recommendations
7370 Pt 3 : 1991	Grounds maintenance. Maintenance of amenity turf

DATA

Topsoil quality to BS 3882: 1994

Topsoil grade	Properties
Premium	Natural topsoil, high fertility, loamy texture, good soil structure, suitable for intensive cultivation
General Purpose	Natural or manufactured topsoil of lesser quality than premium, suitable for agriculture or amenity landscaping; may need fertilizer or soil structure improvement
Economy	Selected subsoil, natural mineral deposit such as river silt or greensand. The grade comprises two subgrades: low clay and high clay, which is more liable to compaction in handling. This grade is suitable for low-production rural land and amenity woodland or conservation planting areas.

Permissible deviations from given levels

Bowling greens	±18 mm from given levels
	±6 mm from finished levels
Sports fields	±75 mm from given levels
	±25 mm from finished levels
Cricket squares	±6 mm from finished levels
Amenity grass	±100 mm from given levels

Typical proprietary seed mixtures

Low maintenance; conservation
13 mm cut; 170–350 kg/ha
40% Creeping red fescue
30% Chewing's fescue
25% Hard fescue
5% Browntop bent

Reclamation; suitable for grazing
No cut; 150–250 kg/ha
43% Perennial ryegrass
20% Flattened meadow grass
15% Creeping red fescue
15% Sheep's fescue
5% Browntop bent
2% White clover

Typical proprietary seed mixtures *continued*

Shaded areas
15 mm cut; 50 g/m²
40% Strong creeping red fescue
20% Smooth-stalked meadow grass
30% Chewing's fescue
10% Browntop bent

Outfield
25 mm cut; 170–350 kg/ha
60% Perennial ryegrass
25% Creeping red fescue
10% Smooth stalked meadow grass
5% Browntop bent

Saline areas
No cut; 150–200 kg/ha
35% Strong creeping red fescue
25% Hard fescue
20% Perennial ryegrass
10% Flattened meadow grass
7% Salt-marsh grass
3% Creeping bent

Low-maintenance amenity
25–50 mm cut; 250–350 kg/ha
60% Perennial ryegrass
35% Slender creeping red fescue
5% Browntop bent

Playing fields
25 mm cut; 180–300 kg/ha
50% Perennial ryegrass
25% Chewing's fescue
15% Smooth-stalked meadow grass
10% Browntop bent

Clover
No cut; 20 kg/ha
17.5% Alsike clover
17.5% Lucerne
17.5% Bird's-foot trefoil
17.5% Sainfoin
15% White clover
15% Red clover

Slopes and banks
13 mm cut; 250–350 kg/ha
65% Strong creeping red fescue
20% Chewing's fescue
10% Smooth-stalked meadow grass
5% Browntop bent

Bowling green
5 mm cut; 35 g/m²
80% Chewing's fescue
20% Browntop bent

Fine lawn
7 mm cut; 35 g/m²
70% Chewing's fescue
20% Creeping red fescue
10% Browntop bent

Grass concrete
30% Chewing's fescue
30% Creeping red fescue
30% Smooth-stalked meadow grass
10% Browntop bent

Typical proprietary pre-seeding NPK fertilizers

Moderately fertile soil   6:9:6 at 7 g/m²
Impoverished soil   10:15:10 at 7 g/m²
Reclamation sites   16:16:16 at 500 kg/ha

OUTLINE SPECIFICATION

Recreation area: existing topsoil

a. *On ground previously cultivated, bring area to given levels, treat with agreed herbicide applied by certified operator at manufacturer's recommended rates, all in accordance with manufacturer's instructions.*

b. *Incorporate spent mushroom compost at 1 m³ to 5 m³ of topsoil, bring seed-bed to a fine tilth, and rake or harrow to final levels; all stones and debris over 50 mm to be removed, the area lightly firmed and the top 25 mm loosened.*

c. *Carry out seeding in two passes at right angles with seed mixture 50% perennial ryegrass, 25% Chewing's fescue, 15% smooth-stalked meadow grass, 10% browntop bent at 200 kg/ha. Harrow lightly and roll lightly.*

d. *Permitted deviation from given levels ±75 mm over 30 m.*

As above, but spread with imported topsoil to BS 3882 : 1994 General Purpose grade 75 mm thick.

for cultivation	*D20*
for topsoil, fertilizers, and weedkillers	*Planting generally*
for turf: Turf and turfing in	*Q30*
for land drainage	*R13*
for chestnut paling: Protective fencing in	*Q40*

Erosion control mats

a. *Lay 20 mm thick open-texture erosion control mat with 100 mm laps, fixed with 8 × 400 mm steel pegs at 1.0 m centres.*

b. *Sow with low-maintenance grass 30% Chewing's fescue, 30% creeping red fescue, 30% smooth-stalked meadow grass, 10% browntop bent at 35 g/m².*

c. *Spread imported topsoil to BS 3882 General Purpose grade 25 mm thick incorporating medium grade peat substitute at 3 kg/m² and Growmore fertilizer at 70 g/m²; water lightly.*

d. *Slope to be watered regularly in dry weather until vegetation is established.*

for cultivation	*D20*
for topsoil, fertilizers, and weedkillers	*Planting generally*
for land drainage	*R13*
for chestnut paling: Protective fencing in	*Q40*

Ornamental lawn

a. *On ground previously cultivated, bring area to given levels, treat with agreed herbicide applied by certified operator at manufacturer's recommended rates, all in accordance with manufacturer's instructions.*

b. *Incorporate to 150 mm depth: washed sand to BS 882 grade F at 50 kg/m²; peat substitute or loose sphagnum peat to BS 4156 at 0.3 m³/m²; NPK 6:9:6 fertilizer at 70 g/m². Bring seed-bed to a fine tilth, and rake to final levels; all stones and debris over 25 mm to be removed, the area lightly firmed and the top 25 mm loosened.*

c. *Lay turf edging 300 mm wide along all edges; turf grass species to be similar to grass seed mixture.*

d. *Carry out seeding in two passes at right angles with seed mixture 70% Chewing's fescue, 20% creeping red fescue, 10% browntop bent at 35 g/m². Rake lightly and roll lightly. Permitted deviation from given levels ±25 mm over 30 m.*

for erosion control	*D11*
for cultivation	*D20*
for topsoil, fertilizers, and weedkillers	*Planting generally*
for turf: Turf and turfing *in*	*Q30*
for land drainage	*R13*
for chestnut paling: Protective Fencing in	*Q40*

Hydroseeding

Typical hydroseeding mix for 12% grassed slope:

Slurry	*5000 kg/ha*
NPK 10:15:10	*250 kg/ha*
Seaweed extract	*12 l/ha*
Biodegradable adhesive	*60 kg/ha*
Soil conditioner	*75 kg/ha*
Seed mixture	*60 kg/ha*

   45% perennial ryegrass
   20% Chewing's fescue
   25% slender creeping red fescue
   5% browntop bent
   5% white clover

Q30 Turf and turfing

GENERAL GUIDANCE

This section covers the treatment of existing turf, the lifting and care of turf, and the final ground preparation, pre-turfing treatment, laying turf and contract period maintenance. All grassed areas must be adequately drained; land drainage is covered in section R13. Topsoil, soil amelioration, fertilizers, herbicides and pesticides are covered in *Planting generally*.

Turf may be natural meadow turf, which may contain weeds and undesirable soil organisms, or cultivated turf, which is grown from selected species and cultivars to give a particular performance. Traditionally, salt-marsh turf produced very fine-quality turves, but such areas are now protected, and the turf is not readily available. Commercial turf supplied to BS 3969: 1990 *Recommendation for turf for general landscape purposes* is classed as:

- fine sports or ornamental turf without ryegrass;
- general-purpose utility turf without perennial ryegrass;
- general-purpose utility turf with perennial ryegrass.

Commercial turf is usually supplied in 1 yd² rolls from 18 mm to 30 mm thick; thin turf becomes established quicker, but very thin turves will dry out and break up in handling, while very thick turves are slow to establish and heavy to handle.

Industrial turf is grown on a mesh; this enables it to be lifted in large rolls, which can be laid by machine. It is thinner than natural turf, and may need more after-care to get it properly established. It should be laid within 24 hours of lifting, with a maximum delay of 48 hours in cool weather.

Reinforced turf is formed by spreading a layer of plastic mesh clippings mixed with a growing medium and then seeding the area in the normal way so that the roots are interlocked with the mesh. This is designed to be more resilient and harder wearing than normal turf.

Existing turf

Existing turf may be treated to bring it into better condition, either by means of chemical treatment and fertilizing, or by physical amelioration, or both. Chemical treatment may include locally or generally applied selective weed-killers, fungicides, mosskillers and insecticides, while top-dressing with lime, sand, fertilizers and trace elements can replace soil deficiencies and alter the pH to a limited extent. Physical treatment to improve the turf structure may include spiking and core removal, sand-slitting, scarifying, raking or harrowing to remove thatch and leaf litter.

Lifting turf for reuse

Turf should be lifted by hand or machine and stacked immediately, not higher than 1.0 m; the turves may be rolled or laid flat but not folded, as this creates a crack, which spoils the turf structure. Turf should be re-laid within 24 hours of lifting, but if this cannot be done the turf must be protected from frost in winter and kept watered in dry weather; the turves should be laid out flat (grass down) and watered.

Preparation of turf-bed

After previous cultivation and bringing to levels (see D20), the final levels may be adjusted with a blade grader. The turf-bed should be raked or harrowed, lightly rolled and treated with a non-residual weedkiller; all stones and debris over 50 mm must be removed.

Laying turf

Turf is best laid in autumn or early winter if the weather is open, and it should not be laid in very dry, waterlogged, frosted, cold or drying wind conditions.

In summer, turf should be saturated on delivery and kept lightly watered during the day until temperatures fall; turf left exposed can die in 3–4 hours in hot weather. In winter, turf should be unrolled and laid out to prevent yellowing if it cannot be laid immediately. Before laying, the turves should be laid on a hard surface and boxed; this means that they must be placed in a wooden former and cut to the same size and thickness so that the laid turf does not show gaps or lumps. Turves are laid in stretcher bond, closely laid, and only whole turves should be used at edges, with all making up of odd shapes done at least one course in from the edge. The turfed area should be designed to avoid sharp and awkward angles. Newly laid turf must not be rolled, but tamped down with a wooden turf-beater as the work proceeds, because hard compacting prevents proper root establishment. Turf must on no account be beaten down to correct faulty levels, but the offending turf must be lifted and the turf-bed corrected. After laying, the joints in the turf should be filled with a fine sand/peat/soil mixture and the surplus brushed off. All work on newly laid turf must be carried out from boards laid on the surface; no walking or machinery should be allowed on the turf itself. The finished level should be about 50 mm above the adjoining paving to allow for settlement.

Turf laid on slopes steeper than 30° should be laid diagonally and pinned with steel hairpins 200 mm long. Traditionally very steep slopes were laid with two layers of turf, one upside down and one right way up, but current practice is to lay erosion control mats on large steep slopes, either pre-seeded or site seeded, which are more likely to remain stable. Diagonal turf is more likely to be used on small shallow slopes.

Temporary protection

The newly laid turf should be protected by chestnut paling, stock fencing, or light wire netting according to the level of protection needed (see Q40 *Fencing*). Proper access for watering and making good must be provided.

After-care during maintenance period

Newly laid turf should be given light regular watering until it is established. The first cut for fine grass should be taken when the grass is about 50 mm high; before cutting, all stones and debris over 25 mm that have accumulated should be removed and the turf lightly rolled. A five- or seven-bladed cylinder mower should be used, and the cuttings boxed off. During the contract maintenance period the turf should be kept cut to the specified height and watered if necessary. Coarse turf on slopes and verges may be cut when 100 mm high with a rotary mower or power scythe.

Depending on weather conditions and the quality of turf required, some work may have to be done to the turf during the maintenance period. The operations that may be specified include watering, stone picking, spiking, scarifying, rolling, harrowing, leaf-litter sweeping, and the application of selective weedkillers and fertilizers. Damaged areas should be scraped out down to firm ground, filled with sieved topsoil and seeded to match turf.

BRITISH STANDARDS

1831 : 1969 & AMD	Recommended common names for pesticides. This standard is continuously updated
3882: 1994	Specification for topsoil (bought and sold)
3969 : 1990	Recommendation for turf for general landscape purposes
4156 : 1990	Peat for horticultural and landscape work
4428 : 1989 & AMD	Code of practice for general landscape operations
7370 Pt 1 : 1991	Grounds maintenance. General recommendations
7370 Pt 3 : 1991	Grounds maintenance. Maintenance of amenity turf

DATA

Turf specified to be to BS 3969 must comply with certain requirements:

- *Either* the seed mixture used to grow the turf must be declared, and 95% of the species present in the turf should consist of the declared species, *or* all species present in the turf at time of sale must be declared, and the percentage of each declared to the nearest 10% if over 5%, and to the nearest 1–5% if under 5%.
- Certain grasses must be declared if present: *Holcus lanatus, Holcus mollis, Dactylis glomerata, Lolium perenne* and *Lolium* spp.
- The turf grass shall have been cut regularly to not more than 35 mm at time of sale.
- The turf shall be visibly free of diseases and pests.
- The turf soil shall not be more than 40% clay, and shall be free of stones over 15 mm; it should match the site soil unless otherwise specified.
- Turf should have no more than 10 mm thatch and 7–18 mm soil thickness below the thatch.
- Turves should be rectangular, and not less than 300 mm in the shorter dimension.

Typical proprietary industrial turf

Industrial turf is supplied either in small squares for hand laying, or in long rolls for machine laying.

Fine lawns	Low maintenance	Recreation areas
Browntop bent	Browntop bent	Browntop bent
Creeping bent	Creeping bent	Chewing's fescue
Chewing's fescue	Chewing's fescue	Slender creeping fescue
Slender creeping fescue	Slender creeping fescue	Perennial ryegrass
	Smooth-stalked meadow grass	

Typical commercial traditional turf

Commercial turf has to have a strong root structure to withstand handling.

Fine turf	Recreation turf
40% Chewing's fescue	40% Perennial ryegrass
30% Strong creeping red fescue	30% Strong creeping red fescue
20% Slender creeping red fescue	20% Chewing's fescue
10% Browntop bent	10% Browntop bent

Typical proprietary NPK fertilizer applications

Fine turf	Sports turf	Reclamation areas
Autumn and winter	Autumn and winter	General Purpose
3:10:5 at 70 g/m²	3:12:12 at 700 kg/ha	20:10:10 at 500 kg/ha
5:5:15 at 35 g/m²	10:10:15 at 350 kg/ha	
slow release	slow release	
Spring and summer	Spring and summer	
11:5:5 at 35 g/m²	9:7:7 at 350 kg/ha	
14:2:4 at 35 g/m²	11:6:9 at 35 g/m²	
15:0:5 at 35 g/m²	slow release	
slow release		

OUTLINE SPECIFICATION

Lifting turf

Lift designated area of turf by hand in turves 100 mm × 300 mm × 40 mm thick minimum and stack on site not higher than 1.0 m; turves may be rolled or laid flat but not folded. Turves to be laid within 24 hours of lifting
or
Turves to be laid flat grass down, and watered.

Laying traditional turf

a. *After previous cultivation, bring to given final levels with blade grader, remove all stones and debris over 50 mm, rake turf-bed to fine tilth, roll lightly, treat with agreed non-residual weedkiller applied by certified operator at manufacturer's recommended rates and in accordance with manufacturer's instructions.*

b. *Supply turf Type XXX from specified supplier, to be lifted not more than 24 hours before laying, lay turves closely in stretcher bond across gradient, whole turves only at edges, tamp down turf with wooden turf beater as the work proceeds, any error in levels to be corrected in turf-bed.*

c. *Fill joints with fine sand/peat/soil mixture and surplus brushed off.*

d. *All work on newly laid turf must be carried out from boards laid on the surface; no walking or machinery allowed on laid turf.*

for topsoil, fertilizers, and weedkillers   Planting generally

Laying proprietary industrial turf

a. *After previous cultivation, bring to given final levels with blade grader, remove all stones and debris over 50 mm, incorporate to 150 mm depth washed sand to BS 882 grade F at 50 mkg/m²; loose peat substitute or sphagnum peat to BS 4156 at 0.3 m³/m²; NPK 6:9:6 fertilizer at 70 g/m², rake turf-bed to fine tilth, roll lightly, treat with agreed non-residual weedkiller applied by certified operator at manufacturer's recommended rates and in accordance with manufacturer's instructions.*

b. *Supply turf rolls Type XXX from specified supplier, to be lifted not more than 24 hours in hot weather and 48 hours in cool weather before laying; lay turf roll closely across gradient with joints staggered, and roll lightly.*

c. *Fill joints with fine sand/peat/soil mixture and surplus brushed off. Any error in levels to be corrected in turf-bed layer. Immediately after laying, saturate turf with sprinkler-applied water.*

d. *All work on newly laid turf must be carried out from boards laid on the surface; no walking or machinery allowed on laid turf.*

for topsoil, fertilizers, and weedkillers   Planting generally

After-care

a. *Cut grass to 35 mm high when the growth is about 50 mm high and reduce height of cut at each mowing until specified height is reached; remove all stones and debris over 25 mm and roll turf lightly before cutting.*

b. *Use five- or seven-blade cylinder mower and box off cuttings. Keep turf cut to 18 mm and water with sprinkler in dry weather until end of maintenance period.*

c. *Apply agreed selective weedkiller once, apply specified NPK fertilizer once.*

d. *Roll once, and make good damaged areas, which must be scraped out down to firm ground, filled with sieved topsoil and seeded to match turf.*

for seeding see Seeding grassed areas *in*   Q30
for topsoil, fertilizers, and weedkillers   Planting generally

Q31 Planting: trees

GENERAL GUIDANCE

This section deals with the planting of trees, including whips and transplants, standards, advanced nursery stock, root-balled trees, semi-mature trees, and forestry planting. Cultivation, soil amelioration and the use of pesticides and herbicides are covered in *Planting generally*.

BS 3936 Pt 1 : 1992 contains useful definitions of plants:

- **Tree**: Woody perennial with a distinct stem or distinct stems.
- **Conifer**: Tree or shrub of the order Coniferales, Taxales or Ginkgoales.
- **Container-grown plants**: Plants that have been grown in any type of container (e.g. pot, plastic bag, peat block) for sufficient time for root growth to have substantially filled the container, but that are not root bound (the size of the container should be in proportion to the size of the plant).
- **Containerized plants**: Plants in containers that have not been in the container long enough to have made substantial new root growth.
- **Bare-root plant**: A plant the root system of which has been grown in the ground, and which on transplanting does not require soil to be attached. (Most deciduous trees and shrubs come into this category.)
- **Root-balled plant**: A plant the root system of which has been grown in the ground, and which on transplanting requires soil to be firmly attached and supported with a suitable porous material such as hessian. (Most evergreen trees, shrubs and conifers come into this category; conifers must be either root balled or container grown.) Conifer transplants for forestry can be bare rooted.
- **Root wrapped**: A term applied to a plant, the root system of which is enclosed to avoid drying out. (This applies to bare-root subjects only.)
- **Country of origin**: The country where the plant has grown for the latter half of the most recent growing season.

Specifying trees

The most suitable tree species and cultivars for landscape work are listed in the *JCLI Plant List*; the landscape designer should refer to this document before specifying trees. Individual trees may be preselected (reserved) and tagged in the nursery. Trees can be obtained in any size from transplants 20 cm high (nurserymen use centimetres, not millimetres) to semi-mature trees 15 m high. The stock selected should have been in the supplier's nursery for at least one year, although this includes stock previously imported from other countries which can be described as UK-grown stock. The source of the plant material should be compatible with the proposed planting soil and climate, and should be as close to the site as possible, though other criteria such as quality and availability of species are more important. All trees must be free from disease or pests, and imported stock must have a phyto-sanitary

Fig. 4.41 *Tilia euchlora* 45–50 cm girth, semi-mature tree with square-clipped crown. (Lorenz von Ehren Nurseries)

certificate. Common species of trees of standard size are usually supplied bare rooted as dug from the nursery; difficult and rare species and conifers are container grown. Container-grown trees are listed by height or girth and the size of the container in litres. Container-grown trees can be planted at any time, but bare-rooted trees must be planted during the dormant season.

Seedlings are supplied as one-, two-, three- or four-year-old plants, which may have been transplanted or undercut during their growing period. Undercutting is the severing of roots below the soil surface to encourage fibrous root growth without transplanting. Transplants are the cheapest form of bulk tree planting, and can be used on reclamation sites where mass planting of larger trees would be too expensive. They can also be used for rural hedge planting where rapid maturity of the hedge is not important, though they will need protection. Seedlings can be supplied as 'cell-grown' stock grown in cell blocks of 12–15 plants 10–60 cm high according to species. These trees can be planted at any time, as they are a type of miniature container-grown stock.

Half-standards, standards and selected standards are usually bare rooted, but they can be supplied root balled, while trees intended to be moved when mature can be grown in root balls ready for moving.

Fig. 4.42 *Tilia platyphyllos* 'Prince's Street'. (Lorenz von Ehren Nurseries)

Semi-mature trees are supplied grown in root balls or prepared and root balled before lifting. The root ball should have a diameter 2.5 times the girth. Such trees are often transplanted several times, or containerized before sale, as this tends to produce a dense root system capable of being transplanted without severe damage. They should be healthy, well grown, vigorous, self-supporting (tree stakes should have been removed at least a year earlier), well formed, true to the species shape and size, and with a well-balanced branch structure. The landscape designer's specification should include:

- age of the tree;
- girth at 1.0 m above ground;
- overall height;
- spread and shape;
- clear stem height;
- number of transplantings;
- root condition: root ball size or container size;
- method of lifting: hand or machine dug and type of root ball formed;
- wrapping method and materials: biodegradable, disposable or reusable;
- crown form: standard crown will be supplied unless otherwise specified;
- branches: tied in for transport unless otherwise specified;
- antidesiccants: specify for certain species and extreme conditions;
- trunk wrapping: specify to prevent damage and in extreme conditions.

Fig. 4.43 *Quercus rubra* 70–80 cm girth. (Lorenz von Ehren Nurseries)

Note: All trees must be protected from heat, cold and wind during transport. Planting should take place within a specified time of lifting.

Girth is the main criterion on which semi-mature trees are offered. Landscape designers should satisfy themselves as to the adequacy of other aspects, but specifying them all without knowledge of their availability from a particular source would probably result in specifying the impossible.

Pre-planting

A semi-mature tree must be able to root into the surrounding ground; it will eventually require access to approximately 100 m³ of accessible soil when it is a mature tree with a fully developed root system. Compacted ground should be subsoiled to well below the root ball with a 150 mm layer of coarse gravel laid in the bottom of the pit to assist drainage, but if the soil tends to water-logging, the tree pit may be drained by means of land drains laid at the bottom of the pit. Conversely, trees in dry areas should have a 75 mm perfo-rated flexible plastic pipe laid round the root ball (without constricting the roots) and connected to irrigation points. If trees cannot be planted as soon as delivered, they must be protected against drying winds, hot sun and frost. Bare-rooted trees should be heeled in a trench with the roots completely covered; root-balled trees should be packed closely together and covered with peat or sand and kept moist. Equally, trees should not be stored where they can become waterlogged or contaminated by site materials.

Planting trees: traditional method

'Easy transplanting' trees should be planted in October to March, while 'difficult transplanting' trees and conifers should be planted in early spring. Root-balled and container trees can be planted in summer, although constant irrigation will be needed. Tree pits should be dug immediately before planting to the full size specified with straight sides, scarified to allow root penetration, and the bottom 250 mm should be loosened. In poor soils the fill should have well-rotted stable manure (not FYM) incorporated at a rate of 1 m³ manure to 5 m³ of soil. Bare-rooted trees should have their roots spread out evenly. Trees with badly damaged roots should be rejected. Root-balled and root-wrapped trees with biodegradable wrapping should be left intact; wire mesh is normally ungalvanized and is left in position. A board should be laid across the pit so that the original ground level of the tree trunk can be maintained (check that the level is that of the final ground level).

Tree pits may be backfilled using the excavated soil if this is good enough to encourage root development, and if it is compatible with the root-ball soil. The pit should be backfilled with a mixture of topsoil, sand, peat (or peat substitute) and bonemeal, the proportions varying with the nature of the existing soil, and it should be as dense as or denser than the surrounding soil. Extra fertilizer may be needed, which should be low nitrogen. Backfill should be placed in 150 mm layers with all air spaces completely filled round the roots, and the tree gently vibrated to settle the soil. The tree and the surrounding soil should be watered to field capacity after backfilling, and a 75 mm mulch of bark, peat or other organic material should be applied, leaving the stem clear.

Each tree should be labelled with a weatherproof label, and if a permanent identification label is required this should be included in the specification.

Formative pruning

Good nurseries carry out formative pruning prior to supply; formative pruning may be regarded by a contractor as an alternative to supplying quality trees. Formative pruning is mainly a continuing maintenance task. Usual pruning actions are:

- Cut out diseased and dead wood.
- Cut out crossing branches.
- Select trees with a distinct strong leader *or* prune to leave only one leader as most trees are apically dominant.
- Shorten to a lateral all branches threatening to compete with the leader.
- Remove basal shoots from the root stock.
- The crown should not be pruned to compensate for die-back; wait until next season to determine what wood needs to be removed.

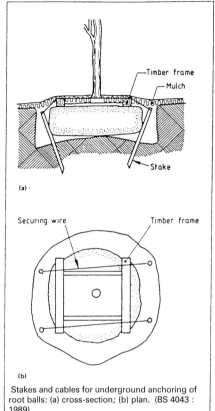

Stakes and cables for underground anchoring of root balls: (a) cross-section; (b) plan. (BS 4043 : 1989)

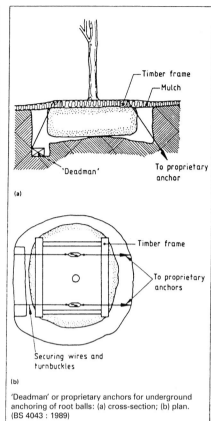

'Deadman' or proprietary anchors for underground anchoring of root balls: (a) cross-section; (b) plan. (BS 4043 : 1989)

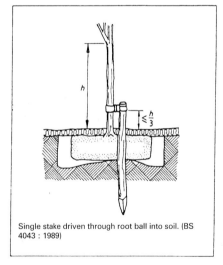

Single stake driven through root ball into soil. (BS 4043 : 1989)

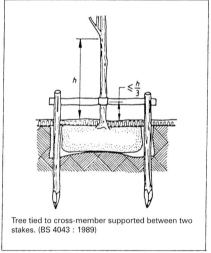

Tree tied to cross-member supported between two stakes. (BS 4043 : 1989)

Fig. 4.44 Underground anchoring

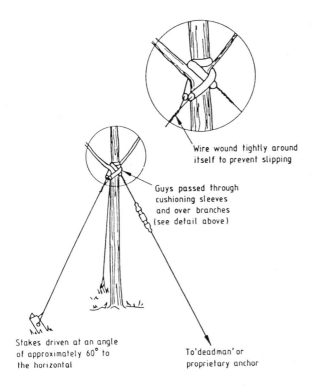

Wire wound tightly around
itself to prevent slipping

Guys passed through
cushioning sleeves
and over branches
(see detail above)

Stakes driven at an angle
of approximately 60° to
the horizontal

To 'deadman' or
proprietary anchor

Fig. 4.45 Overhead anchoring. (BS 4043 : 1989)

Pruning actions during the establishment period are:

● Prune out dead and diseased wood.
● Shorten and eventually remove potentially weak structures.
● Prune to favour a single central leader.
● Never cut out the leading shoot.

Some species, such as *Juglans regia*, can be pruned only at certain times of the year; large cuts at the wrong time can cause permanent severe damage, or even cause the tree to bleed to death. All wounds must be cleaned with a suitable disinfectant, and inspected regularly until healing has taken place.

Planting trees: Amsterdam Tree Soil method

This method has been developed particularly for urban trees in paved areas, where root-balled semi-mature trees are normally used. These trees need to form a good root system rapidly in order to become well established, and a feeding space of 15-20 m³ is required to develop a full root system. A tree pit at least 2000 mm × 2000 mm square (or a trench 2000 mm wide) and 750–900 mm deep is excavated, and a filter-wrapped land drain laid in a trench dug in the bottom of the pit (see *Land drainage* in section R13). The tree is placed in position and the pit backfilled with Amsterdam Tree Soil, a proprietary sand/soil/nutrient mixture. This material is designed to be compacted in layers so that paving may be laid on it without risk of settlement, while still providing an open soil structure containing nutrients, which will encourage rapid root development. Initially the tree and the surrounding soil should be watered to field capacity, while the land drain allows surplus run-off to be removed to prevent waterlogging.

Tree stakes and guards

Small sturdy trees do not need staking unless they are liable to damage, but those with a slender stem habit and those in public areas will need support, particularly if the backfill and surrounding soil are soft. The recommended height of tree stakes varies considerably. BS 3936 suggests that the stake should be cut off just below the first crotch; other authorities recommend stakes two thirds of the clear stem height or one third of the tree height. Long stakes prevent the tree from being broken close to the ground, but short stakes allow the tree to develop a stronger stem structure. Short stakes are often used as markers in rough planting to prevent damage from machinery.

The tree stake should be inserted on the windward side of the tree before backfilling and the tree secured in an upright position. Trees can be secured with one or two tree stakes, underground frames, tensioned underground anchors, props, or guy wires and anchors. Single stakes are normally used for lighter bare-rooted trees in open ground and two stakes for root-balled trees, while guying or props are preferred for semi-mature trees. Underground frames or tensioned anchors are more suitable for semi-mature trees in paved areas where stakes are not desirable and where guys would be dangerous. Trees must be tied to stakes with flexible ties. Ties should be adjusted as the stem increases, but this is very unlikely to be carried out; even when the constriction is not visible, irreversible damage can be done. Although the British Standard specifies treated stakes, it may be preferable to use untreated ones, which will rot by the time the tree is established if it is likely that the stakes will not be removed at the proper time. Removal at the start of the second growing season should have given the tree time to form a self-supporting root system, and five years is the maximum permissible time for supports and ties to remain.

Underground frames consist of timber baulks laid crosswise over the roots close to the trunk and held in place by the backfill. Underground tension wires and anchors are constructed by placing ungalvanized wire netting and tension wires across the root ball and tensioning them to untreated timber ground anchors; this enables the angle of the tree trunk to be corrected by loosening or tightening the anchors.

Guys consist of three steel wires fixed to anchors at a suitable distance from the tree, with the wire protected where it is looped round the tree. To prevent accidents, the lower 2.4 m of the wire must have high-visibility sleeving where the public have access. Propping by three stakes keeps the tree rigid, but care is necessary in fixing the top of the prop to the tree.

After-care

Temporary tree shelters of plastic tube or slotted plastic mesh must be placed round young trees to protect them against damage by rabbits, voles and deer, and to provide a microclimate round the trunk that promotes growth. Where stock are run in newly planted areas, a guard of timber rails and posts plus wire netting must be constructed to prevent browsing and rubbing. In urban areas, permanent tree guards of welded wire, steel railing, or ornamental cast iron are normally used; these are often fixed to the tree grille (see *Tree grilles* in section Q25). It is bad practice to strim or mow grass around young trees, because this only increases the competition for soil moisture.

The after-care of semi-mature trees is extremely important, as it can take many years for a tree to recover from the trauma of being moved. Many nurseries now offer a supply-and-care service, which continues until the tree is safely established; this obviates the large losses due to post-contract lack of care.

For important specimen trees, if the tree is likely to suffer from drying winds or hot sun the trunk and the main branches can be treated with an antidesiccant and wrapped with hessian or special paper (not plastic) and sprayed to prevent foliage loss until the root system can cope. An irrigation system may be installed for individual or groups of trees, because watering may have to be carried out regularly for some years, especially in dry summers. The maintenance of a mulch round the tree will help to prevent weed growth and water loss, because weeds must be controlled for at least two years after planting. Trees may suffer less trauma after transplanting if the foliage is reduced by cutting back the crown by 20–30%; the amount and form of pruning will depend on the shape of the tree.

Forestry and woodland planting

Forestry stock is specified in BS 3936 Pt 4, which includes recommended sources for conifer seed, and it is important to choose the correct species for the soil type in the proposed plantation, because many species will not develop properly in inappropriate soils. The minimum height for open ground stock ranges from 75 mm for Corsican pine to 300 mm for sycamore, and except for *Abies* spp. the seedlings must not be more than four years old when planted. Container-grown forestry stock is obtainable, but more expensive, and the planting height ranges from 250 to 600 mm, while the containers range from 1 litre to 2.35 litres.

Urban woodland planting sites vary considerably in soil condition, climate and vulnerability, and a performance specification may be the best way to achieve successful establishment. The Forestry Authority (the regulatory, grant-aiding and research arm of the Forestry Commission) recommends

5–10 species for a 1 ha urban woodland using small bare-rooted stock, mainly of short-lived tolerant species, as initial planting. Compacted ground must be subsoiled and cultivated, and an application of low-nitrogen fertilizers initially with increased nitrogen in the second season may be necessary. Weed control, irrigation and especially fencing must be provided.

Commercial forestry planting is usually carried out in rough country, where fine cultivation is neither economic nor feasible. The drainage of the planting area is very important; it may be necessary to provide cut-off drains along the contour if the planting is on wet land, and the plantation itself must be drained to remove heavy rainfall without drying out the soil. The Forestry Commission should be consulted on the principles to be followed in relation to the particular type of soil and climate, because any alterations to badly designed drains are impossible after planting has been carried out.

The ground is first cleared of unwanted vegetation, and if it is peaty or compacted it is then ploughed 300 mm deep at 2–3 m apart; this gives a ridge of inverted turf for planting. On very hard soils, subsoiling may be necessary. The ridge-and-furrow system allows surplus water to drain along the furrows to a main ditch until the trees are mature enough to take up the water. While in transit or temporary storage the seedlings must be kept moist and shaded and well heeled in, with the bundles of transplants opened out, because the roots can dry out irrevocably in a very short time. In open ground the seedlings may be planted by screefing off surface vegetation and cutting a planting hole with a mattock, inserting the seedling and backfilling. In ploughed areas the seedlings can be notch planted by cutting a T- or L-shaped slot in the ground with a sharp spade, inserting the seedling, and pressing the turf back; alternatively, the turf can be cut out, inverted, and the seedling planted in a notch cut in the inverted turf.

Protection of newly planted trees can be by either physical or chemical means. Chemical repellents have a limited effect and a short life, so unless there is very good supervision it is better to provide physical barriers. Voles can cause considerable damage, and the best protection is to maintain bare ground for 1.0 m round the tree. Rabbits and deer are a problem in forestry planting, because small deer are increasing rapidly throughout Britain. Large plantations can be fenced (see rabbit and deer fencing under Q40 *Wire fencing*), while for small numbers of trees individual wire netting or plastic rabbit/deer guards can be used. The comparative costs of fencing and guarding can be calculated by the formula:

$$\text{Critical area index} = \frac{\text{Cost of fencing per metre} \times \text{Length of fencing in metres}}{\text{Cost of one tree guard} \times \text{Number of trees}}$$

If the critical area index is less than 1, it is cheaper to fence; if over 1 it will be cheaper to guard. Spiral guards are less satisfactory than rigid guards, and the type must be selected with regard to the kind of attack expected. Rigid solid guards provide a microclimate and protect the tree against herbicide sprays, while being resistant to most type of browsing. All types of guard may need support from stakes.

The contract will normally require young trees to be weeded and fertilized at least once during their first season, and to 'beat up' or replace the failed seedlings; up to 10% loss is customarily allowed in the contract. Machine weeding in post-contract maintenance is feasible where the seedlings are widely spaced, otherwise selective weedkillers can be used, and weeding should continue until the trees are at least 150–200 cm high.

Tree work

The law relating to trees and planning procedures is mainly contained in the Town and Country Planning Act 1990 c.8, Part VIII Special Controls, Chapter 1 Trees, while a useful guide to TPOs is given in the Department of the Environment publication *Tree Preservation Orders: a guide to law and good practice*. A TPO can be applied to a tree with at least a diameter of 75 mm at 1.5 m above ground, though some species are exempt, and special trees can be given protection. General guidance on tree work is given in BS 3998 : 1989 AMD 1990 *Recommendations for tree work*; it is important to give a clear specification of the work to be done, and this is best prepared by an arboriculturist. The specification should include safety requirements and professional competence standards such as those issued by the National Proficiency Test Council, and the Electricity Companies' Code of Practice for work near power lines. The principal operations are:

- cleaning out holes, water pockets and forks;
- removing climbers and epicormic shoots;
- cutting out dead and diseased wood;
- pruning crossing branches;
- crown lifting or reduction;
- crown thinning.

If trees are coppiced or pollarded, they will need different treatment.

Trees in relation to construction

Guidance on safe planting distances from buildings and roads is given in BS 5837 : 1991 *Code of practice for trees in relation to construction*, which also includes guidelines for existing tree assessment on development sites. The Standard covers protection of services and fencing round trees during construction. Trees are graded into four categories:

(a) **Retention most desirable**: healthy, visually attractive, enhance the development, specimen trees;
(b) **Retention moderately desirable**: healthy but not perfect, immature but with potential;
(c) **Retention possible**: sound but need attention, immature, not visually attractive;
(d) **Removal desirable**: dead, diseased, structurally unsafe, poor root-hold, potentially dangerous.

Table 4.2   Preparations for wound treatment: uses and limitations
(BS: 8990 : 1989)

Preparations	Uses and limitations
(a) 3% mercuric oxide paint	A very wide-spectrum protectant with some eradicant properties against early infections, but EC legislation* restricts the use to pomaceous trees, after harvesting and before budding.
(b) 6% thiophanate methyl paint	Protective against bark pathogens only; not for use where silver-leaf is a problem but exceptionally good for stimulating callus growth.
(c) 3% dichlofluanid + 3% active ingredient captan, in water, applied by brush	A wide-spectrum protectant.
(d) 2.4% carbendazim + 5% triadimefon in 4% sodium alginate solution†	Effective against the main wood and bark pathogens, with good systemic properties and hence some eradicant action against early infections, but it is essential that users should mix the fungicide with alginate solution.
(e) Trichoderma spore suspension in water, plus latex-based sealant as an after-treatment	Protective against wood pathogens and possibly bark pathogens, but the viability of spores should be checked after storage.

* Council Directive 79/117/EEC: Section A1 of the Annex.
† Long Ashton formulation.

Although the formulations in the table have proved effective in research, they have not yet been approved for use as tree wound dressings under the Control of Pesticides Regulations (SI 1510) 1986. Advice should therefore be sought from the manufacturers before using these products. *Note*: Some of the preparations shown in this table were not generally available at the time this standard was published and may have to be formulated.

Guidance on the relationship of trees to buildings is also given in the National House-Building Council Standards Ch.4.2.

Tree roots can damage drains and pipe runs; rigid jointed pipes can be penetrated or broken, and flexible pipes can be constricted. Any site services that run close to tree roots should be sleeved with impervious piping or fully surrounded with concrete to beyond the final root spread of the tree. Where services must be run under existing trees the pipe trench should be formed below or between main roots with as little disturbance as possible. Existing trees should be fenced off at a safe distance and no contaminants allowed to run off over the root area, and bonfires must be forbidden within a given distance from the tree.

Trees can interfere with overhead cables and with high vehicles; the future development of roads and services should be taken into account when specifying trees on development sites, and existing trees may have to be crown-lifted if high vehicles will be travelling close to the trees.

Fig. 4.46 Reducing the crown.
Note: Branches pruned back to a suitable outward-pointing bud or small branch. (BS 3998 : 1989)

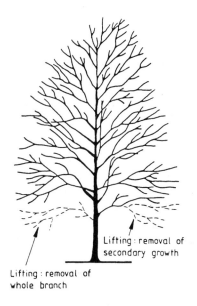

Fig. 4.47 Crown lifting. (BS 3998 : 1989)

Table 4.3   Protection of trees: minimum distances for protective fencing around trees (BS 5837:1991)

Tree age	Tree vigour	Trunk diameter (mm)	Minimum distance (m)
Young trees (age less than 1/3 life expectancy)	Normal vigour	<200	2.0
		200 to 400	3.0
		>400	4.0
Young trees	Low vigour	<200	3.0
		200 to 400	4.5
		>400	6.0
Middle age trees (1/3 to 2/3 expectancy)	Normal vigour	<250	3.0
		250 to 500	4.5
		>500	6.0
Middle age trees	Low vigour	<250	5.0
		250 to 500	7.5
		>500	10.0
Mature trees	Normal vigour	<350	4.0
		350 to 750	6.0
		>750	8.0
Mature trees and overmature trees	Low vigour	<350	6.0
		350 to 750	9.0
		>750	12.0

Note (1) It should be emphasized that this table relates to distances from centre of tree to protective fencing. Other considerations, particularly the need to provide adequate space around the tree including allowances for future growth, and also working space, will usually indicate that structures should be further away.

Note (2) With appropriate precautions, temporary site works can occur within the protected area, e.g. for access or scaffolding.

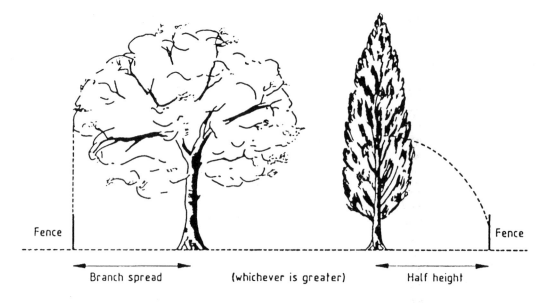

Fence   Fence

Branch spread   (whichever is greater)   Half height

Fig. 4.48 Alternative location for protective fencing. (BS 5837:1991)

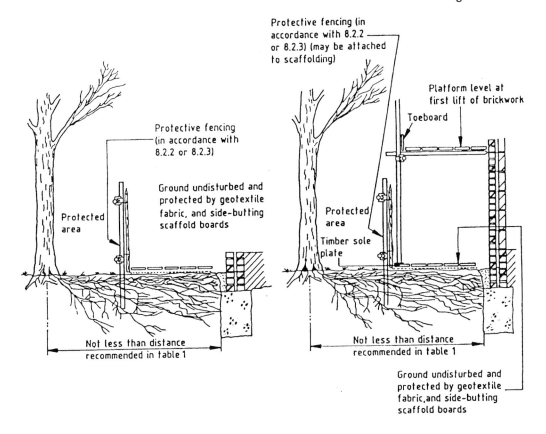

Fig. 4.49 Scaffolding within a protected area: (a) initial stage; (b) secondary stage with scaffolding. (BS 5837 : 1991)

BRITISH STANDARDS

1831 : 1969 & AMD	Recommended common names for pesticides. This standard is continuously updated
3882: 1994	Specification for topsoil (bought and sold)
3936 Pt 1 : 1992	Nursery stock. Trees and shrubs
3936 Pt 4 : 1984	Nursery stock. Forest trees for timber production
3936 Pt 5 : 1985	Nursery stock. Poplars and willows for shelter belts, amenity planting, and timber production
3998 : 1989 & AMD	Recommendations for tree work
4043 : 1989	Recommendations for transplanting root-balled trees
4156 : 1990	Peat for horticultural and landscape work
4428 : 1989 & AMD	Code of practice for general landscape operations
5837 : 1991	Code of practice for trees in relation to construction
7370 Pt 1 : 1991	Grounds maintenance. General recommendations
7370 Pt 4 : 1993	Grounds maintenance. Maintenance of soft landscape (other than amenity turf)

Table 4.4 Minimum dimensions for open ground stock (BS 3936 : Pt 4 : 1984)

Species	Minimum height (mm)	Minimum collar diameter for heights not less than								
		75 mm	100 mm	150 mm	200 mm	250 mm	300 mm	400 mm	500 mm	600 mm
Alder										
Common alder	300	–	–	–	–		5.0	6.0	7.5	9.0
Grey alder	300	–	–	–	–	–	4.5	5.0	6.0	7.0
Italian alder	300	–	–	–	–	–	4.5	5.0	6.0	7.0
Ash	200	–	–	–	5.0	5.5	6.5	8.0	9.5	11.0
Beech	200	–	–	–	4.0	4.5	5.0	6.0	7.5	9.0
Birch										
Common silver birch	200	–	–	–	3.0	3.5	4.0	4.5	5.5	6.5
Chestnut										
Sweet chestnut	200	–	–	–	5.0	5.5	6.5	8.0	9.5	11.0
Cypress										
Lawson cypress	200	–	–	–	4.0	4.5	5.0	6.5	7.5	8.5
Fir										
Douglas fir	200	–	–	–	3.0	3.5	4.0	5.0	6.0	–
Grand fir	100	–	4.5	4.5	5.0	6.0	6.4	8.0	9.5	–
Noble fir	100	–	5.0	5.0	5.0	6.0	6.5	8.0	9.5	–
Gean	200	–	–	–	5.0	5.5	6.5	8.0	9.5	11.0
Hemlock										
Western hemlock	200	–	–	–	2.5	3.0	3.5	4.0	5.0	5.5
Larch										
European larch	200	–	–	–	2.5	3.5	4.0	5.0	6.0	6.5
Hybrid larch	200	–	–	–	2.5	3.5	4.0	5.0	6.0	6.5
Japanese larch	200	–	–	–	2.5	3.5	4.0	5.0	6.0	6.5
Lime										
Large leaf lime	200	–	–	–	5.0	5.5	6.5	8.0	9.5	11.0
Small leaf lime	200	–	–	–	5.0	5.5	6.5	8.0	9.5	11.0
Maple										
Norway maple	200	–	–	–	3.5	4.0	4.5	5.0	6.0	7.0
Sycamore	300	–	–	–	–	–	4.5	5.0	6.0	7.0
Oak										
Sessile	200	–	–	–	5.0	5.5	6.5	8.0	9.5	11.0
Pine										
Bishop pine	100	–	3.0	4.0	4.5	5.5	6.5	8.0	9.5	–
Corsican pine	75	3.0	3.0	4.0	4.5	5.5	6.5	8.0	9.5	–
Lodgepole pine	100	–	3.0	4.0	4.5	5.5	6.5	8.0	9.5	–
Scots pine	100	–	3.0	4.0	4.5	5.5	6.5	8.0	9.5	–
Shore pine	100	–	3.0	4.0	4.5	5.5	6.5	8.0	9.5	–
Southern beech										
Rauli	200	–	–	–	4.5	5.0	5.5	6.0	7.5	9.0
Roble	200	–	–	–	4.5	5.0	5.5	6.0	7.5	9.0
Spruce										
Norway spruce	150	–	–	3.0	4.0	4.5	5.0	6.5	8.0	9.5
Sitka spruce	150	–	–	2.5	3.0	3.5	4.0	5.0	6.0	7.0
Sycamore (see under Maple)										
Western red cedar	200	–	–	–	4.0	4.5	5.0	6.5	7.5	8.5

Table 4.5 Minimum dimensions for container-grown stock (BS 3936 : Pt 4 :1984)

Species	Minimum height (mm)	Minimum collar diameter (mm)	Minimum container size* (ml)
Alder			
Common alder	600	6	125
Grey alder	600	6	125
Italian alder	600	6	125
Ash	600	6	235
Beech	300	5	235
Birch			
Common silver birch	600	6	125
Chestnut			
Sweet chestnut	600	6	235
Fir			
Douglas fir	200	3	100
Limes			
Large leaf lime	600	6	235
Small leaf lime	600	6	235
Maple			
Norway maple	600	6	235
Sycamore	600	6	235
Oak			
Sessile	300	6	235
Pine			
Corsican pine	120	2	100
Lodgepole pine	200	4	100
Scots pine	200	4	100
Southern beech			
Rauli	600	6	125
Roble	600	6	125
Spruce			
Norway spruce	200	4	100
Sitka spruce	200	3	100
Sycamore (see under Maple)	–	–	–

* if the plants are grown in Japanese paper pots (JPPs) the following size designations for containers are appropriate:

100 ml F408
125 ml F508
235 ml F515

(BS 5236 : 1975 *Advanced nursery stock* has been withdrawn and replaced by: BS 3936 Pt 1 : 1990 *Trees and shrubs* and BS 4043 : 1989 *Recommendations for transplanting root-balled trees.*)

(BS 3975 withdrawn.)

DATA

Forms of trees to BS 3936: 1992

- Standards: shall be clear with substantially straight stems. Grafted and budded trees shall have no more than a slight bend at the union. Standards shall be designated as half, extra light, light, standard, selected standard, heavy, and extra heavy.

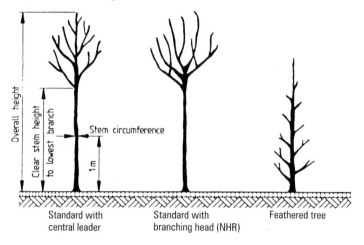

Standard with   Standard with   Feathered tree
central leader   branching head (NHR)

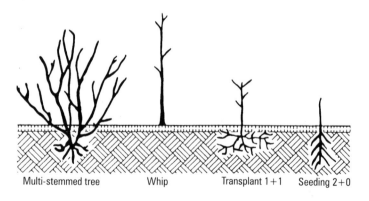

Multi-stemmed tree   Whip   Transplant 1+1   Seeding 2+0

Fig. 4.50 Examples of forms of trees.
Note: These outline diagrams are intended only to indicate the different forms of trees. They are not drawn to scale or intended to convey the number of branches. (BS 3936 Pt 1 : 1992)

- Semi-mature trees: between 6.0 m and 12.0 m tall with a girth of 20–75 cm at 1.0 m above ground.
- Feathered trees: shall have a defined upright central leader, with stem furnished with evenly spread and balance lateral shoots down to or near the ground.
- Whips: shall be without significant feather growth as determined by visual inspection.
- Multi-stemmed trees: Shall have two or more main stems at, near, above or below ground.

Seedlings

Seedlings grown from seed and not transplanted shall be specified when ordered for sale as:

1 + 0   one-year-old seedling
2 + 0   two-year-old seedling
1 + 1   one year seed-bed, one year transplanted   = two-year-old seedling
1 + 2   one year seed-bed, two years transplanted   = three-year-old
        seedling
2 + 1   two year seed-bed, one year transplanted   = three-year-old seedling
1 u 1   two years seed-bed, undercut after 1 year   = two-year-old seedling
2 u 2   four years seed-bed, undercut after 2 years   = four-year-old
        seedling

Cuttings

The age of cuttings (plants grown from shoots, stems, or roots of the mother plant) shall be specified when ordered for sale. The height of transplants and undercut seedlings/cuttings (which have been transplanted or undercut at least once) shall be stated in centimetres. The number of growing seasons before and after transplanting or undercutting shall be stated.

0 + 1   one-year cutting
0 + 2   two-year cutting
0 + 1 + 1   one year cutting bed, one year transplanted
            = two-year-old seedling
0 + 1 + 2   one year cutting bed, two years transplanted
            = three-year-old seedling

Sizes of standards

Heavy standard	12–14 cm girth × 3.50–5.00 m high
Extra heavy standard	14–16 cm girth × 4.25–5.00 m high
Extra heavy standard	16–18 cm girth × 4.25–6.00 m high
Extra heavy standard	18–20 cm girth × 5.00–6.00 m high

Transplanting semi-mature trees

Easy transplanting	Medium transplanting	Difficult transplanting
Acer platanoides	*Alnus*	*Abies*
Acer pseudoplatanus	*Betula*	*Carpinus*
Malus	*Buxus*	*Cedrus*
Platanus	*Crataegus*	*Chamaecyparis*
Sorbus	*Fraxinus*	*Fagus*
Tilia	*Liquidambar*	*Ilex*
	Populus	*Juglans*
	Prunus	*Larix*
	Pyrus	*Liriodendron*
	Quercus	*Magnolia*
	Robinia	*Pinus*
	Salix	*Picea*
		Thuja

Conifers generally do not transplant well.
Container size should be related to girth:

25–30 cm girth	200 l container
30–35	600
40–45	1000

Sizes of tree pits

Bare rooted

Whips, small feathered trees	600 mm dia. × 450 mm deep
Light standard, large feathered trees	900 mm dia. × 600 mm deep
Tall standards	1200 mm dia. × 750 mm deep

Root balled

1500 mm × 1500 mm × 1000 mm deep for root ball up to 1000 mm dia. × 750 mm deep

2000 mm × 2000 mm × 1000 mm deep for root ball 1100–1500 mm dia. × 750 mm deep

500 mm wider than root ball and 250 mm deeper than root ball for root ball over 1000 mm dia. × 750 mm deep

Backfilling mixtures

Good-quality existing or imported topsoil to BS 3882 General Purpose grade.

Add to light topsoil:	30% peat, 600 g/m^3 bonemeal
Add to medium light topsoil:	20% peat, 600 g/m^3 bonemeal

Add to medium topsoil: 10% peat, 600 g/m³ bonemeal
Add to medium heavy topsoil: 20% peat, 10% sand, 600 g/m³ bonemeal
Add to heavy topsoil: 25% peat, 15% sand, 600 g/m³ bonemeal

OUTLINE SPECIFICATION

Planting whips

a. *In previously cultivated ground dig planting pit 600 mm dia. × 450 mm deep, fork over bottom of pit to 250 mm.*

b. *Supply and plant whips of the species and at the spacing shown on planting plan, mix excavated topsoil with spent mushroom compost or other equal and approved organic manure at the rate of 1 m³ manure to 3 m³ of topsoil incorporating bonemeal at 3 kg/m³, and backfill in layers not exceeding 150 mm, soil to fill all air spaces round roots.*

c. *Water to field capacity, mulch with 75 mm of conifer bark 20–40 mm particles.*

for excavation	*D20*
for topsoil, fertilizers, and weedkillers	Planting generally
for fencing see rabbit and deer fencing in	*Q40*

Planting bare-rooted standards

a. *In previously cultivated ground excavate planting pit 1200 mm dia. × 750 mm deep, fork over bottom of pit to 250 mm.*

b. *Supply and plant 10 no. bare-rooted heavy standard trees of the species and in the locations shown on planting plan, maintain orientation and keep original soil mark level with adjoining finished ground level. Mix excavated topsoil with proprietary tree and shrub planting compost Ref: XXX at the rate of 1 m³ compost to 1 m³ excavated topsoil, fix tree stake 100 mm dia. 300 mm into firm ground below tree pit before planting, set tree upright, tie with two flexible tree ties, and backfill with screened excavated topsoil to cover roots in layers not exceeding 150 mm, soil to completely fill all air spaces round roots, backfill to final ground level with tree and shrub planting compost:topsoil mixture.*

c. *Cut stake to 100 mm below first branch or crotch, water thoroughly, mulch with 75 mm of conifer bark 20–40 mm particles.*

for excavation	*D20*
for topsoil, fertilizers, and weedkillers	Planting generally
for fencing see Rabbit and deer fencing *in*	*Q40*

Planting root-balled semi-mature trees in paved areas

a. *Excavate tree pit 2500 mm × 2500 mm × 900 mm deep, excavate drain trench 150 mm × 450 mm deep, supply and lay filter- wrapped*

60 mm dia. perforated plastic pipe to BS 4962 : 1989, backfill drain trench with gravel.

b. *Supply and plant as shown on planting plan 1 no. previously selected semi-mature Quercus rubra 25 cm girth 6.0 m high ref: XXX, maintain orientation and keep original soil mark level with adjoining finished ground level.*

c. *Supply and lay 100 mm plastic irrigation pipe round root ball with access point and cover at finished ground level, anchor with 4 no. untreated timber ground anchors 100 mm dia. and tension wires across root ball, backfill with Amsterdam Tree Soil well compacted in layers not exceeding 150 mm.*

d. *Wrap trunk and main branches with hessian wrapping, water to field capacity; tree grille and paving specified elsewhere.*

for excavation	*D20*
for tree grilles see Tree grilles *in*	*Q25*
for paving	*Q24/25*
for topsoil, fertilizers, and weedkillers	Planting generally
for fencing see Rabbit and deer fencing *in*	*Q40*

Forestry planting

a. *Clear unwanted vegetation, treat area with agreed herbicide applied by certified operator at manufacturer's recommended rates and in accordance with manufacturer's instructions, plough 300 mm deep at 2.0 m centres.*

b. *Supply and plant* Fagus sylvatica, Fraxinus excelsior, Quercus petraea, Betula utilis, Acer campestre *Elsrijk*, Acer saccharum, *1 + 2 as shown on planting plan, notch-plant seedlings.*

c. *Fence area with rabbit fencing specified elsewhere, weed and fertilize once, beat up once.*

for excavation	*D20*
for topsoil, fertilizers, and weedkillers	Planting generally
for fencing see Rabbit and deer fencing *in*	*Q40*

Q31 Planting: shrubs and hedges

GENERAL GUIDANCE

This section covers the final preparation of the planted area, pre-planting treatment, and planting in open ground and planters, planting hedges and the contract period maintenance. Cultivation, soil amelioration, and the use of pesticides and herbicides are covered in *Planting generally*.

Sizes of shrubs

Shrubs can be supplied as bare-rooted transplants for bulk planting when cover must be established over large areas where container-grown shrubs would be too expensive. The source of the plant material should be compatible with the proposed planting soil and climate, and should be as close to the site as possible. Transplants are one-, two-, three- or four-year-old plants, which may have been transplanted or undercut during their growing period. Undercutting is the severing of roots below the soil surface to encourage fibrous root growth without transplanting. Transplants for woodland planting are usually native woodland species such as *Cornus*, *Corylus* and *Crataegus*, which can establish themselves without much maintenance; the younger plants are easier to establish, but need protection more than older stock, and may not suit the client's requirements for rapid results. Sizes of transplants start at 30 cm high, and range from *Salix alba* 0 + 1, 60 cm high to *Quercus robur* 1 + 2, 120 cm high.

Individual shrubs for ornamental landscape work are usually container grown; they can be planted at any season except frost, unlike bare-rooted stock, which must be planted in the dormant season. Shrubs are specified by container size, ranging from 1.5 litre pots of *Eleagnus* 30 cm high to 10 litre pots of *Euonymus* 120 cm high. Low-spreading shrubs such as Ericas are specified by their spread. Specimen shrubs are supplied in 18–80 litre pots, 40–200 cm high, but these are many times more expensive than ordinary container-grown stock. When specifying a plant list, it is advisable to have substitute plants in mind in case of stock failure at the nursery.

Pre-planting operations

Basic cultivation is covered in section D20, and soil amelioration, weedkillers and fertilizers in *Planting generally*. If the existing topsoil can be brought to a healthy state, the top 300 mm is finely cultivated just before planting, and all stones and debris over 50 mm removed. Immediately before planting, the ground is brought to final levels and raked. In very poor soil, such as occurs on reclamation sites, the cultivation can be done on an agricultural scale by chisel ploughing and harrowing, incorporating fertilizers and conditioners, or by hydroseeding with shrub species.

The eventual level of the planted area should be lower than adjoining paving or grass to avoid run-off in heavy rain, and spill from earth and debris loosened in cultivation, so the soil is finished at about 50 mm above the paving or grass to allow for settlement. If the planted area is intended to be higher than the adjoining surface in order to prevent run-off from the paving or road, a kerb and channel must be provided to restrain the soil and to drain the paving.

Planting shrubs

In order to reduce damage from handling and transport, the packaging should be specified to comply with the Code of Practice for Plant Handling available from the Horticultural Trades Association. Delivery of shrubs should take place on the planting date; if this is not feasible the contractor should arrange to collect the plants from the nursery. Bare-rooted plants should be packaged in co-extruded polythene bags, and as soon as they are delivered they should be heeled in under moist shaded conditions until planting. Container shrubs should be kept shaded and watered, because more plants are lost by drying out of roots (which can occur in a very short time) than by damage. Container-grown shrubs may be planted at any time, but bare-rooted stock must be planted during the dormant season, except for conifers, which should be planted at their least active period of growth. Bare-rooted deciduous shrubs may be dipped in a water-retaining dip before planting, and conifers should be dipped or sprayed with antidesiccant in dry weather. The roots should be gently spread out and any damaged roots pruned before planting; stock that has become very pot bound or with distorted tap roots encircling the root ball should be rejected.

Planting holes should be formed 150 mm wider than the roots and at least 400 mm deep, with the bottom 150 mm forked over; tap roots should be accommodated in extra depth. If shrubs of a suckering habit are to be kept under control, a geotextile membrane may be inserted at the furthest edges of the desired root spread. The shrubs should be planted upright with the original soil mark level with the finished ground level, and firmly backfilled with a mixture of topsoil, peat, bonemeal and sand, or a proprietary tree and shrub planting compost, with the backfill packed carefully round roots to fill all air pockets. The shrubs are then pruned to their desired shape and fully watered with a spray. All stock should be watered after planting to field capacity. A closely planted area should be mulched overall with peat, bark, or other mulch, while individual shrubs can be mulched separately for about 450 mm radius round the plant. Standard roses and tall floppy shrubs such as *Buddleia* should be staked with 50 mm dia. stakes and two flexible ties, at least until they are established, and shrub beds may need temporary protection from animals and people by means of cleft chestnut paling. Guards are not usually required for shrubs, though standard roses or clear stem shrubs in grass should be provided with tree tubes to protect them from grass-cutting machinery unless an area at least 400 mm radius is left round the shrub.

Where urban planting is close to open or derelict land there may be a risk of damage from rabbits, badgers, or voles. Heavy wire netting dug into the ground on stakes at 2.0 m centres will keep off rabbits and badgers, while voles and mice can be deterred by a 1 m clear space around shrubs.

Formative pruning

Shrubs may need formative pruning in order to induce them to develop into the desired final shape. Usual pruning actions are:

- Cut out diseased and dead wood.
- Cut out crossing branches.
- Select shrubs with a shape characteristic of the species and cultivar.
- Remove basal shoots from the root stock.
- The leader should never be cut in shrubs with an apical form.

Pruning actions during the establishment period are:

- Prune out dead and diseased wood.
- Shorten and eventually remove potentially weak structures.

All pruning wounds must be treated with disinfectant and inspected regularly until healing has taken place.

Each shrub should be clearly labelled with a weatherproof label, and if a permanent identification label is wanted this should be included in the specification.

After-care

Shrubs must be kept weeded and watered until the completion of the maintenance period; ties should be checked, and mulching maintained. Mulch may have to be replaced if it is scraped away from the shrubs by animals or birds; from this point of view bark is more satisfactory, as it is not so tempting to blackbirds as a compost mulch. An irrigation system may be installed for individual or groups of shrubs, because watering may have to be carried out regularly for some years, especially in dry summers. The maintenance of a mulch round shrubs will help to prevent weed growth and water loss, because weeds must be controlled for at least two years after planting.

Most species have an optimum time for pruning, especially for flowering shrubs, and the cultivars within species may require different pruning. A pruning specification with diagrams of the correct shrub structure should be prepared and handed to the client for reference, especially where light or hard pruning is required at different seasons. Specimen shrubs can be treated individually, but large shrub areas are likely to be trimmed by hedge trimmer or power cutter, and the landscape designer should ensure that all shrubs in such a group can be treated alike. Even in small areas it is advisable to group shrubs that need similar pruning together if possible, as supervision and control will be made easier.

Planting hedges

Hedges may be divided into three types: the trimmed formal hedge, the conservation hedge and the stock-proof farm hedge. Formal hedges are usually composed of species that respond to close cutting and which form a visual barrier, such as beech, hornbeam and privet, while conservation hedges contain a mixture of plant material, which usually includes berry- and nut-bearing species such as hawthorn, bullace, holly, blackthorn, elder and hazel, which also provide good nesting cover. Hedges where stock are run should not contain yew, broom, rhododendron, box or laburnum, as these are not recommended for use where stock can eat them.

Hedging plants are sold as bare-rooted transplants 1 + 1 or 1 + 2, from 40 to 120 cm high. Young transplants can be used for rural hedge planting where rapid maturity of the hedge is not important, although they will need protection against deer and rabbits. Wire netting can be erected each side of the hedge, and removed when the plants are established.

Both types of hedge may be either single row or double staggered row, though conservation hedges may be planted as a double hedge with a narrow overgrown passage down the centre, which offers considerable protection to wild life, and they are often planted on a bank and ditch landform. Traditionally the hedge belongs to the owner of the land on the ditchless side, with the boundary lying on the further edge of the ditch (unless legally stated otherwise), and the landscape designer should bear this in mind when laying out hedge and ditch lines. Regional variations in conservation hedges include hedges on earth banks, hedges on stone-faced banks, and hedge with double-ditch forms.

The hedgerow site should be either ploughed or trenched to a depth of 300 mm and 600 mm wide for a double-row hedge, and the site treated with glyphosate in July–September; planting is done in the dormant season. Unless the soil is fertile topsoil it should be mixed with compost, manure or inorganic fertilizer as for shrubs, though too much nitrogen will induce soft floppy plants. The hedging plants, which are usually whips from $1 + 1 \times 500$ mm up to $2 + 1 \times 900$ mm high, are set 200–300 mm apart for single rows, and with the two rows 400 mm apart staggered for double-row planting. All hedging plants are supplied bare rooted, and as for shrubs and tree seedlings, the hedge plants must be kept moist and shaded before planting. Small plants should be kept together and planted separately from larger plants. The whips are planted to the original soil mark, and the excavated soil is firmed around them. The complete trench is backfilled after all whips have been planted, and if necessary a proportion of peat, sand and bonemeal is added to the backfill. Watering to field capacity should be carried out in dry weather.

Hedges may need some support for the taller whips, and a single or double line of wire on timber posts can be provided (see *Boundary fencing* in section Q40); if there is a risk of rabbit or deer damage a temporary wire netting fence can be set each side of the hedge. If people are likely to take short cuts through the hedge, chestnut paling may be installed to form a barrier until the hedge is fully established; alternatively, permanent stock fencing can be used to prevent gaps from being made in the hedge, because stock fencing wires cannot be easily forced apart. Stock fencing is also more visible to the hedge trimmer operator than line wires.

After-care

Hedges of all types require weed control for at least the first year, and a leaf-mould or compost mulch should be used. The use of herbicides is not recommended if stock are adjacent, and black plastic can get loose and be eaten by cattle, with fatal results. Watering in dry weather should be carried out regularly. Plants that are dead or dying after practical completion of the contract are usually the supplier's responsibility.

Formal hedges are usually cut once a year and often trimmed at intervals if a crisp appearance is required, but conservation and field hedges should only be cut between February and the middle of March so that nests and fruiting are not disturbed. Herbicides should not be used on rural hedges or the adjoining headlands in order to maintain a diverse habitat. The hedge shape may be vertical and rectangular for formal hedges or less rigid for rural hedges; if no undergrowth is wanted an 'A' shape is satisfactory, but a rectangular shape allows bottom vegetation to develop. Evergreen hedges should have pointed or rounded tops to prevent heavy snow from breaking into the hedge.

BRITISH STANDARDS

1831 : 1969 & AMD	Recommended common names for pesticides. This standard is continuously updated
3882: 1994	Specification for topsoil (bought and sold)
3936 Pt 1 : 1992	Nursery stock. Trees and shrubs
3936 Pt 2 : 1990	Nursery stock. Roses
3936 Pt 3 : 1990	Nursery stock. Fruit, soft and tree fruit
4156 : 1990	Peat for horticultural and landscape work
4428 : 1989 & AMD	Code of practice for general landscape operations
7370 Pt 1 : 1991	Grounds maintenance. General recommendations
7370 Pt 4 : 1993	Grounds maintenance. Maintenance of soft landscape (other than amenity turf)

(BS 3975 withdrawn.)

DATA

Typical planter-filling mixtures

Good quality existing or imported topsoil to BS 3882 Premium grade

Add to light topsoil:	30% peat, 600 g/m³ bonemeal
Add to medium-light topsoil:	20% peat, 600 g/m³ bonemeal
Add to medium topsoil:	10% peat, 600 g/m³ bonemeal
Add to medium-heavy topsoil:	20% peat, 10% sand, 600 g/m³ bonemeal
Add to heavy topsoil:	25% peat, 15% sand, 600 g/m³ bonemeal

Topsoil quality to BS 3882 : 1994

Topsoil grade	Properties
Premium	Natural topsoil, high fertility, loamy texture, good soil structure, suitable for intensive cultivation
General Purpose	Natural or manufactured topsoil of lesser quality than premium, suitable for agriculture or amenity landscaping, may need fertilizer or soil structure improvement

Topsoil quality to BS 3882 : 1994 *continued*

Topsoil grade	Properties
Economy	Selected subsoil, natural mineral deposit such as river silt or greensand. The grade comprises two subgrades: 'low clay' and 'high clay', which is more liable to compaction in handling. This grade is suitable for low-production agricultural land and amenity woodland or conservation planting areas.

Hedging species

Formal hedges	Conservation hedges	Stock-proof hedges
Fagus sylvatica	*Prunus spinosa*	*Crataegus* spp.
Buxus sempervirens	*Malus sylvestris*	*Prunus spinosa*
Prunus laurocerasus	*Crataegus* spp.	*Ilex aquifolium*
Cupressus spp.	*Ilex aquifolium*	
Carpinus betulus	*Prunus cerasifera*	
Ligustrun ovalifolium	*Hippophae rhamnoides*	
Taxus baccata	*Cornus sanguinea*	
Rosa spp.	*Sambucus nigra*	
	Corylus avellana	
	Tamarix spp. (for saline areas)	

OUTLINE SPECIFICATION

Planting shrubs

a. *After grading and general cultivation specified elsewhere, bring top 300 mm of topsoil to a fine tilth, incorporating 30% peat or peat substitute and 600 g bonemeal per m³ of topsoil, rake and bring to given levels, remove all stones and debris over 50 mm, dig planting holes average 400 mm deep × 450 mm dia., fork over bottom 150 mm of planting hole.*

b. *Supply and plant specified shrubs in accordance with planting plan, reject pot-bound or damaged shrubs, spread out roots, trim any damaged roots, backfill with topsoil mixture as above with original soil mark level with finished ground level.*

c. *Stake shrubs as necessary with 50 mm dia. stakes and two ties, trim damaged branches, water to field capacity and mulch 450 mm round shrub with 75 mm bark chips 20–40 mm size, spray conifers with antidesiccant in hot weather.*

d. *Water and weed regularly until completion of contract and replace failed plants. All bare-rooted shrubs to be supplied in co-extruded polythene bags and heeled in unless planted immediately on delivery; container-grown plants to be kept in shaded moist conditions until planted.*

for topsoil, fertilizers, and weedkillers   Planting generally

Planting formal hedges

a. *After general cultivation previously specified in section D20, mark out line of hedge, excavate trench 600 mm wide × 300 mm deep and store topsoil, fork over bottom 150 mm of trench.*

b. *Supply and plant* Carpinus betulus *whips 2 + 1 × 900 mm high, 200 mm apart with two rows 400 mm apart staggered, spread out roots and trim all damaged roots and shoots, with original soil mark level with finished ground level, backfill with topsoil screened to remove all stones over 40 mm and incorporate 600 g bonemeal per m³ of topsoil. Small plants should be kept together and planted separately from larger plants.*

c. *Water to field capacity after planting, supply and erect temporary chestnut paling 1.0 m high along one side of hedge and remove at completion of contract, water and weed regularly until completion of contract. Replace failed plants. Whips to be supplied in co-extruded polythene bags and heeled in unless planted immediately on delivery.*

Planting conservation hedges

a. *Clear surplus vegetation from area of hedge, mark out line of hedge, excavate two parallel trenches 300 mm wide × 300 mm deep 900 mm apart and store topsoil, fork over bottom 150 mm of trench.*

b. *Supply and plant the following species 2 + 1 × 600 mm high average as shown on planting plan:* Prunus spinosa, Crataegus oxyacantha, Ilex aquifolium, Corylus avellana, Rosa rugosa, Rosa eglanteria, *200 mm apart in each row, spread out roots and trim all damaged roots and shoots, with original soil mark level with finished ground level, backfill with topsoil screened to remove all stones over 40 mm.*

c. *Water to field capacity after planting, supply and erect rabbit-stop fencing 900 mm high specified in section Q40 along both sides of hedge, water and weed regularly until completion of contract. Replace failed plants. Whips to be supplied in co-extruded polythene bags and heeled in unless planted immediately on delivery.*

As above but with stock fencing 1100 mm high as specified in Q40 on both sides of hedge in lieu of rabbit-stop fencing.

for cultivation generally	*D20*
for chestnut paling	*Q40*
for rabbit-stop fencing	*Q40*
for stock fencing	*Q40*

Q31 Planting: ground cover, climbing and herbaceous plants

GENERAL GUIDANCE

This section covers the final preparation of the planted area, pre-planting treatment, and planting of herbaceous plants in open ground and planters, and the contract period maintenance. The term 'ground cover' can also include some shrubby and herbaceous plant material. Topsoil, soil amelioration, fertilizers, herbicides and pesticides are covered in *Planting generally*.

Sizes of herbaceous plants

Herbaceous plants are sold as container-grown plants, but because of their variable shapes and growth the size of the container is not an accurate guide to the size of the plant. Pot sizes are usually 9 cm diameter, but it is advisable to inspect the stock at the nursery before specifying particular sizes. Container-grown plants can be planted at any time, but some protection against hot sun may be needed for the delicate species, and planting should not be carried out if frosty weather is expected. Plants should be well developed and have been at least a year in their container, but plants two years in the container should be checked, although some slow-growing species may be satisfactory.

Climbers

Climbers may be categorized as tendril twiners, which need support, or self-clingers and self-supporting types, which do not need support. The type of climber must be considered when specifying the supports, also whether they are annual or perennial. Climbers are supplied in the same sizes and also in larger pots of 13 cm diameter or 3 and 10 litre pots depending on the species. Large climbers such as *Wisteria* are supplied in 10 litre pots. The plants should have well-developed shoots, branching or upright according to the species, and all suckers and dead wood removed. If there is a 'best aspect' of the plant, this should be set facing away from the wall.

It is essential to provide strong support for climbers, as they can be very heavy in full leaf, and also subject to pressure from wind. The strength and nature of the support should be determined by the type of climber specified. Support wires should be firmly fixed to walls at least one brick thick. Strong vine eyes should be driven into the wall (fixing holes must be drilled first) and the wire passed through them, and fixed with fence strainers so that they can be kept tensioned at all times. Climbers that need support should not be fixed to fences that require regular maintenance. Light wire mesh stock fencing is a good support for soft floppy climbers, as the mesh offers better opportunity for tying in than line wires, and when the climber is fully developed the fence will not be conspicuous.

If painting or other maintenance is needed for the wall or fence behind the climbers in special circumstances, they must be fixed with removable ties so that they can easily be laid out flat during maintenance operations.

Pre-planting treatment

If established weeds are present in planting areas, a herbicide to control growing weeds and emergent weeds should be applied before the topsoil is disturbed. Depending on the soil, pre-planting general fertilizers, compost, manure or lime, and soil conditioners such as peat, leaf-mould or sand may be added at the final stage of cultivation. If the existing topsoil is healthy and fertile, it is necessary only to ensure that the top 300 mm is finely cultivated just before planting, and all stones and debris over 50 mm removed.

Immediately before planting the ground is brought to final levels and raked. The eventual level of the planted area should be lower than adjoining paving or grass to avoid run-off in heavy rain, and spill from earth and debris loosened in cultivation, so the soil should be left about 50 mm above the paving or grass to allow for settlement. If the planted area is intended to be higher than the adjoining surface in order to prevent run-off from the paving or road, a kerb and channel must be provided to restrain the soil and to drain the paving.

Stock delivered to the site before planting must be kept moist and shaded at all times, and watered but not waterlogged. Plants should be inspected before planting, and any pot-bound, diseased or undersized stock rejected.

Planting herbaceous plants

The plants must be kept cool, shaded, moist and protected from frost during temporary storage on site. They should not be removed from their containers until planting time, and then all container material must be removed. Occasionally a smaller pot has inadvertently been left inside the larger container, and this should be checked. A planting hole 150 mm wider and 100 mm deeper than the container is formed, and the plants set with the roots well spread out; if there is a best aspect of the plant this should face the front of the planted area. Plants should be evenly spaced over the allocated area at the rate and in the mix specified, avoiding straight lines or patterns. The hole should be backfilled with crumbled topsoil and ameliorants or fertilizers if necessary (or a top-dressing of general compound fertilizer applied at the manufacturer's recommended rate), and the plant firmed in. Some plants may have different fertilizer and soil pH needs, and any such requirements should be specified; this applies particularly to calcifuges. It is advisable to water to field capacity and then to spread a medium or coarse bark chip mulch over the area. Very light peat or bark mulches are liable to displacement by wind and birds.

Each specimen plant or group of plants should be clearly labelled with a weatherproof label, and if a permanent identification label is wanted this should be included in the specification.

Planting climbers

The general principles of planting are the same as those for herbaceous plants. Planting holes for climbers depend on the species, but generally they should be 300 mm wider than the container and at least 500 mm deep. Climbers that need support, particularly if they are not regularly pruned, can become very heavy, and strong wires and fixings should be provided. Closely spaced wires will help to train climbers into the correct shape.

After-care

Plants must be kept weeded and watered until they are established, and the ties on climbers must be checked and adjusted to keep the plants growing correctly; some light pruning may be necessary. Additional top-dressing with general or special fertilizers may be needed for the first season if the soil is light, or alternatively a slow-release granular or capsule fertilizer may be incorporated in the backfill. These are available for one-, two- or three-year release periods. Plants that are dead or dying after practical completion of the contract are the supplier's responsibility.

BRITISH STANDARDS

1831 : 1969 & AMD	Recommended common names for pesticides. This standard is continuously updated
3882 : 1994	Specification for topsoil (bought and sold)
3936 Pt 7 : 1989	Nursery stock. Bedding plants grown in boxes or trays
3936 Pt 10 : 1990	Nursery stock. Ground cover plants
3936 Pt 11 : 1984	Nursery stock. Container-grown culinary herbs
4156 : 1990	Peat for horticultural and landscape work
4428 : 1989 & AMD	Code of practice for general landscape operations
7370 Pt 1 : 1991	Grounds maintenance. General recommendations
7370 Pt 4 : 1993	Grounds maintenance. Maintenance of soft landscape (other than amenity turf)

(BS 3975 is withdrawn.)

DATA

Typical planter-filling mixtures

Good-quality existing or imported topsoil to BS 3882 Premium grade.

Add to light topsoil:	30% peat
Add to medium-light topsoil:	20% peat
Add to medium topsoil:	10% peat
Add to medium-heavy topsoil:	20% peat, 10% sand
Add to heavy topsoil:	25% peat, 15% sand

Topsoil quality to BS 3882: 1994

Topsoil grade	Properties
Premium	Natural topsoil, high fertility, loamy texture, good soil structure, suitable for intensive cultivation
General Purpose	Natural or manufactured topsoil of lesser quality than Premium, suitable for agriculture or amenity landscaping, may need fertilizer or soil structure improvement
Economy	Selected subsoil, natural mineral deposit such as river silt or greensand. The grade comprises two subgrades: 'low clay' and 'High clay', which is more liable to compaction in handling. This grade is suitable for low-production agricultural land and amenity woodland or conservation planting areas.

OUTLINE SPECIFICATION

Planting herbaceous plants

a. *After grading and general cultivation specified elsewhere, bring top 300 mm of topsoil to a fine tilth, incorporating 20% peat or peat substitute, 10% sand, bring to given levels, remove all stones and debris over 50 mm, dig planting holes 150 mm wider and 100 mm deeper than plant container.*

b. *Supply and plant herbaceous plants of the following species from approved suppliers in accordance with planting plan, all in 1 litre pots: Acanthus mollis, Ajuga reptans, Bergenia cordifolia, Brunnera macrophylla, Echinops ritro, Euphorbia amygdaloides, Hosta lancifolia, Vinca minor blue and white, Waldsteinia ternata, remove all container material, reject pot-bound or damaged plants, spread out roots, trim any damaged roots, and backfill with topsoil mixture as above.*

c. *Water to field capacity, top-dress with 7:7:7 general fertilizer at 35 g/m² and mulch planting area with 75 mm bark chips 10–20 mm size, water and weed regularly until completion of contract. Replace failed plants. Container-grown plants to be kept in shaded moist conditions until planted.*

d. *Supply and erect temporary chestnut paling round planted area and remove at completion of contract.*

for cultivation	*D20*
for brick planters	*F10*
for waterproofing planters	*J*
for topsoil, fertilizers, and weedkillers	*Planting generally*
for wire fencing	*Q40*

Planting climbers

a. *Before planting supply and fix 3 mm plastic-coated tensioned galvanized line wires to wall at 600 mm centres vertically, fixed 50 mm away from wall to galvanized steel vine eyes 150 mm long at 750 mm centres.*

b. *After grading and general cultivation specified elsewhere, bring top 300 mm of topsoil to a fine tilth, incorporating 20% peat or peat substitute, 10% sand, and bring to given levels, remove all stones and debris over 50 mm, dig planting holes 300 mm wider than container and 500 mm deep 150 mm away from the wall or clear of shallow foundations, fork over bottom 150 mm of hole.*

c. *Supply and plant climbers of the following species from approved suppliers in accordance with planting plan, all in 2 litre pots:* Clematis orientalis, Clematis tangutica, Clematis *hybrid Ville de Lyon,* Forsythia suspensa, Jasminium officinale, *remove all container material, reject pot-bound or damaged plants, spread out roots, trim any damaged roots, backfill with topsoil mixture as above, and tie in all shoots to each line wire.*

d. *Water to field capacity, top-dress with 7:7:7 general fertilizer at 35 g/m^2 and mulch planting area with 75 mm bark chips 20–40 mm size. Supply and attach permanent plastic weatherproof identification labels to each plant. Water and weed regularly until completion of contract. Replace failed plants. Container-grown plants to be kept in shaded moist conditions until planted.*

for cultivation	*D20*
for brick planters	*F10*
for waterproofing planters	*J*
for topsoil, fertilizers, and weedkillers	Planting generally
for wire fencing	*Q40*

Q31 Planting: bulbs

GENERAL GUIDANCE

This section covers the final preparation of the planted area, pre-planting treatment, and planting in open ground and planters, and the contract period maintenance. Topsoil, soil amelioration, fertilizers, herbicides and pesticides are covered in *Planting generally*.

Sizes of bulbs

For display planting in planters or raised beds, *Narcissus* spp. should be specified as double-nosed top-size bulbs, and *Tulipa* spp. as top size bulbs, but for mass planting or naturalizing mixed sizes as lifted are satisfactory. Spring crocus are supplied in 7–8 cm, 8 cm up, 9–10 cm, and 10 cm up sizes.

Preparation

In planting areas, the soil should be cultivated to a depth of 300 mm, and the soil should be warm and dry in order to ripen the bulbs after flowering. Soil amelioration should be carried out with peat and gritty sand, and – in very poor soils – a little compost or manure. In wet soils, land drainage or at least sand-slitting will have to be provided. Bulbs do not need a highly fertile spoil, as this can lead to over-production of leaf and lowered resistance to disease: *Liliums* in particular do not like fresh animal manure or acid peat. The soil suitable for herbaceous plants has too high a level of nitrogen to suit bulbs; they may be planted in separate areas or planters.

Planting bulbs

Bulbs in planted areas should be placed in small groups where they will not be disturbed by cultivation of the ground, and their location should be marked by permanent labels. For specimen bulbs, planting holes should be formed 25 mm deeper than the correct planting depth; in heavy soils, gritty sand mixed with peat or leaf-mould and bonemeal should be placed at the bottom of the hole. For larger-scale planting it is more economical to form the hole and place the bulb directly on the soil. Bulbs in shrub-planted areas are less likely to be disturbed than bulbs in herbaceous-planted areas, but extra nutrients may have to be provided as low-nitrogen slow-release fertilizers or bonemeal.

Bulbs in grassed areas are usually laid out randomly by throwing them onto the defined area and planting them where they fall, but better results are obtained if a slight regrouping is carried out during the planting. Bulbs should have a plug of turf lifted with a bulb planter. The topsoil is removed to the correct planting depth plus 25 mm, gritty sand and bonemeal are placed at the bottom of the hole, and the bulb is placed at the correct depth and covered with a leaf-mould/sand or peat/sand mixture. The turf plug is replaced and firmed down but not compacted. Care must be taken to ensure that stones are not left on the turf after planting. For mass planting it may be cheaper to lift patches of turf, plant the bulbs in groups, and replace the turf.

According to the landscape design, wild flora seed can be sown in the grassed bulb areas to conceal the dying foliage of the bulbs. *Narcissus* spp. tend to spread mainly by bulb division, but other species such as snowdrops spread by seeding as well, and will extend the bulb area.

After-care

Bulbs in grass should be left for at least six weeks after flowering before the grass is cut, and the client should be advised of this programme; if grass cutting is not carefully supervised it is advisable to contain naturalized bulbs within a clearly defined area, and to seed the area with low-maintenance grass species. A potash top-dressing will help bulb growth without over-stimulating grass growth.

Generally, bulbs that are normally lifted in the autumn, such as gladioli and hyacinths, are not suitable for landscape work, and the hardier species are usually specified.

BRITISH STANDARDS

1831 : 1969 & AMD	Recommended common names for pesticides. This standard is continuously updated
3882 : 1994	Specification for topsoil (bought and sold)
3936 Pt 9 : 1987	Nursery stock. Bulbs, corms and tubers
4156 : 1990	Peat for horticultural and landscape work
4428 : 1989 & AMD	Code of practice for general landscape operations
7370 Pt 1 : 1991	Grounds maintenance. General recommendations
7370 Pt 4 : 1993	Grounds maintenance. Maintenance of soft landscape (other than amenity turf)

(BS 3975 now withdrawn.)

DATA

Suggested planting depths for bulbs and corms

Roots below soil, corms exposed
Cyclamen spp.
Iris germanica

5 cm deep to top of bulb
Anemone blanda
Crocus spp.
Crocosmia
Galanthus spp.
Iris reticulata
Ornithogalum
Muscari
Scilla

12 cm deep to top of bulb
Darwin tulips
Dutch iris
Tulipa fosteriana
Hyacinth
Narcissus spp.
Tulipa turkestanica

15 cm deep to top of bulb
Lilium spp.

Typical planter-filling mixtures

Good-quality existing or imported topsoil to BS 3882 Premium grade.

Add to light topsoil:	30% peat, 600 g/m³ bonemeal
Add to medium-light topsoil:	20% peat, 600 g/m³ bonemeal
Add to medium topsoil:	10% peat, 600 g/m³ bonemeal
Add to medium-heavy topsoil:	20% peat, 10% sand, 600 g/m³ bonemeal
Add to heavy topsoil:	25% peat, 15% sand, 600 g/m³ bonemeal

Topsoil quality to BS 3882: 1994

Topsoil grade	Properties
Premium	Natural topsoil, high fertility, loamy texture, good soil structure, suitable for intensive cultivation
General Purpose	Natural or manufactured topsoil of lesser quality than Premium, suitable for agriculture or amenity landscaping, may need fertilizer or soil structure improvement
Economy	Selected subsoil, natural mineral deposit such as river silt or greensand. The grade comprises two subgrades: 'low clay' and 'high clay', which is more liable to compaction in handling. This grade is suitable for low-production agricultural land and amenity woodland or conservation planting areas.

OUTLINE SPECIFICATION

Bulbs in grass

a. *Define area to be planted with a line of sand, scatter bulbs casually over area to be planted at average of 20 bulbs/m², remove turf plug with bulb planter, form planting holes, to given depth plus 25 mm, place 25 mm 1:2 gritty sand/peat and bonemeal at 600 g/m³ mixture in hole, plant winter flowering crocus species, miniature daffodils* Narcissus cyclamineus *'Jack Snipe', single* Galanthus nivalis.

b. *Backfill to top of bulb with peat/sand/bonemeal mixture as above, replace turf plug and firm lightly.*

c. *Remove all stones and soil from turf and leave clean. Grass not to be cut until six weeks after end of flowering. Replace failed bulbs.*

for topsoil, fertilizers, and weedkillers   Planting generally

Bulbs in planted areas

a. *Form holes in planted area where shown on planting plan to given depth plus 25 mm, place 25 mm 1:2 gritty sand/peat and bonemeal at 600 g/m³ mixture in hole, plant double-nosed top-size bulbs of* Narcissus *cultivars Division 1a 'King Alfred', 1b 'Magnet', 2b 'Flower Record', Division 7 'Trevithian', Division 9 'Pheasant Eye'.*

b. *Backfill with existing topsoil and firm lightly.*

c. *Supply and insert permanent identification labels. Replace failed bulbs.*

for topsoil, fertilizers, and weedkillers   Planting generally

Bulbs in planters

a. Lay 100 mm of 25 mm coarse neutral aggregate, lay geotextile layer, fill planter with moist medium-light topsoil mixed with 20% peat, 600 g/m³ bonemeal to within 200 mm of finished soil surface.

b. Place bulbs on soil as shown on planting plan and plant Fritillaria imperialis *'Lutea maxima, paeony-flowered tulips 'Angelique',* Hyacinthus orientalis.

c. Backfill planter to finished levels with moist peat/sand/bonemeal mixture as above. Replace failed bulbs as necessary.

for topsoil, fertilizers, and weedkillers   Planting generally

Q31 Planting: aquatics

GENERAL GUIDANCE

This section covers the final preparation of the aquatic area, pre-planting treatment, planting in water and wetland, and the contract period maintenance.

Aquatic plants

Aquatic plants may be classified as **deep submerged plants** (500–1000 mm below surface), **marginal plants** growing at the edge of the water (moist soil, 100 mm deep), **floating plants**, which live freely in the water, and **wetland plants**. They are usually supplied in basket containers, except for floating plants, which are supplied loose. There is no British Standard for aquatic plants, so it is advisable to inspect the supplier's stock before specifying; plants should be at least a year old, healthy, well grown, not pot-bound, and free of disease and pests, with a well-developed root system. On no account should *Crassula helmsii* (*Tillaea recurva*) be planted, as this alien species is a serious pond and lake pest, which will rapidly dominate any pond or wetland.

Marginal plants

These are supplied either in crates or in clumps, and should be of the quality given for aquatics.

Preparation

Although many species are intolerant of changes in the water level, exact levels and seasonal variations cannot always be predicted, such as local water-table variations and reservoirs. Aquatic planting can be done by trial and error, or by selecting plants with a wide tolerance. Some plants, such as watercress, do better in running water, and some prefer still water. The pH and other chemical properties of the water are also important, as plants such as brooklime do not tolerate acid water well, and the water should be analysed before deciding on the species of plants. If fish are to be established in the water it is important to know whether they are plant eaters or carnivorous, as this will affect the whole ecology of the water feature, and if sport fishing is to be part of the use, the types of aquatic plant must be selected accordingly. In particular, reed beds will expand rapidly unless cut regularly. The pond or lake may be a formal design, or a natural ecologically balanced water feature, and the method of planting will vary accordingly.

Planting

The planting season for aquatics and marginal plants is May or June to September; the water will be at its lowest level during this period, and allowance should be made for winter levels.

The plants must be secured in their baskets by permanent mesh or line ties, and baskets in marginal areas must be weighted to ensure that they do not move from their positions. Where wildlife is using the water it is advisable to leave a sloping hard-surfaced access and escape area; this can be laid with heavy cobbles or natural stone setts. For formal flat-bottomed pools, decorative perforated terracotta or clay pipes set on end in groups can be used as planters; the depth of planting can be controlled by the height of the unit and the depth of soil fill. Such planters also serve to control the spread of aggressive species. Some shelter for fish and amphibians should also be provided.

Marginals and wetland plants are planted in notches cut in the soil and carefully firmed into place; great care is necessary to avoid cutting liners or clay pond bases. Fertilizers should not be applied to marginal plants, because they may leach into the water and cause over-production of algae. For this reason high-fertility topsoil should not be used for marginal planting, and Economy grade topsoil to BS 3882 will be suitable if the pH is satisfactory. The choice of 'low clay' or 'high clay' subgrade will depend on the planting design and the layout of the water feature.

BRITISH STANDARDS

3882 : 1994   Specification for imported topsoil
4428 : 1989 & AMD   Code of practice for general landscape operations
7370 Pt 1 : 1991   Grounds maintenance. General recommendations

There is no British Standard specifically for aquatic plants.
(BS 3975 now withdrawn.)

DATA

Typical aquatic and marginal plants

Submerged plants
Callitriche stagnalis
Ranunculus aquatilis
Sagittaria sagittifolia

Free floating plants
Ceratophyllum demersum
Hydrocharis morsus-ranae
Nymphaea alba (rooted)

Emergent plants
Butomus umbellatus
Phragmites australis
Rumex hydrolapathum

Marginal plants
Carex nigra
Iris pseudacorus
Typha angustifolia
Caltha palustris
Geum rivale
Veronica beccabunga

Topsoil quality to BS 3882 : 1994

Topsoil grade	Properties
Economy	Selected subsoil, natural mineral deposit such as river silt or greensand. The grade comprises two subgrades: 'low clay' and 'high clay', which is more liable to compaction in handling. This grade is suitable for low-production agricultural land and amenity woodland or conservation planting areas.

OUTLINE SPECIFICATION

Aquatics

a. *Supply and place 1* Callitriche stagnalis, *2* Sagittaria sagittifolia *and 3* Ranunculus aquatilis *per 10 m² of water surface.*

b. *Supply and plant 1* Ceratophyllum demersum *and 1* Nymphaea alba *per 10 m² of water surface, secure plants in crates with non-biodegradable wire or plastic ties, trim off all damaged roots and shoots, weight crates with stone or mature concrete blocks, place in position according to required depth. Ensure that given water level and pH are maintained during establishment of stock, and replace failed plants.*

Marginals

Supply and plant 10 each of Phragmites australis, Iris pseudacorus, Rumex hydrolapathum, Typha angustifolia, Caltha palustris *and* Acorus calamus *as shown on planting plan, trim off all damaged roots and shoots, form notch in soil and place plants with roots spread out, firm back soil.*

Sharp or pointed tools must not be used to form notches. Ensure that given water level and pH are maintained during establishment of stock, and replace failed plants.

Q40 Fences generally

GENERAL COMMENTS

All types of fences require frequent regular maintenance, and the timing of maintenance operations should be discussed before the landscape designer specifies the fences. They are also more easily damaged than walls, and frequent checks should be made to note and repair any loose tension wires, other wires, cracked boards etc. that could enable animals to escape, or cause damage to people and animals.

Fencing, including gates, railings, and all other types of fencing, in timber, steel, cast iron or concrete or a mixture of two or more of these materials, should be considered as an important part of any project, and the landscape designer should give the client a clear understanding of the uses to which the various types of fence can be put, as well as some indication of their economic life in terms of capital outlay and running costs.

REGULATIONS AND PLANNING REQUIREMENTS

Before starting on a project, check any legal restraints on the height or type of fencing. Planning permission may be needed for boundary fences in National Parks, AONBs and all types of conservation areas, and local planning authorities may impose conditions as part of the 'Reserved Matters' matters conditional on the grant for full planning permission. Highway authorities can impose limitations on fences bordering a highway, in particular where sight-lines at road junctions are involved. Private estates may also impose conditions over the type and height of fences used on their land; this is often so on housing sites developed within land under their overall control.

Gates may well be required for access by emergency vehicles, and the client or the landscape designer should be aware of any requirements imposed by the local authority for access by the fire brigade, ambulance service or police.

Any project that involves keeping wild animals has to be licensed, and is subject to stringent conditions; riding establishments and boarding kennels also have to be licensed by the local authority, and they too are likely to impose conditions on the fencing commensurate with the safety of the public.

SAFETY AND THE DUTY OF CARE

Whether or not the client asks for safety or security fencing, it must be remembered that the Health and Safety at Work etc. Act and the Occupier's Liability Acts place a duty of care on the owner and/or occupier, so that failure to protect the occupants, the workforce or the public at large from foreseeable hazards is an offence. If negligence on the part of the landscape designer can be proved he may find himself in considerable trouble if he has not considered the problems of hazards carefully. The Acts also apply to sites under construction.

Playground fencing and gates, and any fence in grounds housing the mentally handicapped, require particular care in their design, erection and maintenance. Many types of fencing may be used, but they should aim to comply with the following criteria:

- There should be no spikes, sharp edges or arrises where children can cut or impale themselves.
- They should be unclimbable, or so low that children cannot fall off and break or sprain limbs.
- Gaps between wires, boards or rails should be no more than 100 mm, to prevent heads from getting trapped.
- Bolt heads and nuts should be sunk, and preferably removable only with special tools.
- Paint or preservative should be at least non-toxic and non-staining.
- Gates, particularly where very young children play, should not swing or slam so as to trap fingers or toes against posts and walls or between pairs of gates.
- Barriers must be placed between the children's access and the road or footpath to prevent them from running directly into danger.
- Boundary fences next to roads should be unclimbable, and where necessary they should have a mesh capable of stopping golf balls or other small projectiles.

SECURITY

Visibility is an important aspect of security, and in many ways an open wire or cast iron fence provides better security than a wall of equivalent height, as it provides better surveillance of the area. It is of course important to know whether a fence is a barrier intended to keep the public in or out. Where intruders are a mild nuisance a 1.8 m fence with one line of barbed wire and a field gate with a padlock may suffice. Where the installation behind the barrier has to be very secure, the client should be informed that the budget allocation is likely to be high, and that it is advisable to call in a security specialist; they can design and erect a complete electronic security system combined with 3.5 m high-tensile welded fencing and electronically controlled gates.

Where hedges are planted to provide an animal-proof barrier, it is advisable to erect a stock-proof fence before planting the hedge; this gives immediate security, and will not be visible once the hedge has grown to a good solid mass. Where stock hedges are planted, fencing may be necessary to stop stock from eating a new hedge.

BRITISH STANDARDS

British Standards cover fencing in great detail, and all standardized fences have reference numbers, which can be used to specify a fence type without further detail. It is not, however, sufficient to specify a fence 'to be to BS 1722 Pt 3, type XXX', as there are many options in each Standard, and the landscape designer must state which of these he wishes to form part of the contract. There are also many firms producing perfectly adequate fencing that is not exactly to any British Standard; this does not necessarily mean that their products are inferior. Fences constructed to MAFF (Ministry of Agriculture Fisheries and Food) Standards are a suitable alternative for agricultural fences.

Refer to subsections:

for hedges	Q31
for playgrounds	Q50

Q40 Fences and gates: temporary protective fencing

GENERAL GUIDANCE

This type of fencing is used to protect whole construction sites, or specific areas where existing or new trees, shrubs, hedges, grassed areas and similar areas need protection both during and after completion of the contract.

Cleft chestnut fencing

Timber pale fencing is most commonly used for temporary site protection where vandalism is not a serious threat. It is also used as an economical fencing material around allotments, gardens or parkland. The distance between pales is related to the degree of protection required, closer pales and posts giving the greater protection. Cleft chestnut pale fences are specified in BS 1722 Pt 4; the pales are woven with galvanized wire to form a continuous paling fence, which is delivered to site in rolls. Although tension line wires are not normally used, the landscape designer should specify them to support the fence if there is likely to be pressure on it.

Welded steel mesh panels

Welded steel mesh wire panels are now frequently used to protect sites, particularly in urban areas where vandalism is more prevalent. They are also used to encompass areas within country house gardens and so forth where fairs, exhibitions and festivals are held. This proprietary fencing comes from the manufacturer in ready-made panels, which are fitted in heavy precast, pre-holed concrete bases laid out in straight lines or angled to the required shape.

BRITISH STANDARDS

1485 : 1983 (1989)	Zinc coated hexagonal steel wire netting
1722 Pt 2 : 1989	Rectangular wire-mesh and hexagonal wire netting fences
1722 Pt 4 : 1986	Cleft chestnut pale fences
1722 Pt 14: 1992	Open mesh steel panel fences
4102 : 1990	Steel wire and wire products for fences

DATA

Spile fencing

Cleft chestnut spiles 50 mm dia. each driven into the ground at 150 mm centres and secured with two or three rows of 4 mm seven-ply line wire according to height.

Fence posts similar to those used for chestnut pale fencing are used at 3 m intervals.

Cleft chestnut pale fences to BS 1722 Pt 4 : 1986

Pale length (mm)	Pale spacing (mm)	Wire lines	Application
900	75	2	Temporary protection
1050	75 or 100	2	Light protective fences
1200	75	3	Perimeter fences
1350	75	3	Perimeter fences
1500	50	3	Narrow perimeter fences
1800	50	3	Light security fences

Posts for cleft chestnut fencing

Pale length (mm)	Concrete posts (mm)	Timber posts (mm)
900	1570 × 100 × 100	1500 × 190–230 girth
1050	1720 × 100 × 100	1650 × 190–230 girth
1200	1870 × 125 × 125	1800 × 190–230 girth
1350	2020 × 125 × 125	1950 × 190–230 girth
1500	2320 × 125 × 125	2250 × 230–250 girth
1800	2620 × 125 × 125	2550 × 190–230 girth

Welded steel panel fences to BS 1722 Pt 14

Category 1:

Height (mm)	Mesh size centre to centre of wires (mm)
Up to 1800	200 high × 50 wide
	50 × 50
	75 × 25
	50 × 12.5
	12.5 × 50
	12.5 × 12.5

OUTLINE SPECIFICATION

Temporary fencing

a. *Supply and erect temporary protective fencing to BS 1722 Pt 4, 1200 mm high cleft chestnut pale fencing, pales at 75 mm spaces pointed at top on three lines of wire.*

b. *Posts to be larch posts 1800 mm × 200 mm girth driven 600 mm into firm ground at 3.0 m centres, and removed after planting is established. Line wire only to be tensioned.*

Alternative specification

a. *Supply and erect cleft chestnut pale fencing to BS 1722 Pt 4, 1500 mm high, pales at 50 mm spaces pointed at top on three lines of wire.*

b. *Posts to be concrete posts 2320 mm × 125 mm × 125 mm set 600 mm into ST2 concrete base at 3.0 m centres. Line wires only to be tensioned to 2320 mm × 125 mm × 125 mm straining posts and struts at 50.0 m centres.*

Welded steel mesh panel fencing

Supply and erect proprietary welded steel mesh panel fencing complying with category 1 of BS 1722 Pt 14, round areas indicated on drawings, maintain in good condition and remove after completion of contract.

Q40 Fences and gates: wire fencing

GENERAL GUIDANCE

This section covers light wire fencing used for demarcating boundaries in fields, gardens and parks, as well as those used to contain stock and to keep rabbits and deer from gardens and plantations.

Wire fencing

This is the most economical type of light fencing. The common types are: plain post-and-wire (stob-and-wire); chain-link or stock fencing on timber, steel or concrete posts. Included in this category are rabbit-stop, deer-stop fences and the like, used to guard planting. Post centres for normal usage vary from post centres at 3.0 m to 3.5 m, while security fences 2.4 m or more high require posts at a minimum of 1.6 m centres: here the closer centres help to prevent the fence from being ruptured by ramming. 'Droppers' or 'prick posts' are light posts that help to brace the line wire, but are not fixed into the ground; where they are used, post centres for light wire fencing may be extended.

Forestry fencing

Forestry fencing is designed to prevent access of both public and stock to plantations. Rabbit fencing is essential for the protection of newly planted woodland or ornamental shrubs; here it is critical that the wire is buried 150 mm below ground and turned out another 150 mm to prevent burrowing. Deer-stop fences are also required to protect gardens and other planted areas that need protection from roe, fallow and other small deer. In areas such as Exmoor and the Scottish Highlands, where there are red deer, substantial fencing is required, and droppers are essential to prevent their antlers from being trapped between loose wires, as well as to brace the fencing. Keepers will require access to any protected woodland or forest, and deer ladders should be provided for this purpose.

Field fencing or stock fencing

This is a woven or welded square mesh galvanized wire fencing material in a range of heights and mesh sizes. Though not normally used for ordinary urban and commercial landscape contracts, it is used in rural landscape schemes. It should be used where farm animals form part of an attraction in a country park, and the correct type and height of field fencing for the particular animals kept should be specified. Where zoo or dangerous animals are kept, the site or even the garden housing them has to be licensed, and strict controls are required. The owner will be aware of this, and will instruct the landscape designer on the type of fencing stipulated in the licence. The main types of stock fence are cattle fencing 1100 mm high, pig and lamb fencing 800 mm high, and sheep fencing 900 mm high; all have wires closer together at the bottom than at the top. Field fencing also makes an excellent and economic support for young hedging, climbing plants and shrubs.

Chain-link fencing

This is probably the most commonly used wire fencing for housing estates and local authority parks. It is relatively unclimbable, easily and rapidly erected, and readily available from stock. It does, however, have the unfortunate habit of bunching up if the adjacent section is damaged; stretching, bulging and sagging are also faults in chain-link fencing. Minor repairs are rarely entirely successful, and replacement of a whole section between posts is the best option. Tubular or steel angle framed gates with infill panels to match the chain-link specified are available.

Chain-link mesh is specified in BS 4102 : 1986 and as chain-link fencing in BS 1722 Pt 1 : 1986; these British Standards give detailed specifications for the wire itself, its coatings, the sizes and quality of timber, concrete and steel posts, details of all fixings required, and methods of construction. Reference numbers are given for a series of standardized heights and weights of fencing, so that the landscape designer, who is quite certain that he has specified the correct reference, need only quote a number to cover the height and weight of the chosen fence.

Chain-link mesh is now manufactured in an ever-increasing range of gauges, from the standard light 50 mm mesh suitable for most garden boundaries to heavy-duty 25 mm mesh for security fencing. Material used for this can be supplied in galvanized, plastic-coated, or galvanized and plastic-coated forms, mild steel or high-tensile steel for security applications. Light plastic-coated galvanized wire is 2.65/2.0 mm thick, medium is 3.15/2.24mm, heavy is 3.55/2.5mm, extra heavy is 4.0/3.0mm, and super weight is 4.75/3.55mm.

Posts

Posts for all types of boundary wire fencing may be timber, concrete, mild steel or high-tensile steel in angle or tube form. Timber posts are cheap, comparatively short lived but easily erected by driving them into the ground. BS timber posts for boundary fences in treated softwood are normally round, square or sectored; heavier posts to MAFF standard are required for farm and conservation grant contracts.

Precast reinforced concrete posts are the most frequently used alternative, because they are available in a good selection of heights and various finishes, and have preformed line wire holes; unless run into by a vehicle or deliberately smashed, they are reasonably damage resistant.

Mild steel tubular section posts are used mainly for security fencing and tennis courts, and when made in the form of high-tensile posts they are very difficult to cut or break. For very high fencing, steel lattice posts are more satisfactory. They are usually galvanized and/or plastic coated for weather protection. For ordinary boundary fences up to 1500 mm high the landscape designer should specify mild steel rolled angles supplied pre-drilled or notched to take the line wires.

BRITISH STANDARDS

1722 Pt 1 : 1986	Chain-link fences and gates and gate posts for fences up to 1.8 m high
1722 Pt 2 : 1989	Rectangular wire-mesh and hexagonal wire netting fences
1722 Pt 3 : 1986	Strained wire fences
4102 : 1990	Steel wire and wire products for fences

DATA

Chain-link fencing

Gates for proprietary fencing are supplied as part of the fencing contract. Chain-link fencing for a range of applications:

Situation	Type	Mesh (mm)	Height (mm)	Lines
Gardens	Light	50	900	2
	Medium	50		
Playgrounds	Heavy	50	1200	3
	Extra heavy	50		
Highways and factories	Medium,	40/50	1400	3
	Heavy	40/50		
	Extra heavy	50		
Playing fields	Extra heavy	50	1400	3
Industrial and security	Heavy	50	1800	3
	Extra heavy,	50		
	Super weight	50		
Railway	Heavy	40	1800	3
Tennis courts	Light	45	2750/3600	4/5
	Medium	45/50	2750/3600	4/5
	Heavy	50	2750/3600	4/5

Steel posts to BS 1722 Pt 1

Rolled steel angle iron posts for chain-link fencing (all dimensions in mm):

Posts	Fence height	Strut	Straining post
1500 × 40 × 40 × 5	900	1500 × 40 × 40 × 5	1500 × 50 × 50 × 6
1800 × 40 × 40 × 5	1200	1800 × 40 × 40 × 5	1800 × 50 × 50 × 6
2000 × 45 × 45 × 5	1400	2000 × 45 × 45 × 5	2000 × 60 × 60 × 6
2600 × 45 × 45 × 5	1800	2600 × 45 × 45 × 5	2600 × 60 × 60 × 6
3000 × 50 × 50 × 6*	1800	2600 × 45 × 45 × 5	3000 × 60 × 60 × 6

*With arms

Circular hollow section and rectangular hollow section posts are also covered in the British Standard.

Concrete posts to BS 1722 Pt 1

Concrete posts for chain-link fencing (all dimensions in mm):

Posts and straining posts	Fence height	Strut
1570 × 100 × 100	900	1500 × 75 × 75
1870 × 125 × 125	1200	1830 × 100 × 75
2070 × 125 × 125	1400	1980 × 100 × 75
2620 × 125 × 125	1800	2590 × 100 × 85
3040 × 125 × 125*	1800	2590 × 100 × 85

*With arms

Rolled steel angle posts to BS 1722 Pt 2

Rolled steel angle posts for rectangular wire-mesh (field) fencing (all dimensions in mm):

Posts	Fence height	Strut	Straining post
1200 × 40 × 40 × 5	600	1200 × 75 × 75	1350 × 100 × 100
1400 × 40 × 40 × 5	800	1400 × 75 × 75	1550 × 100 × 100
1500 × 40 × 40 × 5	900	1500 × 75 × 75	1650 × 100 × 100
1600 × 40 × 40 × 5	1000	1600 × 75 × 75	1750 × 100 × 100
1750 × 40 × 40 × 5	1150	1750 × 75 × 100	1900 × 125 × 125

Concrete posts to BS 1722 Pt 2

Concrete posts for rectangular wire-mesh (field) fencing (all dimensions in mm):

Posts	Fence height	Straining post	Strut
1270 × 100 × 100	600	1420 × 100 × 100	1200 × 75 × 75
1470 × 100 × 100	800	1620 × 100 × 100	1350 × 75 × 75
1570 × 100 × 100	900	1720 × 100 × 100	1500 × 75 × 75
1670 × 100 × 100	600	1820 × 100 × 100	1650 × 75 × 75
1820 × 125 × 125	150	1970 × 125 × 125	1830 × 75 × 100

Timber posts to BS 1722 Pt 2

Timber posts for wire-mesh and hexagonal wire netting fences.

Round timber for general fences
All dimensions in mm:

Posts	Fence height	Straining post	Strut
1300 × 65 dia.	600	1450 × 100 dia.	1200 × 80 dia.
1500 × 65 dia.	800	1650 × 100 dia.	1400 × 80 dia.
1600 × 65 dia.	900	1750 × 100 dia.	1500 × 80 dia.
1700 × 65 dia.	1050	1850 × 100 dia.	1600 × 80 dia.
1800 × 65 dia.	1150	2000 × 120 dia.	1750 × 80 dia.

Squared timber for general fences
All dimensions in mm:

Posts	Fence height	Straining post	Strut
1300 × 75 × 75	600	1450 × 100 × 100	1200 × 75 × 75
1500 × 75 × 75	800	1650 × 100 × 100	1400 × 75 × 75
1600 × 75 × 75	900	1750 × 100 × 100	1500 × 75 × 75
1700 × 75 × 75	1050	1850 × 100 × 100	1600 × 75 × 75
1800 × 75 × 75	1150	2000 × 125 × 100	1750 × 75 × 75

Round timber for high-tensile fences
All dimensions in mm:

Posts	Fence height	Straining post	Strut
1500 × 50 dia.	750	1750 × 100 dia.	1750 × 80 dia.
1800 × 80 dia.	900	2000 × 100 dia.	2000 × 80 dia.
1800 × 80 dia.	1050	2300 × 100 dia.	2000 × 80 dia.
2500 × 50 dia.	1800	2500 × 100 dia.	2500 × 80 dia.
2600 × 80 dia.	1800	2800 × 120 dia.	2500 × 100 dia.

Concrete mixes to BS 5328 Pt 1 & AMD

Application	Standard mix	Designated mix	Slump (mm)
Mass concrete fill and blinding	ST 2 (C10P)	GEN 1	75
Mass concrete foundations in non-aggressive soils	ST 2 (C10P)	GEN 1	75

Concrete mixes to BS 5328 Pt 1 & AMD *continued*

Application	Standard mix	Designated mix	Slump (mm)
Trench fill foundations in non-aggressive soils	ST 2 (C10P)	GEN 1	125
Foundations			
in Class 2 sulphate soils	–	FND 2	75
in Class 3 sulphate soils	–	FND 3	75
in Class 4A sulphate soils	–	FND 4A	75
in Class 4B sulphate soils	–	FND 4B	75

Traditional concrete mixes

Application	Name of mix	Cement:aggregate:sand (parts by volume)
Setting posts, filling weak spots in ground, backfilling pipe trenches:	one, two, three	1:2:3
	or	or
	one to five all-in	1:5
Paving bases, bedding manholes, setting kerbs, edgings, precast channels:	one, two, four	1:2:4
	or	or
	one to six all in	1:6
In-situ concrete paving:	one, one & half, two & half	1:1.5:2.5

'All-in' is mixed aggregate as dug from the pit.

OUTLINE SPECIFICATION

Chain-link fence to BS 1722 Pt 1

a. *Supply and erect chain-link fencing to BS 1722 Pt 1 900 mm high, plastic-coated galvanized 3 mm gauge wire 51 mm mesh, galvanized line and tying wires.*

b. *Form post holes and erect 1500 mm × 40 mm × 40 mm × 5 mm ms angle-iron posts at 3.0 m centres with stretcher bars and 50 mm × 50 mm × 5 mm straining posts and 40 mm × 40 mm × 5 mm struts at 50 m centres set in C7P concrete.*

All as above but with 1570 mm × 100 mm × 100 mm precast concrete posts and 1500 mm × 75 mm × 75 mm struts at 50 m centres set in C7P concrete.

Chain-link gate

Supply and erect gate 900 mm wide of chain-link fencing to match above, posts of 80 mm × 80 mm × 3.6 mm hollow steel section including one pair galvanized hinges and padbolt field fencing to BS 1722 Pt 2.

Rectangular wire-mesh fence

a. Supply and erect rectangular wire-mesh fencing to BS 1722 Pt 2 1150 mm high medium weight Type C.
b. Form post holes and erect 1820 mm × 125 mm × 125 mm concrete posts and straining posts 1970 mm × 125 mm × 125 mm and struts 1830 mm × 75 mm × 100 mm at 50 m centres set in C7P concrete.

Rabbit-stop fencing

a. Supply and erect rabbit stop fencing 900 mm high of 1200 mm × 31 mm mesh galvanized wire netting on two 2.5 mm line wires, with 150 mm of netting buried below ground and 150 mm turned outwards from protected area.
b. Supply and erect 1600 mm × 75 mm diameter treated softwood posts at 3.5 m centres driven 700 mm into firm ground with 2000 mm × 125 mm × 125 mm straining posts at 30 m centres with struts 2000 mm × 75 mm × 75 mm.

Deer-stop fencing for roe and muntjac deer

Supply and erect high-tensile rectangular mesh fencing 1800 mm high specially manufactured for protection against the smaller deer on 2500 mm × 80 mm treated round softwood posts at 15.0 m centres and 2500 mm × 100 mm dia. straining posts at 60.0 m centres driven into firm ground.

Deer-stop fencing for red deer

Supply and erect high-tensile wire fencing 2000 mm high of 5 no. lines of 2.5 mm galvanized line wire and 5 no. lines of two-ply galvanized barbed wire fixed to 2800 mm × 150 mm dia. treated softwood posts driven into firm ground at 3.0 m centres, with 25 mm × 38 mm timber droppers 1000 mm long stapled to the wire at 1500 mm centres.

Forestry fencing

Supply and erect forestry fencing of three lines of 3 mm plain galvanized wire stapled to 1700 mm × 65 mm dia. treated softwood posts at 5.0 m centres with 1850 mm × 100 mm dia. straining posts and 1600 mm × 80 mm dia. struts at 50.0 m centres driven into firm ground.

Q40 Fences and gates: security fencing

GENERAL GUIDANCE

This section deals with steel and wire fencing intended to protect property against intruders and to prevent access to potentially dangerous areas. These fences are used for industrial and public building contracts, and private estates. The security fencing described here is not intended to be used for areas or installations likely to be targets for terrorists and other dedicated intruders; this requires highly specialized fencing and surveillance equipment, and is contracted out to specialist firms.

Barbed wire or tape

In situations where security is very important, barbed wire or the much more vicious barbed tape with closely spaced razor blades along a flat steel tape can be used, but as the latter in particular can cause serious injury, the landscape designer must specify that it is fixed no lower than 1800 mm above ground level in any area where the public can gain access and it must not project over a public path. Because claims for damages are not uncommon, the landscape designer should ensure that the specification includes a clause requiring barbed tape to be accompanied by warning notices 'DANGER THIS PROPERTY IS PROTECTED BY BARBED TAPE' that are visible at all times (night lighting may be required). An expensive though safe alternative is to provide an outer perimeter fence on the public side of the security fence.

General requirements for security fences and gates

- Narrow mesh that is proof against an intruder's toe or foothold.
- Barbed or jagged tops to the fence can be used to prevent intruders climbing over them; where several strands of barbed wire are used on top of the fence, droppers must to be fitted in order to prevent the wires from being squeezed together. Any barbed wire must be at least 2000 mm above ground level.
- Mesh should be buried or set on a concrete sill at least 125 mm wide × 150 mm deep to prevent the fence from being kicked in or dug under.
- Right angles or obtuse angles should be used whenever the fence changes direction in order to prevent 'chimney' climbing.
- Signs, street furniture, trees or other aids to climbing should be kept away from the fence, and no vehicles should be allowed near it. If necessary, vehicle barriers may have to be used in front of the fence to keep vehicles away. Signboards should not be fixed to the outside of the fence, as they are likely to provide a foothold.
- Electronic security systems require an unobstructed view along the length of fence that they serve.
- All fastenings should be hammered over, or have clutch heads to prevent unfastening.

- Gates must be at least as secure as the fencing; this point is often over-looked.
- Any opening in the fence must be as well protected as the fence itself.

Chain-link

The simplest wire fence that gives protection against casual intruders is an 1800 mm or 2000 mm chain-link fence on concrete or steel posts with straight or cranked arms carrying two or three lines of two-strand galvanized barbed wire. These 'anti-intruder' fences are described in BS 1722 Pt 10. This fence can be penetrated by cutting a flap in the chain link with heavy wire cutters, and greater security is provided by welded mesh.

Welded mesh

Welded mesh panel fences are normally supplied complete as a proprietary fencing system. The four categories of welded mesh panel fencing to BS 1722 Pt 14 are given below. The lighter forms of welded mesh panel fencing can be used to keep small animals and birds from straying, as well as for for temporary protective fencing around sites. No sizes of posts are given in the Standard, but posts of any material must have a moment of inertia not less than 12.5 cm⁴, and the landscape designer should ensure that the fence manufacturer can confirm that the posts supplied comply with this requirement.

Palisade

Palisade security fences provide more serious security, and are covered by BS 1722 Pt 12 : 1990, which classifies them as General Purpose (GP) fences up to 2400 mm high, and Special Purpose (SP) fences 2400, 3000 and 3600 mm high. This section deals with SP fences only. The British Standard covers all fixings and fasteners in great detail.

BRITISH STANDARDS

1722 Pt 10 : 1990	Anti-intruder chain-link fences
1722 Pt 12 : 1990	Steel palisade fences
1722 Pt 14 : 1992	Open mesh steel panel fences

DATA

Anti-intruder chain-link fences to BS 1722 Pt 10 : 1990

Chain-link to be 40 mm or 50 mm mesh.
Welded mesh to be 50 mm × 50 mm × 3 mm or 75 mm × 25 mm × 3 mm vertical pattern.
Droppers are fitted in each bay to the barbed wire on extension arms.
Intermediate posts to be at not more than 3.0 m centres.
Straining posts to be at not more than 69.0 m centres.

Sizes of posts for intermediate weight chain-link fencing

All dimensions in mm:

	Intermediate posts	Straining posts	Struts
Concrete	3150 × 125 × 100	3150 × 150 × 150	3250 × 100 × 100
RSJ	3200 × 50 × 50 × 6	3200 × 60 × 60 × 8	3200 × 45 × 45
Tubular	3200 × 48.3 × 3.2	3200 × 60.3 × 4	3200 × 48.3 × 3.2
Rectangular	3200 × 50 × 50 × 3	3200 × 60 × 60 × 3	3200 × 50 × 50 × 3

Sizes of gate posts

All dimensions in mm:

	Width of gates		
	Up to 4.0 m	Up to 6.0 m	Up to 10.0 m
RSJ	127 × 114	152 × 127	203 × 203
Tubular	139.7 × 5	168.3 × 5	219.1 × 6.3
Rectangular	100 × 100 × 4	150 × 150 × 5	180 × 180 × 6.3

All posts should be set 750 mm deep in concrete:

- Intermediate post holes not less than 75 mm all round the post for square posts and 300 mm dia. for round posts.
- Straining post holes not less than 450 mm × 450 mm for square posts and 450 mm dia. for round posts.
- Gate post holes 450 mm × 450 mm.

Extension arms for barbed wire may be straight or angled at 40–45° to the vertical. They may be integral with the post or separate; separate arms are 45 mm × 45 mm × 5 mm steel angles.

The bottom of the fence shall be either:

- buried 300 mm deep in the ground *or*
- fixed with hairpin staples cast into a concrete sill 125 mm × 225 mm deep *or*
- fixed with hairpin staples grouted into concrete or rock.

Welded mesh panel fences to BS 1722 Pt 14

All fencing parts must resist a load of 700 N for each metre length at a height of 1.25 m above ground level. All dimensions in mm.

	Height	Mesh size centre to centre of wires	
Category 2 (normal security)	2400	75 high	× 25 wide
		50	× 50
Category 3 (high security)	3000	50	× 50
		75	× 12.5
		25	× 75
Category 4 (maximum security)	3000	75	× 12.5
		12.5	× 12.5

Steel palisade fences to BS 1722 Pt 12: 1990

Security palisade fences are made of corrugated thick sheet steel pales fixed to horizontal rails.

Pales to be of steel Grade HR15, 3.0 or 3.5 mm thick.

Pales to be fixed not more than 155 mm centres, bolted, welded or riveted to the rails.

Heads of pales can be straight, rounded, or split to give two or three points, but pointed pales should not be used for fences under 1800 mm high.

An ST2 concrete sill should be laid below the fence 125 mm × 225 mm deep. Longer pales can be embedded in the concrete. Posts to be set in ST2 concrete.

Fence height	RSJ posts	Angle iron rails	Oversail		Footings	
			Upper	Lower	Round	Square
2400	102 × 44	45 × 45 × 6	300	210	450	300 × 300
	UB posts					
3000	127 × 76	50 × 50 × 6	475	380	600	450 × 450
3600	127 × 76	65 × 50 × 6	650	550	600	450 × 450

All dimensions in mm.

The **oversail** is the projection of the pales above and below the rails.

Two rails are required for 1500 mm and 1800 mm fences and three rails for 2100 mm and 2400 mm fences.

Gates are constructed to the same specification as the fences.

Concrete mixes to BS 5328 Pt 1 & AMD

Application	Standard mix	Designated mix	Slump (mm)
Mass concrete fill and blinding	ST 2 (C10P)	GEN 1	75
Mass concrete foundations in non-aggressive soils	ST 2 (C10P)	GEN 1	75
Trench fill foundations in non-aggressive soils	ST 2 (C10P)	GEN 1	125
Foundations			
in Class 2 sulphate soils	–	FND 2	75
in Class 3 sulphate soils	–	FND 3	75
in Class 4A sulphate soils	–	FND 4A	75
in Class 4B sulphate soils	–	FND 4B	75

Traditional concrete mixes

Application	Name of mix	Cement:aggregate:sand (parts by volume)
Setting posts, filling weak spots in ground, backfilling pipe trenches:	one, two, three	1:2:3
	or	or
	one to five all-in	1:5
Paving bases, bedding manholes, setting kerbs, edgings, precast channels:	one, two, four	1:2:4
	or	or
	one to six all in	1:6
In-situ concrete paving:	one, one & half, two & half	1:1.5:2.5

'All-in' is mixed aggregate as dug from the pit.

OUTLINE SPECIFICATION

Anti-intruder chain-link fence

a. *Supply and erect chain-link fence 2400 mm high, 40 mm mesh, with line wires and stretcher bars bolted to concrete posts 3150 mm × 125 mm × 100 mm at 3.0 m centres and straining posts 3150 mm × 150 mm × 150 mm with struts 3250 mm × 100 mm × 100 mm.*

b. *Posts set in FND 2 concrete 450 mm × 450 mm × 750 mm deep. Fit straight extension arms of 45 mm × 45 mm × 5 mm steel angle with three lines of barbed wire and droppers.*

c. *Base of fence to be fixed with hairpin staples cast into ground beam of GEN4 concrete 125 mm × 225 mm deep.*

d. *All metalwork to be factory hot-dip galvanized for painting on site.*

   for protective coatings   *M60*

Gates

Supply and erect pair of gates 2.0 m wide overall constructed as above hung on RSJ gate posts 127 mm × 114 mm set in concrete as above, complete with welded hinges and lock.

Palisade fence

a. *Supply and erect palisade security fence 3000 mm high to BS 1722 Pt 12 with corrugated steel pales 3.5 mm thick at 155 mm centres on three 50 mm × 50 mm × 6 mm rails.*

b. *Rails bolted to 254 mm × 254 mm × 73 kg universal beam posts set in FND3 concrete 450 mm × 450 mm × 750 mm deep at 2750 mm centres, tops of pales to be split two prongs.*

c. *Construct GEN4 concrete ground beam 125 mm × 225 mm deep, clearance between pales and beam not more than 50 mm.*

d. *All metalwork to be hot-dip factory galvanized for painting on site.*

   for protective coatings   *M60*

Gate

Supply and erect single gate 1000 mm wide to match above complete with welded hinges and lock.

Welded steel mesh panel fencing

a. *Supply and erect proprietary welded steel mesh panel fencing to BS 1722 Pt 14 Category 4 on steel angle posts with a moment of inertia not less than 12.5 cm⁴, set 750 mm deep in FND4 concrete foundations 600 mm × 600 mm.*

b. *Supply and erect proprietary single gate 2.0 m wide to match fencing.*

   for protective coatings   *M60*

Q40 Fences and gates: trip-rails

GENERAL GUIDANCE

Trip-rails are low rails with a maximum height of 600 mm above the ground. There is no British Standard for trip-rails. They must not (as their name would imply) be so positioned that people can actually trip over them, but they are useful to demarcate and to some extent deter pedestrians, skaters, cyclists and vehicles from crossing grassed or planted areas without being too obtrusive. In recreation areas they should be used only where very young children can be expected to be accompanied by an adult; where older children are likely to enjoy jumping over them, it is advisable to provide a space between the fence and the area they are intended to protect.

The main materials used are timber and metal with timber or concrete posts; the choice will depend upon where they are to be erected. White or brightly coloured paint should be used where they are to be clearly visible.

BRITISH STANDARDS

As trip-rails are constructed on site from timber or metal with concrete or steel posts to the landscape designers specification, there is no British Standard applicable for the whole construction.

4848 Pt 2 : 1991 Steel: hot-finished hollow sections

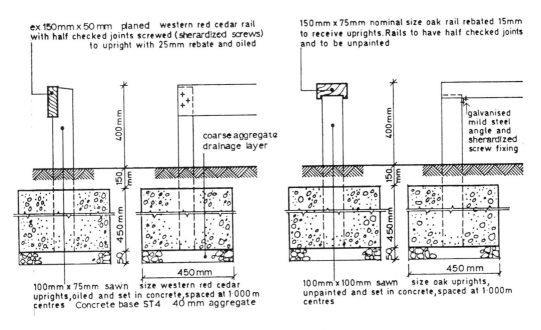

ex 150mm x 50 mm planed western red cedar rail with half checked joints screwed (sherardized screws) to upright with 25mm rebate and oiled

coarse aggregate drainage layer

100mm x 75mm sawn size western red cedar uprights, oiled and set in concrete, spaced at 1·000 m centres Concrete base ST4 40 mm aggregate

150mm x 75mm nominal size oak rail rebated 15mm to receive uprights. Rails to have half checked joints and to be unpainted

galvanised mild steel angle and sherardized screw fixing

100mm x 100mm sawn size oak uprights, unpainted and set in concrete, spaced at 1·000m centres

Fig. 4.51 Low rails: timber, hardwood.

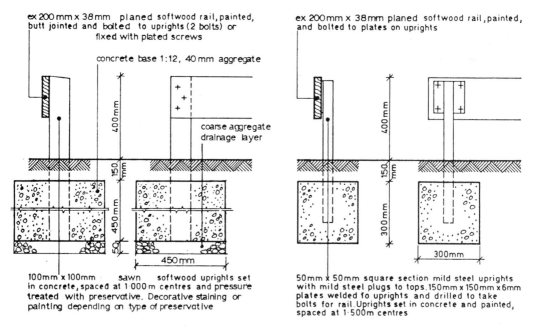

ex 200mm x 38mm planed softwood rail, painted,
butt jointed and bolted to uprights (2 bolts) or
fixed with plated screws

concrete base 1:12, 40mm aggregate

400mm

coarse aggregate
drainage layer

150mm

450mm

50

450mm

100mm x 100mm   sawn   softwood uprights set
in concrete, spaced at 1·000m centres and pressure
treated with preservative. Decorative staining or
painting depending on type of preservative

ex 200mm x 38mm planed softwood rail, painted,
and bolted to plates on uprights

400mm

150mm

300mm

300mm

50mm x 50mm square section mild steel uprights
with mild steel plugs to tops. 150mm x 150mm x 6mm
plates welded to uprights and drilled to take
bolts for rail. Uprights set in concrete and painted,
spaced at 1·500m centres

Fig. 4.52 Low rails: timber, softwood.

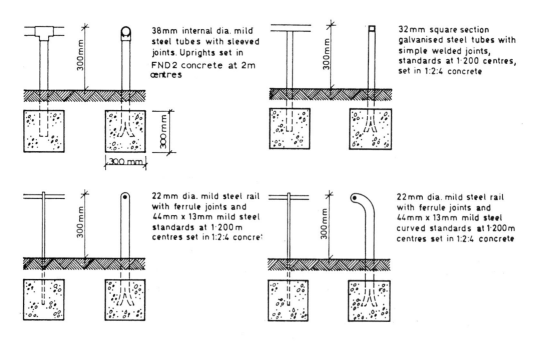

300mm

300mm

300 mm

38mm internal dia. mild
steel tubes with sleeved
joints. Uprights set in
FND2 concrete at 2m
centres

300mm

32mm square section
galvanised steel tubes with
simple welded joints,
standards at 1·200 centres,
set in 1:2:4 concrete

300mm

22mm dia. mild steel rail
with ferrule joints and
44mm x 13mm mild steel
standards at 1·200m
centres set in 1:2:4 concrete

300mm

22mm dia. mild steel rail
with ferrule joints and
44mm x 13mm mild steel
curved standards at 1·200m
centres set in 1:2:4 concrete

Fig. 4.53 Low rails: metal.

DATA

Steel hollow sections to BS 4848 Pt 2 : 1991

Steel: hot-finished hollow sections (all dimensions in mm)

Square sections		Rectangular sections		Circular sections (outside diameter)	
Size	Thick	Size	Thick	Size	Thick
20 × 20	2.0	50 × 25	2.0	21.3	3.2
	2.5		2.5	26.9	3.2
	3.2		3.2	33.7	2.6
25 × 25	2.0	50 × 30	2.5		3.2
	2.5		3.0		4.0
	3.2		3.2	42.4	2.6
30 × 30	2.6		4.0		3.2
	3.2	60 × 40	2.5		4.0
40 × 40	2.6		3.0	48.3	3.2
	3.2		3.2		4.0
	4.0		4.0		5.0
50 × 50	2.5	80 × 40	3.0	60.3	3.2
	3.0		3.2		4.0
	3.2		4.0		5.0
	4.0		5.0	76.1	3.2
	5.0				4.0
60 × 60	3.0				5.0
	3.2			88.9	3.2
	4.0				4.0
	5.0				5.0
80 × 80	3.0				
	3.6				
	5.0				

All shapes of hollow steel sections are available in larger sizes, and there are thicker sections available in the larger sizes.

Timber rails

Sizes of hardwood depend greatly on what species and sizes are available at the time; generally 75 mm × 75 mm is the smallest size practicable for light trip-rails.

Stock sizes of sawn softwood suitable for trip-rails are:
75 mm × 75 mm, 75 mm × 100 mm, 75 mm × 150 mm
100 mm × 100 mm, 100 mm × 150 mm

Concrete mixes to BS 5328 Pt 1 & AMD

Application	Standard mix	Designated mix	Slump (mm)
Mass concrete fill and blinding	ST 2 (C10P)	GEN 1	75
Mass concrete foundations in non-aggressive soils	ST 2 (C10P)	GEN 1	75
Trench fill foundations in non-aggressive soils	ST 2 (C10P)	GEN 1	125
Foundations			
in Class 2 sulpahte soils	–	FND 2	75
in Class 3 sulphate soils	–	FND 3	75
in Class 4A sulphate soils	–	FND 4A	75
in Class 4B sulphate soils	–	FND 4B	75
Kerb bedding and haunching	ST 1 (C7.5P)	GEN 0	10
Drain pipe support in non-aggressive soils	ST 2 (C10P)	GEN 1	10
Manholes etc. in non-aggressive soils	ST 2 (C10P)	GEN 1	50
In-situ concrete paving:			
pedestrian or light vehicle	–	PAV 1	75
heavy vehicle	–	PAV 2	50

GEN concretes with low cement content may not be suitable for good cast and direct finished surfaces.

Traditional concrete mixes

Application	Name of mix	Cement:aggregate:sand (parts by volume)
Setting posts, filling weak spots in ground, backfilling pipe trenches:	one, two, three	1:2:3
	or	or
	one to five all-in	1:5
Paving bases, bedding manholes, setting kerbs, edgings, precast channels:	one, two, four	1:2:4
	or	or
	one to six all in	1:6
In-situ concrete paving:	one, one & half, two & half	1:1.5:2.5

'All-in' is mixed aggregate as dug from the pit.

OUTLINE SPECIFICATION

Timber trip-rail

a. *Supply and erect trip-rail 400 mm high of treated softwood rail 75 mm × 100 mm.*

b. *Rail screwed with non-corroding screws to 100 mm × 100 mm × 1000 mm long oak posts, posts set 500 mm into ground in FND2 concrete with central hole at 2.0 m centres.*

c. *Top of rails and posts twice weathered, 5 mm gap between ends of rails. Rails stained to approved colour after fixing.*

for paint and protective treatment   *M60*

Alternative specification

a. *Supply and erect trip-rail 400 mm high of treated softwood rail 75 mm × 75 mm set at 45°.*

b. *Rail fixed with galvanized clips and nails to pretreated notched softwood posts 100 mm × 100 mm × 900 mm long, posts set 500 mm into ground in FND2 concrete with central hole at 2.0 m centres.*

c. *5 mm gap between ends of rails. Rails and posts stained to approved colour after fixing.*

for paint and protective treatment   *M60*

Metal trip-rail

a. *Supply and erect trip-rail of galvanized steel tube 33.7 mm internal dia. with sleeved tee-joint fixed to 48.3 mm dia. galvanized steel posts 800 mm long, set in ST1 concrete at 2.0 m centres.*

b. *Metalwork primed and painted two coats metal preservative paint.*

Alternative specification

a. *Supply and erect trip-rail 400 mm high of galvanized steel tube 50 mm × 50 mm × 3.0 mm with sleeved flush joints at posts, passing through quarry pitched selected natural stone posts 1000 mm × 200 mm × 200 mm approx. with squared tops, drilled with one 50 mm hole for tube at 2.0 m centres set 500 mm into firm ground.*

b. *Rail to be painted two coats gloss paint after assembly. Rail to be cemented into end posts in 1:3 masonry cement:sand mortar.*

for natural stone   *F20*
for paints   *M60*

Q40 Fences and gates: timber fencing

GENERAL GUIDANCE

This section covers all timber fences, whether made up on site or prefabricated; they include post and rail, hit-and-miss, close boarded, panels, and picket fencing.

Open plank or pickets do not give very good visual privacy, but as the gaps allow air to pass through them, they are not so susceptible to damage in high winds. Timber fences are suitable for a wide range of design applications.

Solid fences, particularly when 1.5 m or more high, do give excellent privacy, though unless very strongly constructed they are vulnerable to high winds. The type and treatment of timber specified will enable them to be either treated with clear or coloured preservative, primed and painted, or left in their natural state. Matching gates are available to match most types of timber fencing. Temporary timber fencing and trip-rails are covered in separate sections.

Post and rail fences

These types of fence have riven rails, which make them highly suitable for rural settings, or sawn rails more suitable for town or country parks.

High-class and therefore more expensive work has morticed posts with the rails set in the centre of the posts, while the cheaper versions have the

Fig. 4.54 Timber absorbing environmental noise barriers. (Stenoak Fencing)

Fig. 4.55 Timber absorbing environmental noise barriers.
(Stenoak Fencing)

rails nailed to one side of the posts; these have either rectangular or sectored rails. When nailed work is used, the rails should be fixed on the side most likely to be leaned upon by people or stock. Sawn and nailed softwood is most familiar as highway boundary fencing through the countryside.

In morticed work, the rails are tapered off at each end to form a tenon that fits into pre-morticed posts. As the fence has to be erected as a single entity, the rails cannot be removed to allow the entry of vehicles or machinery. The landscape designer must provide gates to take the maximum size of vehicles the client is likely to use.

Cleft rail fences

This is a relatively uncommon and expensive fence used in areas that retain the tradition of riven oak structures. It is composed of rails of cleft or riven oak, and it provides a durable and very attractive fence for rural settings. Rails vary in length between 2.5 and 3 m long, and may be bought barked or unbarked. Each end of the rail is tenoned to fit into pre-morticed posts; because of the difficulty of morticing hardwood, posts are often made from softwood. Though the rails are naturally irregular, they must not be so uneven as to leave over-wide gaps between the rails. Two adjoining rails may be set into one mortise, but it is better practice to form two separate mortices, one for each rail.

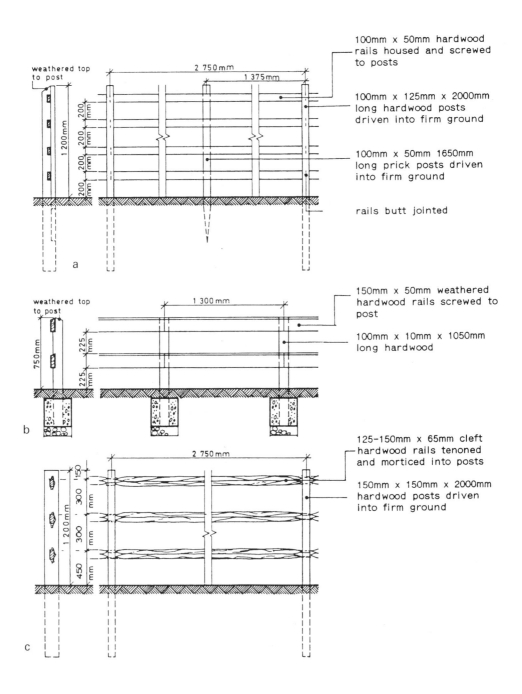

Fig. 4.56 Fences (timber): (a) horizontal rail fence; (b) low horizontal rail fence; (c) cleft rail fence.

Plank fences

This is the generic name for fences made from individual planks fixed on site to posts, as opposed to being made up in preformed panels. In general, they form a sturdier fence than the close-boarded preformed feather-edged panel fence. The boards can be nailed, screwed or bolted to softwood posts, the latter giving the strongest and most lasting fence. They can be open spaced, hit-and-miss, diagonal, and any other patterns designed by the landscape designer. They are most commonly used in residential areas and in public spaces where vandalism is not a serious threat, but where a certain amount of privacy and restricted access is called for. Depending on the type and finish of timber used, these fences can be either treated with coloured preservative stain or primed and painted.

Vertical hit-and-miss fences

Though still open spaced, this type of fence has staggered boards fixed on either face of the rails; this still allows the air to pass through, but makes it difficult to see through. Because of the greater privacy provided by this fence it is frequently used to divide the gardens or patios of terrace or town houses, and to mark the boundary between them and communal areas.

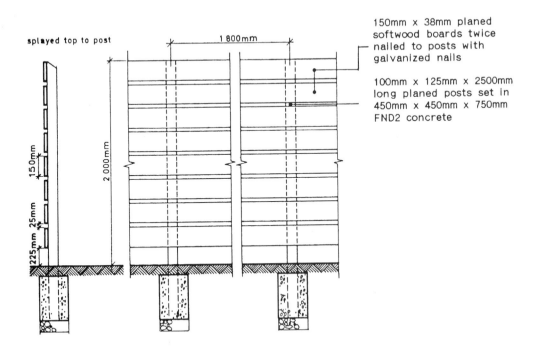

splayed top to post

1 800mm

150mm x 38mm planed softwood boards twice nailed to posts with galvanized nails

100mm x 125mm x 2500mm long planed posts set in 450mm x 450mm x 750mm FND2 concrete

150mm

2 000mm

225mm 25mm 150mm

Fig. 4.57 Fences (timber): horizontal board fence.

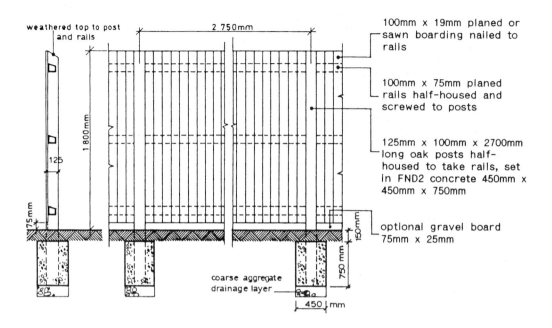

Fig. 4.58 Fences (timber): close-boarded fence. Matching gates available.

Close-boarded or feather-edged fencing

This type of fencing comes in either preformed panels of varying widths and heights, or as individual boards; these can also be used to make up the short-fall in standard panels. BS 1722 Pt 5 deals with the boards, rails, posts and fixings. It is a suitable fence for residential areas where privacy is important, and where vandalism or casual damage is unlikely to occur, but because they are lighter in construction than plank fences, they are vulnerable to strong winds and should not be used in exposed situations. Gates to match are ledged, braced and framed on each side and at the foot; they are normally supplied with galvanized band hinges, locking latches or pad bolts.

Palisade and picket fences

These consist of light dressed timber pales with tops finished to various profiles, normally round or pointed, though pointed ones should not be specified around playgrounds or other play areas as children could damage themselves on them. Both pales and pickets are usually of primed and painted softwood nailed to horizontal rails, which are in turn fixed to supporting posts. The term 'palisade' refers to the higher of the two fences, which is a more serious deterrent. Gates are available to match, with a choice of ironmongery.

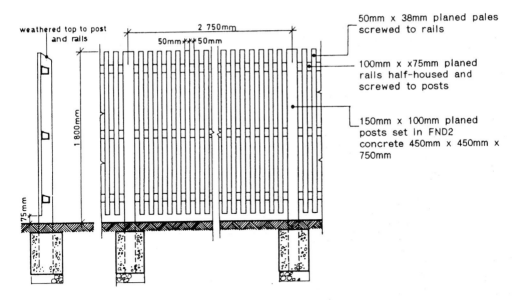

Fig. 4.59 Fences (timber): vertical pale fence.

BRITISH STANDARDS

1722 Pt 5 : 1986	Close-boarded fences 1.05 to 1.8 m high and wooden or reinforced concrete posts
1722 Pt 6 1986	Wooden palisade fences
1722 Pt 7 1986	Wooden post and rail fences
1722 Pt 11 1986	Woven-wood fences and wooden or reinforced concrete posts
3470 : 1975	Field gates and posts in timber and steel
4072 Pt 2 : 1987	Wood preservation: copper/chromium, arsenic compositions
4072 Pt.1: 1987 & AMD	Wood preservation: copper/chromium, arsenic compositions
5082 : 1993	Water-borne priming paints for woodwork
5358 : 1993	Solvent-borne priming paints for woodwork: low lead content
5589 : 1989	Code of practice for preservation of timber
6952 Pt 1 : 1988	Exterior wood coating systems: guide to classification and selection

DATA

Softwood

Stock sizes of sawn softwood suitable for fencing are:

$19 \times 38, 50, 75, 100, 125$ mm

$25 \times 38, 50, 75, 100, 150, 175, 200$ mm

38 × 38, 50, 75, 100, 150, 175, 200, 225 mm
50 × 50, 75, 100, 125, 150, 175, 200, 225 mm
75 × 75, 100, 150, 175, 200, 225 mm
100 × 100, 150, 200 mm
The most commonly available types of timber suitable for fencing are:

Name	Natural durability	Treatment	Hardness	Colour
Western red cedar*	15–25 yrs	None	Soft	Grey
Sweet chestnut	15–25 yrs	None	Medium	Brown
American oak	15–35 yrs	None	Hard	Grey/brown
European oak	15–25 yrs	None	Hard	Grey/brown
Robinia	15–25 yrs	None	Hard	Yellow/brown
Yew	15–25 yrs	None	Tough	Yellow
Douglas fir	10–15 yrs	Treat	Soft	White
Larch	10–15 yrs	Treat	Soft	White
Turkey oak	10–15 yrs	Treat	Medium	Brown
Birch	5–10 yrs	Treat	Soft	White
Dutch elm	5–10 yrs	Treat	Medium	Brown
Fir (most species)	5–10 yrs	Treat	Soft	White
American red oak	5–10 yrs	Treat	Medium	Reddish
Pine (most species)	5–10 yrs	Treat	Soft	Yellow/white
Grey poplar	5–10 yrs	Treat	Soft	White
Spruce (most species)	5–10 yrs	Treat	Soft	White

*British-grown Western red cedar is not so durable.

Close-boarded fences to BS 1722 Pt 5 : 1986

Close-boarded fences 1.05–1.8 m high.
Type BCR (recessed) or BCM (morticed) with concrete posts 140 mm × 115 mm tapered and Type BW with timber posts.

Palisade fences to BS 1722 Pt 6 : 1986

Wooden palisade fences
Type WPC with concrete posts 140 mm × 115 mm tapered
and Type WPW with timber posts.
For both types of fence:
Height of fence 1050 mm: two rails:
    timber posts 100 mm × 100 mm × 1650 mm long, or concrete posts 1600 mm long
Height of fence 1200 mm: two rails:
    timber posts 100 mm × 125 mm × 1800 mm long, or concrete posts 1750 mm long

Height of fence 1500 mm: three rails:
   timber posts 100 mm × 125 mm × 2250 mm long, or concrete posts
   2200 mm long
Height of fence 1650 mm: three rails:
   timber posts 100 mm × 125 mm × 2400 mm long, or concrete posts
   2350 mm long
Height of fence 1800 mm: three rails:
   timber posts 100 mm × 125 mm × 2550 mm long, or concrete posts
   2500 mm long
Height of fence 1800 mm: three rails:
   timber posts 100 mm × 150 mm × 2550 mm long, or concrete posts
   2500 mm long

For heavy-duty fences use 100 mm × 50 mm arris rails.
For posts set in concrete, post holes shall be 300 mm min dia.
For fences up to 1500 mm high, posts to be set 600 mm minimum into the
   ground.
For fences over 1500 mm high, posts to be set 750 mm minimum into the
   ground.
Concrete mix shall be not weaker than 1:10 cement:all-in aggregate with 40
   mm max. aggregate

Post and rail fences to BS 1722 Pt 7

Wooden post and rail fences
Type MPR 11/3 morticed rails and Type SPR 11/3 nailed rails
Height to top of rail 1100 mm
Rail spacing 325 mm and 275 mm from top
Main posts 75 mm × 150 mm × 1800 mm long
Prick posts 87 mm × 38 mm × 1600 mm long
Rails: three rails 87 mm × 38 mm

Type MPR 11/4 morticed rails and Type SPR 11/4 nailed rails
Height to top of rail 1100 mm
Rail spacing 225 mm, 200 mm and 175 mm from top
Main posts 75 mm × 150 mm × 1800 mm long
Prick posts 87 mm × 38 mm × 1600 mm long
Rails: four rails 87 mm × 38 mm

Type MPR 13/4 morticed rails and Type SPR 13/4 nailed rails
Height to top of rail 1300 mm
Rail spacing 250 mm, 250 mm and 225 mm from top
Main posts 75 mm × 150 mm × 2100 mm long
Prick posts 87 mm × 38 mm × 1800 mm long
Rails: four rails 87 mm × 38 mm

For morticed fences the mortice must be the full depth of the rail.
For fence 1100 mm high, post holes shall be 600 mm deep.
For fence 1300 mm high, post holes shall be 700 mm deep.
For posts set in concrete, post holes shall be 300 mm min dia.
Concrete mix shall be not weaker than 1:10 cement:all-in aggregate with 40 mm max. aggregate.

Fences around playgrounds must meet certain criteria: See *General introduction* to fences.

Types of preservative to BS 5589 : 1989

Creosote (tar oil) can be 'factory' applied
    by pressure to BS 144 Pts 1 and 2
    by immersion to BS 144 Pt 1
    by hot and cold open tank to BS 144 Pts 1 and 2
Copper/chromium/arsenic (CCA)
    by full-cell process to BS 4072 Pts 1 and 2
Organic solvent (OS)
    by double vacuum (vacvac) to BS 5707 Pts 1 and 3
    by immersion to BS 5057 Pts 1 and 3
Pentachlorophenol (PCP)
    by heavy oil double vacuum to BS 5705 Pts 2 and 3

Boron diffusion process (treated with disodium octaborate to BWPA Manual 1986)
Note: Boron is used on green timber at source, and the timber is suppled dry.

Concrete mixes to BS 5328 Pt 1 & AMD

Application	Standard mix	Designated mix	Slump (mm)
Mass concrete fill and blinding	ST 2 (C10P)	GEN 1	75
Mass concrete foundations in non-aggressive soils	ST 2 (C10P)	GEN 1	75
Trench fill foundations in non-aggressive soils	ST 2 (C10P)	GEN 1	125
Foundations			
in Class 2 sulphate soils	–	FND 2	75
in Class 3 sulphate soils	–	FND 3	75
in Class 4A sulphate soils	–	FND 4A	75
in Class 4B sulphate soils	–	FND 4B	75

Traditional concrete mixes

Application	Name of mix	Cement:aggregate:sand (parts by volume)
Setting posts, filling weak spots in ground, backfilling pipe trenches:	one, two, three or one to five all-in	1:2:3 or 1:5

OUTLINE SPECIFICATION

Palisade (or picket) fence to BS 1722 Pt 6

a. *Supply and erect traditional picket fence 1500 mm high to BS 1722 Pt 6 with 75 mm × 20 mm treated softwood vertical pickets with pointed tops 75 mm apart, double nailed to two ex 75 mm × 75 mm triangular treated softwood arris rails spaced 225 mm from top and bottom of pales.*

b. *Rails housed into 2250 mm × 100 mm × 125 mm treated softwood posts, at 2.50 m centres set in ST2 concrete bases 300 mm dia. Centre stump to be set under centre of bottom rail.*

for finishes to timber components   *M60*

Post and rail fence to BS 1722 Pt 7

a. *Supply and erect post and rail fence to BS 1722 Pt 7 Type MPR 13/4 1300 mm high with four rails 87 mm × 38 mm.*

b. *Rails morticed into treated softwood posts 2100 mm × 75 mm × 150 mm at 3.0 m centres set 700 mm into firm ground with intermediate prick posts 1800 mm × 87 mm × 38 mm.*

for finishes to timber components   *M60*

Cleft rail fence

a. *Supply and erect cleft oak rail fence, with rails 300 mm minimum girth tenoned both ends.*

b. *125 mm × 100 mm treated softwood posts double morticed for rails, corner posts 125 mm × 125 mm, driven into firm ground at 2.5 m centres.*

for finishes to timber components   *M60*

Close-boarded fence to BS 1722 Pt 5

a. *Supply and erect close-boarded fence to BS 1722 Pt 5 1800 mm high with treated softwood feather-edge boards ex 100 mm × 22 mm lapped*

18 mm, on three ex 75 mm × 75 mm triangular treated softwood rails spaced 225 mm from top and bottom of boards.

b. Rails housed into treated softwood posts 2550 mm × 100 mm × 125 mm at 2.5 m centres set 750 mm into ST2 concrete bases 450 mm × 450 mm × 750 mm.

c. Gravel boards to be treated softwood 150 mm × 32 mm with centre stump set under centre of bottom rail.

for finishes to timber components   *M60*

Vertical hit-and-miss fence

a. Supply and erect vertical hit-and-miss fence 1500 mm high with 150 mm × 25 mm treated softwood sawn pales arranged alternately on opposite sides of the rails and cross-nailed with two galvanized nails, treated softwood rails 150 mm × 38 mm.

b. Rails nailed to 2250 mm × 100 mm × 125 mm treated softwood posts at 1.8 m centres set into 450 mm × 450 mm × 600 mm deep ST2 concrete.

for finishes to timber components   *M60*

Plank fence

a. Supply and erect plank fence 1225 mm high of 125 mm × 19 mm treated softwood planks 25 mm apart.

b. Planks cross-nailed with galvanized nails to 100 mm × 100 mm treated softwood posts at 2.0 m centres set into 450 mm × 450 mm × 600 mm deep ST2 concrete base. Gap at bottom of fence 50 mm.

for finishes to timber components   *M60*

Q40 Fences and gates: concrete fencing

GENERAL GUIDANCE

Concrete panel fencing is mainly used as a protective barrier or sound insulation screen, because with the exception of a masonry wall it provides greater sound attenuation than any other boundary wall. These fences can be faced with a vertical board fence where appearance or environmental conditions make this desirable. The panels, which are available in various finishes, are slotted into precast reinforced posts, which have grooves on either side to take them. They present a sturdy (more or less unclimbable) fence; if the panels are damaged, they are easily replaced by sliding in new ones.

The landscape designer should ensure that the contractor stores both posts and panels on a clean level surface, and protects them from oil, creosote, preservative, paint, cement and so forth. Panels must be fitted with care, as broken nibs on the posts should not be accepted.

BRITISH STANDARDS

Precast concrete fencing is supplied and erected as proprietary systems, and therefore there is no British Standard applicable.

DATA

Site control

- Posts and panels must be stacked on clean dry level surfaces.
- Neither posts nor panels should be dropped or thrown, because hair cracks may not show up until weathering takes place.
- Posts and panels should be protected from oil, creosote, cement, mud and other pollutants.
- Because post-and-panel fencing is a dry construction, frost precautions are not required for the fence itself, but concrete footings should not be poured in freezing conditions.
- Posts must be correctly centred, and panels must not be so loose that they can be removed, or so tight that they are forced into too small a space.
- The aggregate face is easily damaged by machinery, so the panel erection should be left as late as possible in the contract.

Typical post-and-panel concrete fencing

125 mm × 100 mm posts 1565 mm long, 600 mm in ground with 915 mm high fence panels
125 mm × 100 mm posts 2335 mm long, 760 mm in ground with 1525 mm high fence panels
150 mm × 100 mm posts 3250 mm long, 760 mm in ground with 2440 mm high fence panels
175 mm × 125 mm posts 4000 mm long, 900 mm in ground with 3050 mm high fence panels

OUTLINE SPECIFICATION

Precast concrete post-and-panel fence

Supply and erect precast concrete post-and-panel fence in 2 m bays, fence 1000 mm high, panels to be shiplap profile, aggregate faced one side, posts set 600 mm in ground in ST2 concrete.

As above but with close-boarded fence 1000 mm high on one side fixed to posts.

for close-boarded fencing	*Q40 (timber fences)*
for concrete mixes	*E10*

Q40 Fences and gates: ball-stop fencing

GENERAL GUIDANCE

Ball-stop fencing is specified for tennis courts, sports fields and playgrounds where ball games are played, and it is usually made of chain-link or welded mesh fencing on specially tall tubular steel posts. Ball-stop fencing is most commonly specified complete with doors and fixings that are supplied complete by specialist manufacturers. Where extra high fencing is needed, another 2–3 m of 125 mm × 125 mm 3 mm nylon netting may be used for the upper part of the fence, to reduce loading on the posts. For small children's playgrounds a ball-stop fence made entirely of nylon netting should be adequate, though the landscape designer must bear in mind that if vandalism is prevalent in the area nylon can be burnt or slashed.

Tennis court surrounds are described in BS 1722 Pt 13. They are always supplied complete with doors and all fixings, and are specified to enclose one, two or three standard-size courts (competition courts are constructed complete by specialist contractors).

A typical ball-stop fence suitable for controlled playgrounds would be constructed of light-gauge 45 mm chain-link mesh 3000 mm high on 4.75 mm straining wires fixed to tapered tubular steel posts set in concrete at 3 m centres.

Where heavy and possibly rough ball game usage is expected, such as football or cricket practice, it is better to use welded steel 62 mm × 66 mm mesh panels on steel posts at 2 m centres, and if there is much risk of balls travelling over the fence it may be necessary to provide cranked arms to the posts, with the mesh extending 1 m into the play area. In very confined sites such as rooftop sports areas, or where high-speed roads are adjacent to the ball game area, it may even be necessary to roof in the space with light welded mesh panels or nylon ball-stop netting on alloy framing, and in this case a structural engineer should be consulted.

BRITISH STANDARDS

1722 Pt 1 : 1986 Chain-link fences and gates and gate posts for fences up to 1.8 m high
1722 Pt 2 : 1989 Rectangular wire-mesh and hexagonal wire netting fences

DATA

Ball-stop netting

25 mm mesh high-density polyethylene in green or black

Tennis courts

Chain-link fencing for tennis courts

Type	Mesh (mm)	Height (mm)	Lines
Light	45	2750/3600	4/5
Medium	45/50	2750/3600	4/5
Heavy	50	2750/3600	4/5

Concrete mixes to BS 5328 Pt 1 & AMD

Application	Standard mix	Designated mix	Slump (mm)
Mass concrete fill and blinding	ST 2 (CIOP)	GEN 1	75
Mass concrete foundations in non-aggressive soils	ST 2 (CIOP)	GEN 1	75
Trench fill foundations in non-aggressive soils	ST 2 (CIOP)	GEN 1	125
Foundations			
in Class 2 sulphate soils	–	FND 2	75
in Class 3 sulphate soils	–	FND 3	75
in Class 4A sulphate soils	–	FND 4A	75
in Class 4B sulphate soils	–	FND 4B	75

Traditional concrete mixes

Application	Name of mix	Cement:aggregate:sand (parts by volume)
Setting posts, filling weak spots in ground, backfilling pipe trenches:	one, two, three	1:2:3
	or	or
	one to five all-in	1:5

'All-in' is mixed aggregate as dug from the pit.

OUTLINE SPECIFICATION

Ball court

a. *Supply and erect plastic-coated light chain-link fencing 45 mm mesh 2750 mm high on 60.3 mm dia. nylon-coated tubular steel posts at 3.0 m centres and 60.3 mm dia. straining posts with struts at 60 m centres set 750 mm into FND2 concrete footings 300 mm × 300 mm × 600 mm deep.*

b. *Include framed chain-link gate 900 mm × 1800 mm high to match, complete with hinges and locking latch.*

   for concrete   *E20*

Junior play area

Supply and erect ball-stop fencing where shown on drawing 1800 mm high with 25 mm mesh high-density polyethylene ball-stop netting on 60.3 mm dia. nylon-coated tubular steel posts at 3.0 m centres and 60.3 mm dia. straining posts with struts at 60 m centres set 750 mm into FND2 concrete footings 300 mm × 300 mm × 600 mm deep.

   for concrete mixes   *E10*

Q40 Fences and gates: railings

GENERAL GUIDANCE

This section covers traditional wrought iron railings and prefabricated steel railings; there is a separate section for pedestrian guard rails.

Cast iron railings

Although expensive, these are used in conservation areas and so forth, where grants for their replacement are often available. These railings are now available from manufacturers in standard patterns or purpose-made patterns for a specific contract. The landscape designer should not specify the traditional sharp-pointed railings, because people can impale themselves upon the sharp spikes. There is no British Standard for wrought or cast iron railings, and the manufacturer's technical department should be asked to produce specifications.

Rail and stanchion railings

Rail and stanchion railings come in standard lengths with horizontal bars that pass through stanchions; these range from plain tubular to the ornate, and the landscape designer will find many styles in the manufacturers' catalogues. Heavy iron railings with vertical bars provide greater security, as they are difficult to climb; they have two horizontal rails, with the bars welded to the bottom rail, and these can either stop beneath the top rail or pierce it.

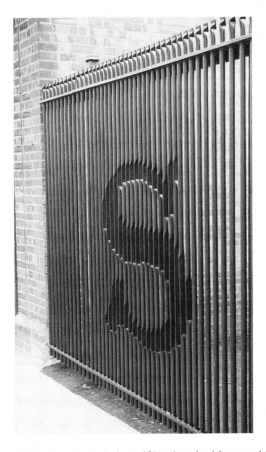

Fig. 4.60 Railings, Sainsbury's, Dulwich. (Claydon Architectural Metalwork)

Gaps between bars should not be greater than 100 mm to prevent children from getting their heads trapped in them.

Stanchions with infill panels

Welded steel mesh panels within a frame are the alternative to using rails. They are useful if small animals have to be kept within bounds, and are difficult to climb. Individual panels normally have lugs welded on for fixing to the stanchions.

BRITISH STANDARDS

729 : 1971   Hot-dip galvanized coatings for iron and steel
1722 Pt 8 : 1992   Mild steel (low carbon steel) continuous bar fences
1722 Pt 9 : 1992   Mild steel fences: round or square verticals; flat standards and horizontals
1722 Pt 14 : 1992 Open mesh steel panels
4921 : 1988   Sheradized coatings for iron and steel

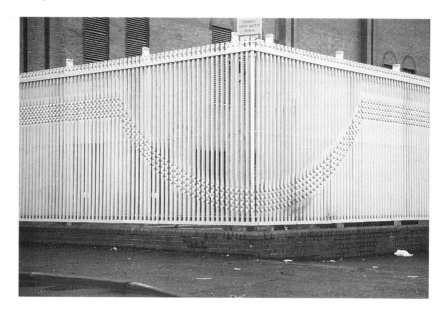

Fig. 4.61 Railings, Sewage Treatment Works, Liverpool.
(Claydon Architectural Metalwork)

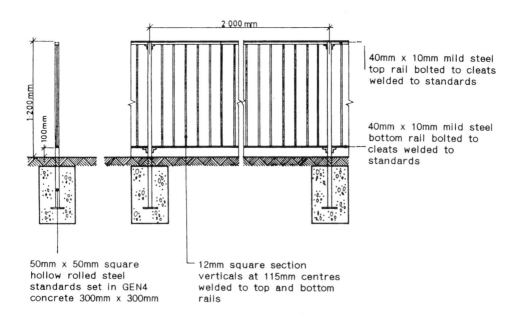

40mm x 10mm mild steel top rail bolted to cleats welded to standards

40mm x 10mm mild steel bottom rail bolted to cleats welded to standards

1 200 mm

100mm

2 000 mm

50mm x 50mm square hollow rolled steel standards set in GEN4 concrete 300mm x 300mm

12mm square section verticals at 115mm centres welded to top and bottom rails

Fig. 4.62 Fences (metal): mild steel bar railing, flat topped.

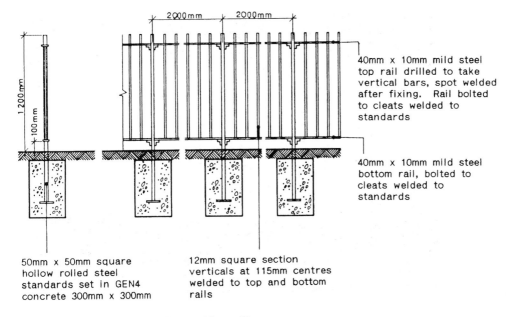

50mm x 50mm square
hollow rolled steel
standards set in GEN4
concrete 300mm x 300mm

12mm square section
verticals at 115mm centres
welded to top and bottom
rails

40mm x 10mm mild steel
top rail drilled to take
vertical bars, spot welded
after fixing. Rail bolted
to cleats welded to
standards

40mm x 10mm mild steel
bottom rail, bolted to
cleats welded to
standards

Fig. 4.63 Fences (metal): mild steel bar railing.

DATA

Steel bar fences to BS 1722 Pt 8 : 1992

Mild steel (low carbon steel) continuous bar fences on posts with two pronged feet. (These are the traditional deer-park railings; they are expensive, but may be used for conservation work.) All dimensions in mm.

Fence height	Top rail	Lower rails	Flat posts
1000	16 dia. round	4 no 25 × 6	30 × 10
1200	20 dia. round	5 no 25 × 6	40 × 10
1200	22 dia. round	5 no 30 × 6	50 × 10
1200	22 dia. round	6 no 30 × 6	50 × 10
1400	20 dia. round	6 no 25 × 6	40 × 10
1400	22 dia. round	6 no 30 × 6	50 × 10
All straining posts:	square hollow section	50 × 50 × 3	
	round hollow section	60.3 × 4	
	rolled steel angles	50 × 50 × 6	

Steel fences to BS 1722 Pt 9 : 1992

Mild steel fences: round or square verticals; flat standards and horizontals.

Tops of vertical bars may be bow-top, blunt, or pointed (note that pointed-top railings are not acceptable where people may become impaled).

Round or square bar railings
All dimensions are in mm:

	Fence height	Top/bottom rails and flat posts		Vertical bars
Light	1000	40 × 10	450 in ground	12 dia. at 115 cs
	1200	40 × 10	550 in ground	12 dia. at 115 cs
	1400	40 × 10	550 in ground	12 dia. at 115 cs
Light	1000	40 × 10	450 in ground	16 dia. at 120 cs
	1200	40 × 10	550 in ground	16 dia. at 120 cs
	1400	40 × 10	550 in ground	16 dia. at 120 cs
Medium	1200	50 × 10	550 in ground	20 dia. at 125 cs
	1400	50 × 10	550 in ground	20 dia. at 125 cs
	1600	50 × 10	600 in ground	22 dia. at 145 cs
	1800	50 × 10	600 in ground	22 dia. at 145 cs
Heavy	1600	50 × 10	600 in ground	22 dia. at 145 cs
	1800	50 × 10	600 in ground	22 dia. at 145 cs
	2000	50 × 10	600 in ground	22 dia. at 145 cs
	2200	50 × 10	600 in ground	22 dia. at 145 cs

Bow top railings
Lighter railings are available. All dimensions are in mm.

	Fence height	Top/bottom rails and flat posts		Vertical bars
Light	1000	30 × 10	450 in ground	12 dia. at 110 cs
	1200	30 × 10	550 in ground	12 dia. at 110 cs
	1400	30 × 10	550 in ground	12 dia. at 110 cs
Medium	1000	40 × 10	450 in ground	16 dia. at 120 cs
	1200	40 × 10	550 in ground	16 dia. at 120 cs
	1400	40 × 10	550 in ground	16 dia. at 120 cs
Heavy	1200	50 × 10	550 in ground	20 dia. at 131 cs
	1400	50 × 10	550 in ground	20 dia. at 131 cs
	1600	50 × 10	600 in ground	20 dia. at 131 cs
	1800	50 × 10	600 in ground	20 dia. at 131 cs

Rolled steel joist end posts are 102 mm × 44 mm for all railings.

Posts for railings up to 1400 mm high:
- Round hollow steel posts are 60.3 mm o/d
- Square hollow steel posts are 50 mm × 50 mm × 3 mm
- Rolled steel angle sections are 50 mm × 50 mm × 6 mm

Concrete mixes to BS 5328 Pt 1 & AMD

Application	Standard mix	Designated mix	Slump (mm)
Mass concrete fill and blinding	ST 2 (C10P)	GEN 1	75
Mass concrete foundations in non-aggressive soils	ST 2 (C10P)	GEN 1	75
Trench fill foundations in non-aggressive soils	ST 2 (C10P)	GEN 1	125
Foundations			
in Class 2 sulphate soils	–	FND 2	75
in Class 3 sulphate soils	–	FND 3	75
in Class 4A sulphate soils	–	FND 4A	75
in Class 4B sulphate soils	–	FND 4B	75

Traditional concrete mixes

Application	Name of mix	Cement:aggregate:sand (parts by volume)
Setting posts, filling weak spots in ground, backfilling pipe trenches:	one, two, three or one to five all-in	1:2:3 or 1:5

'All-in' is mixed aggregate as dug from the pit.

OUTLINE SPECIFICATION

Railings

a. *Supply and erect mild steel vertical bar railing to BS 1722 Pt 9, 1800 mm high with 50 mm × 10 mm flat posts set 600 mm in ground, blunt-top vertical bars 22 mm dia. at 145 mm centres, horizontal rails 50 mm × 10 mm, at 200 mm from top and bottom of railing.*

b. *End and corner posts 102 mm × 44 mm RSJs set 600 mm into concrete base of ST 4 concrete, vertical bars welded to horizontals top and bottom.*

c. *All metalwork hot-dip galvanized after manufacture, and painted two coats low-lead protective paint.*

Alternative specification

As above but panel infilled with mild steel mesh 51 mm square welded to frame, all metalwork galvanized after fabrication.

   for protective paintwork   *M60*

Gate

Supply and erect single gate to match above, 900 mm wide complete with one pair galvanized hinges and padbolt, metal work galvanized after fabrication.

   for protective paintwork   *M60*

Handrail

a. *Supply and erect tubular steel handrailing 1000 mm high with ball joints, 3 no. rails 48.33 mm dia.*
b. *Stanchions 48.3 mm dia. at 2000 mm centres with circular bases bolted to existing paving.*
c. *All metalwork factory primed for painting on site.*

   for protective paintwork   *M60*

Q40 Fences and gates: gates

GENERAL GUIDANCE

Garden and field gates that are not part of a fencing system are included here, together with stiles, bridle and kissing-gates. Chain-link fencing, close-boarded fencing, picket fencing, iron and steel railings, pedestrian guard rails, and security fencing are the most common types of fencing to be supplied complete with matching gates. As mentioned before, the landscape designer must ascertain the type and size of vehicles or machinery requiring access, and whether gate catches or an automated gate system is required.

Formal gates

These are available with or without their own posts; either way they are usually built into masonry walls, and are suitable for courtyards and walled gardens. They come in standard sizes or are purpose made, and are available in hard or softwood, mild steel, wrought iron, and (for high class) with cast iron components. As these gates are designed to fit snugly into the opening, they must be carefully specified. Solid timber gates 2.0 m high, close boarded in a vertical or horizontal pattern and fitted with heavy hook, hook-and-band or reversible hinges, drop bolts and lockable padbolts or handles and locks provide excellent security. Even well-designed metal gates of a similar height can be effective as barriers provided that horizontal rails do not offer foothold.

Informal entrance gates

These follow the traditional pattern of barred and braced gates; the heavier gates have a high hanging stile with a brace running from it down to the foot of the clapping side. These gates can be wide enough to take large vehicles and machinery, and often have pedestrian side-gates to match. This is particularly useful where there is a public right of way through private grounds; in this case the vehicle gate is usually kept locked.

Field gates

These are also suitable for entrance gates to estate roads, allotments and so forth where large vehicles and machinery require access. They are normally made to standard sizes, and are made of treated softwood or tubular steel; heavier section gates are available for use where heavy stock such as bullocks or horses are kept. Hook or hook-and-band hinges and spring catches are normally factory fitted, but lockable bolts may be required.

Bridle or hunting gates

These are made of similar construction to field gates, but are higher (normally 1.3–1.4 m) to enable the rider to bend down to open them without

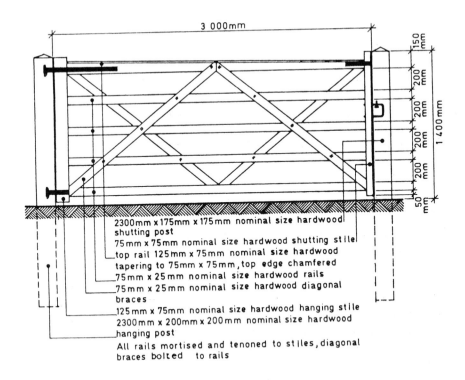

Fig. 4.64 Timber field gate.

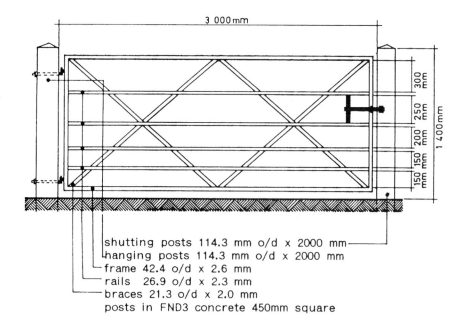

shutting posts 114.3 mm o/d x 2000 mm
hanging posts 114.3 mm o/d x 2000 mm
frame 42.4 o/d x 2.6 mm
rails 26.9 o/d x 2.3 mm
braces 21.3 o/d x 2.0 mm
posts in FND3 concrete 450mm square

Fig. 4.65 Tubular steel field gate.

dismounting; for this they require either an automatic self-closing catch or a loop-over catch to the gate as well as hook or hook-and-band hinges. As their name implies, they are not intended for harness horses, and are used in hunting country and on bridleways.

Kissing-gates

These have three-sided or V-shaped enclosures and a gate hung to clap between the open ends of the enclosure. They are usually made of timber, but standard tubular metal gates are made to match the adjacent fence, when the enclosure can be U-shaped. Wrought iron or mild steel gates can be made to a specific design for conservation work. They are intended to prevent the passage of stock while allowing pedestrians through, and they are useful on public paths because there is no risk of gates being left open. For wheelchair access the enclosure needs to be large enough for easy manoeuvring.

Stiles

Stiles are mainly used on public footpaths where stock have to be kept within their enclosure. Timber stiles are made of post and rail fencing with steps set at an angle to the fence. Metal stiles are of tubular steel and designed like ladders with one high post for holding on to. Where paths are suitable for handicapped people, alternative access should be considered for the elderly, wheelchairs and for prams. In drystone wall regions the stiles are either a ladder type or made of projecting stones.

BRITISH STANDARDS

3470 : 1975	Field gates and posts in timber and steel
4072 Pt 1 : 1987 & AMD	Wood preservation: copper/chromium, arsenic compositions
4072 Pt 2 : 1987	Wood preservation: copper/chromium, arsenic compositions
4092 Pt 1 : 1966	Domestic front entrance gates: metal gates
4092 Pt 2 : 1966	Domestic front entrance gates: wooden gates
5450 : 1977	Sawn hardwoods: sizes of hardwoods and methods of measurement
5707 Pt 1 : 1979	Wood preservatives: general purposes including timber to be painted
5709: 1979	Stiles, bridle gates and kissing-gates

DATA

Timber field gates to BS 3470 : 1975

Gates made to this standard are designed to open one way only.
All timber gates are 1100 mm high.
Width over stiles 2400, 2700, 3000, 3300, 3600 and 4200 mm.
Gates over 4200 mm should be made in two leaves.
Rails are tenoned into the stiles, and braces are bolted to rails.
Timber hanging posts are 2300 mm × 200 mm × 200 mm set into ground, or 2100 mm × 200 mm × 200 mm set in concrete.
Timber shutting posts are 2300 mm × 175 mm × 175 mm set into ground, or 2100 mm × 175 mm × 175 mm set in concrete.

Steel field gates to BS 3470: 1975

Heavy duty: width over stiles 2400, 3000, 3600 and 4500 mm
Light duty: width over stiles 2400, 3000 and 3600 mm
All steel gates are 1100 mm high.

Heavy-duty gates, tubular steel
Frame: 42.4 mm o/d × 2.6 mm
Rails: 26.9 mm o/d × 2.3 mm
Braces: 21.3 mm o/d × 2.0 mm
Hanging posts for 2400 mm and 3000 mm gates 88.9 mm o/d × 3.2 mm × 2000 mm long.
Hanging posts for 3600 mm and 4500 mm gates 114.3 mm o/d × 3.6 mm × 2000 mm long.
Shutting posts 88.9 mm o/d × 3.2 mm × 2000 mm long.

Light duty gates, tubular steel
Frame: 26.9 mm o/d × 2.3 mm
Rails: 21.3 mm o/d × 2.0 mm
Braces: 21.3 mm o/d × 2.0 mm
Hanging posts for 2400 mm and 3000 mm gates 76.1 mm o/d × 3.2 mm × 2000 mm long.
Hanging posts for 3600 mm gates 88.9 mm o/d × 3.2 mm × 2000 mm long.
Shutting posts 88.9 mm o/d × 3.2 mm × 2000 mm long.

Domestic front entrance gates to BS 4092 Pt 1 : 1966

Metal gates
Single gates are 900 mm high minimum, 900 mm, 1000 mm and 1100 mm wide.

Domestic front entrance gates to BS 4092 Pt 2 : 1966

Wooden gates
All rails shall be tenoned into the stiles.
Single gates are 840 mm high minimum, 801 mm and 1020 mm wide.
Double gates are 840 mm high minimum, 2130 mm, 2340 mm and 2640 mm wide.

Timber stiles to BS 5709 : 1979

Minimum width between posts	1000 mm
Maximum height of top rail above ground	900 mm
Maximum height of top rail above top step:	
one-step stile	450 mm
multi-step stile	600 mm
Maximum height of bottom step above ground	300 mm
Maximum gap between rails	300 mm

Supports set 500 mm in ground or in concrete.
Single steps not less than 45° to fence line.
Double steps at right angles to each other.
Stile posts set 750 mm into ground or in concrete.
Stile post must be high enough to give firm handhold above top rail.
All rails and posts to be chamfered on top and sides.
All rails to be mortised into posts.

Timber bridle gates to BS 5709 : 1979 (horse or hunting gates)

Gates open one way only.
Minimum width between posts: 1525 mm
Minimum height: 1100 mm

Timber shutting post: 125 mm × 125 mm
Concrete posts: 125 mm × 125 mm
Tubular hollow steel posts: 88.9 mm o/d × 4 mm
Square hollow steel posts: 80 mm × 80 mm × 3.6 mm
RSJ posts: 127 mm × 76 mm
Rails to be morticed to stiles.
Braces to be bolted to rails and birdsmouthed to bottom rail.
Posts set into ground 750 mm.
Top hinge 600 mm minimum.

Timber kissing-gates to BS 5709 : 1979

Minimum width: 700 mm
Minimum height: 1000 mm
Minimum distance between shutting posts: 600 mm
Minimum clearance at mid-point: 600 mm
Sizes of timbers and construction are the same as those for bridle gates.

Metal kissing-gates to BS 5709 : 1979

Sizes and spacing of rails and posts are the same as those for timber kissing-gates.

Maximum gaps between rails: 120 mm

Note: The dimensions of these kissing-gates are very restrictive, and wider spacings are needed when elderly people, pushchairs and physically handicapped people must use the gates.

Concrete mixes to BS 5328 Pt 1 & AMD

Application	Standard mix	Designated mix	Slump (mm)
Mass concrete fill and blinding	ST 2 (C10P)	GEN 1	75
Mass concrete foundations in non-aggressive soils	ST 2 (C10P)	GEN 1	75
Trench fill foundations in non-aggressive soils	St 2 (C10P)	GEN 1	125
Foundations			
in Class 2 sulphate soils	–	FND 2	75
in Class 3 sulphate soils	–	FND 3	75
in Class 4A sulphate soils	–	FND 4A	75
in Class 4B sulphate soils	–	FND 4B	75

Traditional concrete mixes

Application	Name of mix	Cement:aggregate:sand (parts by volume)
Setting posts, filling weak spots in ground, backfilling pipe trenches:	one, two, three or one to five all-in	1:2:3 or 1:5

'All-in' is mixed aggregate as dug from the pit.

OUTLINE SPECIFICATION

Timber field gate

a. *Supply and erect five-bar diamond-braced treated softwood timber field gate to BS 4370, 1100 mm high × 2700 mm wide.*

b. *Hang on 2300 mm × 200 mm × 200 mm treated softwood hanging post and 2300 mm × 175 mm × 175 mm treated softwood shutting post set 750 mm into firm ground.*

c. *Fittings to be galvanized steel hook-and-band hinges, spring catch and padbolt.*

   for timber treatment   *M60*

Steel field gate

a. *Supply and erect tubular mild steel heavy-duty diamond-braced field gate to BS 4370, 1100 mm high × 2400 mm wide.*

b. *Hang on 88.9 mm o/d × 3.2 mm × 2000 mm long capped tubular steel posts set in ST2 concrete 450 mm × 450 mm × 600 mm deep.*

c. *Fittings including hinges, automatic catch and drop bolt, gate, posts and fittings to be hot-dip galvanized for painting on site.*

   for metal treatment   *M60*

Timber garden gate

a. *Supply and erect timber garden gate to BS 4092 Pt 2, 900 mm high × 1020 mm wide, stiles 95 mm × 45 mm, rails 70 mm × 45 mm, vertical rails 70 mm × 45 mm, all in selected hardwood.*

b. *Hang on 1500 mm × 100 mm × 100 mm hardwood posts set into ST2 concrete 300 mm × 300 mm × 600 mm deep.*

c. *Fittings to be wrought iron tee hinges, drop latch and drop bolt with boat socket. Gate to open one way only.*

for concrete mixes	*E10*
for timber treatment	*M60*

Steel garden gate

a. *Supply and erect mild steel garden gate to BS 4092 Pt 1, 1000 mm high × 1100 mm wide, square section frame 15.9 mm × 15.9 mm with vertical solid mild steel bars 12.7 mm × 12.7 mm.*

b. *Hang on 1600 mm × 100 mm × 100 mm selected hardwood posts set in ST2 concrete 300 mm × 300 mm × 600 mm deep.*

c. *Fittings to be welded hook-and-band hinges, mild steel drop latch and drop bolt with boat socket. All metalwork factory primed for painting on site. Gate to open one way only.*

for concrete mixes	*E10*
for metal treatment	*M60*

Timber stile

a. *Supply and erect selected hardwood stile to BS 5709, 1100 mm wide between posts, top rail 100 mm × 50 mm set 900 mm above ground with two rails 87 mm × 48 mm, all rails mortised into posts.*

b. *2 no. steps 900 mm × 200 mm × 50 mm set at right angles to each other and at 45° to the fence line on step supports 150 mm × 100 mm set 50 mm back from end of step and set 500 mm into firm ground.*

c. *Stile posts 100 mm × 100 mm set 750 mm into firm ground, one post to be 750 mm above top rail, all rails and posts to be chamfered on top and sides*

for timber treatment	*M60*

Kissing-gate

a. *Supply and erect treated softwood kissing-gate and guard rails to BS 5709, 1100 mm high, gate 1000 mm wide, clearance at midpoint 750 mm.*

b. *Gate to be ex 100 mm × 100 mm × 75 mm top rail, 4 no. 75 mm × 25 mm rails and brace. Rails mortised into 100 mm × 75 mm hanging stile and 75 mm × 75 mm swinging stile.*

c. *Hang on 2000 mm × 150 mm × 150 mm post set 750 mm into firm ground, with one pair galvanized mild steel hook-and-band hinges.*

d. *Guard rails; top guard rail 100 mm × 75 mm, 4 no. 75 mm × 25 mm rails housed and nailed to 3 no. 2000 mm × 150 mm × 150 mm posts set 750 mm into firm ground.*

for timber treatment	*M60*

Q40 Fences and gates: pedestrian guard rails

GENERAL GUIDANCE

Standard pedestrian guard rails are used to separate pedestrians from traffic in urban areas, supermarket car parks and other potentially dangerous areas or situations. The guard rails described here are not intended for use in sports stadia or for crowd control where large numbers congregate for festivals, pop concerts and the like. The design and installation of all guard rails is strictly controlled by British Standards, and the specifications given should not be varied. Gates in pedestrian guard rails are made to match the railings.

Guard rails are manufactured of plain round mild steel bars welded top and bottom to flat rails to form panels bolted to plain square or round tubular steel posts. The panels must be firmly fixed to withstand vehicle impact, but they may have to be removed for road works or replacement after traffic accidents. Heavy ground anchors or sockets with locks or sunk-head bolts are suitable for fixing posts. Welded mesh panels may be used where there is a risk of objects falling or being thrown through the bars to the danger of the traffic, but there must be good visibility through them.

Gates in pedestrian barriers are designed to the same specification as the barriers themselves, and are usually the same height as the barrier panel. Single gates provide maintenance access to the pedestrian area; double gates should be specified where plant and vehicles need access, and they should be wide enough for fire engines and rescue vehicles. No gate may open onto the carriageway. The fire brigade, the ambulance service and the police will advise on the most suitable types of gates and locks for public barriers.

BRITISH STANDARDS

729 : 1971	Hot-dip galvanized coatings for iron and steel
1722 Pt 14 1992	Open mesh steel panels
3049 : 1976	Pedestrian guard rails
4921 : 1988	Sheradized coatings for iron and steel

DATA

Categories of pedestrian guard rail to BS 3049 : 1976

Class A for normal use:
the rails must withstand	700 N/m
and the infill panels	500 N/m

Class B where vandalism is expected:
the rails must withstand	700 N/m
and the infill panels	1000 N/m

Class C where crowd pressure is likely:
rails must withstand	1400 N/m
and the infill panels	1000 N/m

Fig. 4.66 Pedestrian guard rail, Wandsworth riverside, incorporates a wave pattern. (Claydon Architectural Metalwork)

Dimensions of guard rails

Distance between top rail and bottom rail overall: 900 mm
Gap between top rail and intermediate rail: 200 mm
Gap between bottom rail and ground: 100 mm
Height from ground to top of top rail: 1000 mm
Posts to be not further apart than: 2000 mm
Minimum depth of posts in ground: 300 mm
No gap in infill or round infill greater than: 100 mm
Mesh aperture perimeter not greater than: 200 mm

Gates must not open onto the highway.
All components to be marked with BS number.

Components of guard rails

All dimensions in mm:

	Steel	Aluminium	Other materials
Minimum thickness of rails:		2.0	4.0
sealed hollow section	2.5		
other sections	4.0		
Minimum thickness of posts:		2.0	5.0
sealed hollow section	3.0		
other sections	5.0		
Minimum thickness of infill:		1.5	
sealed hollow section	2.0		2.0
other sections (and mesh)	3.0		3.0

OUTLINE SPECIFICATION

Pedestrian guard rail

a. *Supply and erect mild steel pedestrian guard rail Class A to BS 3049, panels 1000 mm high × 2000 mm wide with 100 mm toe space and 200 mm visibility gap at top, rails to be rectangular hollow sealed section 50 mm × 30 mm × 2.5 mm, vertical support 25 mm × 19 mm central between intermediate and top rail.*

b. *End and gate posts to be rectangular hollow sealed section 60 mm × 40 mm × 3.0 mm, infill rectangular solid bars 25 mm × 19 mm at 100 mm centres. Posts to be set 300 mm into paving base.*

c. *All components factory welded and factory primed for painting on site.*

for painting   M60

Gates

Supply and erect single gates to match above with self-locking catches and retaining lockable bolts. Gate to be one-way opening away from highway.

Supply and erect double gates to match above with self-locking catches and retaining lockable bolts. Gates to be one-way opening away from highway.

for painting   M60

Q50 Street furniture generally

This section covers most of the fittings and fixtures associated with landscape work: litter bins, outdoor seats, plant containers, cycle holders, flagpoles and also playground equipment and vehicle barriers used in urban areas, as well as the bases for these items.

The vast variety in street furniture of all types leaves the landscape designer with many difficult decisions to make; however, the choice of one particular manufacturer's items, or the use of a specific material to give a coherent appearance, reduces the selection to a reasonable size. This is further reduced by taking into account the area in which the street furniture is to be used, the degree of use expected over its economic/useful life, and the type of people using it. Obviously the requirements of the elderly and/or disabled are not the same as those of healthy young students on a campus, nor are the requirements for seating where people can be expected to sit on it for a few minutes the same as those where people are likely to want to sit and enjoy the scenery for an hour or so at a time. Few of the ergonomic requirements for these situations are compatible, and they must be taken into consideration before a final choice is made. There are also areas where people may wish to sit or lie, but for public order reasons they must be discouraged from doing so.

LOCATION

Street furniture must be located where it does not cause an obstruction to the footway, and allowance must be made for wheelchairs to pass; seats and planters should not obstruct drivers' sight-lines at corners and junctions. The local authority and the highway authority are responsible for the control of street furniture in public roads and urban areas. Most street furniture in public areas requires planning permission before it can be erected. It is preferable not to place street furniture over a service run, as it will have to be moved for repairs, and may not be replaced. Access for emergency vehicles must be kept clear. In order not to weaken the kerb, all ground fixings for street furniture should be kept 450 mm in from the kerb. Signs and markers are often placed in grassed areas to keep them away from the road, and where this is unavoidable they should be set in a concrete or paved base to enable mowing to be carried out easily.

SAFETY

It is perhaps in the area of street furniture and playground equipment in particular that accidents occur, and the designer of the equipment, the landscape designer and the contractor have a 'duty of care' to protect the public at large from foreseeable hazards. It is of course of great importance that the client/owner is fully aware of his responsibilities after completion of the final hand-over of the project, as it is then up to them to ensure that the entire area is well maintained.

Playground fencing and gates have already been dealt with in Q25, but some of the general safety requirements are applicable to street furniture in general as well as play equipment, and will be repeated here:

- There should be no spikes, sharp edges or arrises where children can cut or impale themselves.
- Gaps between rails, boards or planks should be no more than 100 mm to prevent heads from getting trapped.
- Any mesh material used for seats or infill panels must be small to prevent small fingers getting trapped.
- Bolt heads and nuts should be sunk and preferably removable only with special tools.
- Paint or any other finishings should be non-toxic and non-staining.
- Equipment, particularly where toddlers play, should not swing, spring or make any other movement likely to trap fingers or toes.
- Some type of barrier should be placed in pedestrian areas or play areas next to roads where children are liable to run out into danger.
- It is the responsibility of the equipment supplier to state the maximum potential free-fall height (MPFFH), and the responsibility of the ground surfacing manufacturer to specify the actual surface.

Any equipment requiring an electrical supply must of course be wired by a qualified electrician, and all access points must be kept well covered or out of the way of children.

ELDERLY AND DISABLED USERS

Wheelchair users, elderly, physically disabled, poorly sighted or blind people must be considered, as they are general users of all types of street furniture; where necessary their special needs must be considered. Wheelchair users cannot reach comfortably below 300 mm or above 1400 mm, and park seating should have a level space on at least one side of 1000 mm wide so that they can sit comfortably beside their friends. The approach to seating should also be level and wide enough so that manoeuvring of wheelchairs and wheeled walking aids (or 'zimmers') is made safe and easy.

SHELTER

At least some seating should be provided with shelter/shade for both adults and small children, particularly where they can be expected to sit for a considerable time; this applies to parks, playgrounds, picnic areas and so forth. Refer to subsections:

for fences	Q40
for paving	Q10–25

Q50 Site and street furniture: vehicle barriers

GENERAL GUIDANCE

This section includes cattle grids, vehicle barriers, vehicle crash barriers, and bollards.

Cattle grids

These are used mainly in country parks and open forestry areas where cars and service vehicles need constant access, but where stock must be prevented from straying. They consist of a set of heavy round or square steel bars laid on dwarf brick or concrete walls in a pit; the spacing is far enough apart to prevent large animals from walking across them. They are built on site to order. The British Standard gives recommendations for spacing and sizes of grid; the strength of the grid varies according to the type of vehicle expected, but two-way traffic is not usually allowed except on busy public roads such as those in the New Forest. It is usual to construct a field gate alongside for the use of very heavy vehicles, horse-drawn carts and riders; this should be self-closing or locked with authorized entry only, and guard rails must be provided to both sides of the pit. Drainage to the grid pit must be provided, and a crawlway should be provided to enable small mammals and reptiles to escape from the pit. Warning notices at a suitable distance are required on the road at each side of the cattle grid. Maintenance of the grid is essential to prevent build-up of soil in the pit, which would allow stock to escape.

Vehicle barriers

Protective railings and pedestrian guard rails are covered in Q40; this section deals with control barriers only. The massive barriers used on highways are not usually required in landscape contracts, though they may be necessary where heavy lorries have to be kept away from other traffic, or where a change in ground level involves a raised roadway, which could be a hazard for vehicles out of control. Vehicle barriers and traffic control will usually call for expert engineering design input. The most usual purposes of vehicle barriers are to keep pedestrians and cars separate, to control car movement within the site, and to prevent unauthorized or illegal parking.

Barriers may be simple poles operated by hand, drop barriers that prevent people passing underneath, key- or card-operated poles to stop unauthorized entry, or complex electronically controlled security barriers designed to stop ram-raiding. In all types, there must be provision for emergency services to gain access; usually the emergency services and sometimes the local authority have keys to standard padlocks. Otherwise the emergency services must be able to lift off the poles or cut the fastenings on the barrier; they should be consulted before a specification is prepared. 'Flow plates' are small steel plates that can be driven over in one direction while springing back into a vertical barrier to prevent traffic from passing in the other direction.

Bollards

Bollards are most often used where pedestrians, trolleys and bicycles may pass freely, but cars are restricted. Spacing of bollards must allow for the passage of wheelchairs, prams and hand-operated cleaning and mowing equipment, but prevent cars. It is very difficult to prevent motor cycles from getting through bollards; although some types have been designed with a bar at handlebar height, this is equally inconvenient for wheelchairs.

Bollards can be made of almost any material – timber, concrete, stone, steel or cast iron – and they can be made removable but locked by means of locking pins or by sheer weight. Bollards can be illuminated; these are covered in V41.

Timber bollards are usually hardwood stained or painted; they are not suitable for traffic control where there is a risk of crashing.

Precast reinforced concrete bollards are supplied plain, coloured, or aggregate faced. They are made in a wide range of sizes and shapes, from low drum shapes used as seats to heavy crash-proof posts.

Natural stone bollards are available in granite and other hard igneous rocks; sandstone and limestone are suitable only for boulders which can serve the purpose of bollards.

Steel bollards are usually made from circular or square hollow steel sections with ornamental cappings and mouldings applied. They are easy to fit with locking devices, but more susceptible to damage. They must be galvanized, powder coated or painted to prevent corrosion.

Cast or ductile iron bollards are most often used in civic centres, because they can be cast with logos, badges or crests, and can be painted or gilded. They are subject to theft, and should be very firmly fixed; most thefts occur by simply lifting them out of the ground with levers.

BRITISH STANDARDS

3049 : 1976	Pedestrian guard rails
4008 : 1991	Cattle grids on private roads
5696 Pt 1 : 1986 & AMD	Play equipment intended for permanent installations outdoors: methods of test
5696 Pt 2 : 1986 & AMD	Play equipment: construction and performance
5696 Pt 3 : 1979 & AMD	Play equipment: code of practice for installation and maintenance
6180 : 1995	Code of practice for protective barriers in and about buildings, and vehicle barriers up to 10 mph
6571 Pt 1 : 1989	Coin operated clockwork parking meters
6571 Pt 2 : 1989 & AMD	Electrically powered parking meters (coins, tokens, cards)
6571 Pt 3 : 1989	Pay and display equipment
6571 Pt 4 : 1989	Barrier type parking control
6579 Pt 1 : 1988 & AMD	Safety fences and barriers for highways: tensioned corrugated beam on Z posts
6579 Pt 3 : 1988	Safety fences and barriers for highways: tensioned rectangular hollow section beam
6579 Pt 5 : 1986 & AMD	Safety fences and barriers for highways: open box beam, single height
6579 Pt 6 : 1988	Safety fences and barriers for highways: open box beam, double height
6579 Pt 7 : 1988	Safety fences and barriers for highways: untensioned corrugated beam
6579 Pt 8 : 1987	Safety fences and barriers for highways: concrete safety barriers
6779 Pt 1 : 1992	Parapets for vehicle containment on highways: parapets of metal construction
6779 Pt 3 : 1994	Parapets: vehicle containment on highways: combined metal/concrete construction

DATA

Cattle grids

Grids not to be within 18 m of a road junction.
Transverse gradient not more than 1:40.
Pit length for stock (except deer): 2.6 m min.
Pit length for deer: 4.0 m min.
Pit width: 2.75 m min.
Pit depth: 250 mm min., 450 mm max.
Clear space between rails 130–150 mm, transverse rails to be clear space dimension +80 mm max.
Loading shall be appropriate to the road loadings.
Guard rails shall be on both sides of the pit and extend 2.4 m beyond the pit on the stock side and 600 mm on the other.
Escape ramps 150–200 mm wide × 1:2 slope with rough surface. Dividing walls shall have 150 mm × 150 mm openings.

Vehicle barriers

Parking posts
Hinged post with top padlock.
Lift-out post with bottom padlock.
Hinged space reserver with top lock.

Pole barriers
Manually operated up to 7.0 m, lockable closed and open (allow room for the counterweight).
Electrically operated up to 6.0 m long × 1.0 m high, lockable open and closed, manual override.
Electrically operated, heavy duty up to 9.0 m long × 1.0 m high with lattice skirt, warning light, stop notice, impact safety switch, lockable open and closed, manual override.

Controls for pole barriers and flow plates

- Pass card reader
- Key
- Security guard
- Remote radio control
- Coins and tokens
- Timed ticket
- Traffic lights
- CCTV or voice communication
- Vehicle number counters
- Pressure pads

Bollards

Timber

Typical sizes and shapes (all dimensions in mm):

	high	long	round	square
Hardwood	500	750	125–250	125–300
	600	1000	–	150
	820	1220	100	–
Softwood	900	1500	120–190	–

Cast iron

Typical sizes and shapes (all dimensions in mm):

920 high	1210 long	150 round	fluted
1000	1300	160	cannon profile
1080	1480	250 square	ribbed
1100	1350	125	slender, ball top
1131	1285	180	ornamented
1280	1780	450	fluted
250	bolted down	160 round	traditional wharf bollard

Concrete

Typical sizes and shapes (all dimensions in mm):

355 high	405 long	500 round	exposed aggregate
400	450	220–260	verge marker
500	900	280	exposed aggregate
525	950	300	exposed aggregate
750	1200	250	plain concrete
850	1372	220	plain concrete
910	1220	200–230	exposed aggregate

OUTLINE SPECIFICATION

Cattle grid

a. *Excavate pit for cattle grid, spread spoil round pit to 20 m radius, construct 4 no. one-brick dwarf sleeper walls 1.4 mm and 2.0 mm apart in engineering brick Class B laid in Group 1 cement:sand mortar 1:3, opening 2 bricks high × 1 brick wide in each wall.*

b. *Construct concrete escape ramp 200 mm wide at 1:3 slope to one side of pit.*

c. *Excavate trench and lay 110 mm perforated plastic land drain to BS 4962, bedding Type 4 to outfall as shown on drawings.*

d. *Supply and install proprietary cattle grid 2370 mm long × 2900 mm wide in accordance with manufacturer's instructions. Loading 10 tonnes.*

e. *Supply and erect cleft timber post and rail fencing to BS 1722 Pt 7, Type MPR 11/3, 6.0 m long with timber posts set into ground to both sides of grid.*

f. *Supply and erect timber field gate to BS 3470 3300 mm wide with posts set into ground.*

for excavation	*D20*
for timber fences	*Q40*
for timber gates	*Q40*
for land drainage	*R13*

Vehicle barrier

Supply and install as shown on drawings 2 no. proprietary vehicle barriers Type XXX 3500 mm long, electrically operated by vehicle number count, with STOP warning, impact safety switch, lockable open and closed, safety manual override and CCTV to control point.

Parking posts

Supply and install 100 no. proprietary powder-coated steel parking posts and bases Type XXX to specified colour as shown on drawing, fold-down top-locking type, complete with keys and instructions. Supply and install 99. no. powder-coated steel fixed posts to match, posts and bases set in ST1 concrete footing 200 mm × 200 mm × 300 mm deep.

Bollards

Supply and fix 300 no. proprietary hardwood bollards Type XXX as shown on car park drawings, 600 mm high × 150 mm square set 400 mm into ST1 concrete footing 400 mm × 400 mm × 500 mm deep. Bollards to be factory stained to specified colour, one coat stain to be applied on site.
or
Supply and install 50 no. proprietary ductile iron bollards Type XXX 1080 mm high × 250 mm square set 400 mm into ST1 concrete footing 500 mm × 500 mm × 600 mm deep. Bollards to have raised cast lettering as shown on drawing on one face only and to be supplied primed ready for painting and gilding on site.
or
Supply and install 10 no. proprietary precast concrete Type XXX 'mill-stone' bollards 355 mm high above ground × 405 mm deep, 500 mm dia. where shown on drawing, white spar exposed aggregate finish, bedded in ST1 concrete base 750 mm dia. × 300 mm deep.

Q50 Site and street furniture: playground equipment

GENERAL GUIDANCE

This section includes climbing equipment, swings, slides and seesaws.

There are many ready-made pieces of play equipment on the market, but the landscape designer can design pieces to complement these as long as thought is given to the safety aspect in relation to the children of different ages using the equipment, as set out in BS 5696. The standard covers the manufacture and installation of the equipment, and also sets out minimum clear safety dimensions round each type of equipment. BS 7188 sets standards of safety for the playground surface itself; this is covered in section Q26. In practice the safest course is to allocate an area for the playground, to specify the type of equipment to be provided, and then to obtain quotes for combined surface and equipment supply and installation. The safety surface should be designed to match the equipment selected. Both supply and installation must be certified by the contractor to comply with all safety regulations and standards.

The Standards require that the surfacing shall prevent critical head injury if children fall from equipment. This is the **maximum potential free-fall height** (MPFFH), and is calculated as the height of any access point or platform. Children may climb over guard rails as well as slipping off a platform, and it would be preferable to specify the top of the guard rails as the MPFFH. These heights should be obtained from the playground equipment manufacturer. The usual procedure is to select equipment, obtain the MPFFH figures for each item, and then to specify a playground surface that meets those criteria.

Playgrounds generally require seating with shelter/shade for both adults and small children; the type of equipment used should provide children with an opportunity to balance, climb, jump, slide, swing, rock, rotate, and crawl.

All play equipment should be under regular surveillance, and that designed for toddlers should be within an enclosed space surrounded by dog-proof fencing.

Equipment

All playground equipment must be surrounded by the statutory safety zones and be fixed on a safety surface compatible with the height of the equipment. Swings in particular should be sited away from other items, with unclimbable barriers erected to prevent children from running into the path of the swing. Embankment or contour slides can be built into natural or artificial slopes; they comprise an entry section (sit down), a slide at 30°, a transitional (slowing down) section, a run-out chute, and for high slides an additional level run-out chute. Steps should be provided at least 1.0 m minimum away from the slide. The surfaces adjoining the slide should not contain any projections.

Most playground equipment manufacturers supply compound installations of 'multi-play' equipment designed to create a complete exercise course, some-

Fig. 4.67 Action pack play system for 7-14 year olds, modular. (SMP (Playgrounds) Ltd)

Fig. 4.68 Horizon play system for 4-12 year olds, modular. (SMP (Playgrounds) Ltd)

times fancifully constructed and decorated, but individual pieces of equipment are obtainable, which can be coordinated by the landscape designer.

BRITISH STANDARDS

5696 Pt 2 : 1986 & AMD   Playground equipment intended for permanent installation outdoors: specification for construction and performance

5696 Pt 3 : 1979 & AMD   Playground equipment intended for permanent installation outdoors: code of practice for installation and maintenance

7188 : 1969   Methods of test for impact absorbing playground surfaces

Some imported equipment is to the German Standard:
DIN 7926   Play equipment for children

BS 5956: Pt2: 1986 & AMD includes very detailed requirements for safety in play equipment. All play equipment to be certified by the supplier and marked as conforming with the BS.

DATA

Suitable equipment for a toddlers' play area

Dimensions (all in mm) do not include safety zone.

Swings with cradle seats:			
single	2200 wide	× 1900 long	× 2200 high
double	3500	1900	2200
Slides, freestanding:			
single	460	3500	2400
double	3500	1000	2400
Slides, embankment:	460	4200	2800
	460	3900	3300
Playhouse:	700	600	1400
	2000	2000	1900
Spring equipment:	600 dia.	to 1000 dia.	
Seesaw:			
single	400	3000	500
double	1000	3000	550
Small climber:	1000	900	1000
Agility bars:	500	2000	300
Sand play (supervised):	2000	2000	400
Water play (supervised):	500	1000	150
(bed swings are available for handicapped children)			

Note: All must be to BS 5696 or equivalent standard and of a suitable scale for toddlers; preference should be given to an impact-absorbent surface.

Suitable equipment for a junior play area

Dimensions do not include safety zone.

Swings, single or in groups:			
double	2490 mm wide	2000 mm long	2400 high
four	5180	2000	2400
Log cabin, seats, slide:	2580	4180	2430
Composite equipment for climbing and balancing:			
	1000	1200	1200
Combined equipment area:	8.0 m	13.0 m	
Ball game area:	10.0 m	12.0 m	
Wheeled toy area:	6.0 m	10.0 m	

Note: All must be to BS 5696 or equivalent standard and of a suitable scale for juniors.

Suitable equipment for neighbourhood parks

Dimensions do not include safety zone.
For older children provide:

Larger groups of swings:			
six seater	7110 mm wide	2000 mm long	3000 mm high
Composite structures for agility:	300 mm wide	4560 mm long	2000 mm high
Combined equipment area	12.0 m	10.0 m	
Slide:			
freestanding	450 mm	4000 mm	1500 mm high
embankment made to order any length continuous steel trough at 30°			

Note: All must be to BS 5696 or equivalent standard and of a suitable scale for older children. The contractor should be required to certify that it has been erected in accordance with BS 5696.

All equipment purchased should provide the following information:

- life expectancy and guarantees;
- compliance with the current edition of BS 5696;
- manufacturer's name and date of manufacture;
- where applicable, a parts list and assembly instructions;
- minimum use zones;
- suggested surfacing requirement;
- inspection and maintenance details in accordance with BS 5696 Pt 3.

Handrails and guarding rails giving access to equipment

Provide guard rails for:

- All equipment (except climbing frames) where access to equipment exceeds 500 mm above the ground or level below.
- Vertical rails only to be used in order to prevent children climbing on them. Where vertical bars are used, they must not have a clear gap greater than 100 mm between them.
- Top rails to be between 18 and 40 mm and between 500 and 900 mm above the pitch of the steps.
- Where perforated infill panels are used, no hole should be greater than 25 mm in any direction.

Playground surfaces to BS 7188

Impact-resisting surface must be a minimum of 1.75 m beyond outside edge of static equipment, and beyond the limit of travel of moving equipment.

Maximum potential free-fall height (MPFFH)

Height of any access point or platform; heights obtained from playground equipment manufacturer.

Safety of playground equipment and surfaces

- There should be no spikes, sharp edges or arrises where children can cut or impale themselves.
- Bolt heads and nuts should be sunk and preferably removable only with special tools.
- Paint or preservative must be non-toxic and non-staining.

OUTLINE SPECIFICATION

Toddlers' play area

a. *Supply and erect combined units on safety surface 7600 mm × 7600 mm specified in Q26, including steps, slide, platform and enclosed play area.*

b. *Supply and erect two double swing units 3500 mm wide × 1900 mm long × 2200 mm high on safety surface specified in Q26, frames set into concrete foundation, all in accordance with manufacturer's instructions.*

c. *Supply and erect free-standing slide 460 mm wide × 3500 mm long × 2400 mm high on safety surface specified in Q26, frames set into concrete foundation, all in accordance with manufacturer's instructions.*

d. *Supply and erect safety seesaw 400 mm wide × 3000 mm long × 500 mm high on safety surface specified in Q26, frames set into concrete foundation, all in accordance with manufacturer's instructions.*

for playground surfaces   *Q26*

Senior play area

a. *Supply and erect combined units on safety surface 12.0 m × 10.0 m specified in Q26, including fixed trapeze, suspension bridge, platform, slide and steps.*

b. *Supply and erect six-seater swings 7110 mm wide × 2000 mm long × 3000 mm high on safety surface specified in Q26, steel frames set into concrete foundation, all in accordance with manufacturer's instructions.*

for playground surfaces   *Q26*

Q50 Site and street furniture

GENERAL GUIDANCE

Street furniture comprises all the items that are an important part of the street scene, including those used in parks and gardens, with the exception of lighting fittings and columns, which are covered in section V41, and bollards, which are covered in Q50 *Vehicle barriers*. Railings and pedestrian guard rails are covered in section Q40. The safety factors to be considered and the degree of vandalism likely to be encountered are further problems to be addressed; these are discussed in *Street furniture generally*.

The main street furniture items likely to be specified in a landscape contract include seats and benches, tables, picnic benches, litter and salt bins, signs, cycle stands, flag-poles, and prefabricated planters.

Materials

All street furniture must be weatherproof, vandal resistant for both knives and graffiti, ergonomically satisfactory, capable of standing up to misuse, firmly fixed, safe to use, and clearly visible to the partially sighted. Most urban furniture is therefore made of steel, concrete, cast (ductile) iron, heavy-duty plastic, GRC, or a combination of these materials, though hardwood can be used as infill material or as complete units in protected areas.

Seats and benches

Generally, seats have backs and benches are backless. Most types are constructed on a steel or concrete frame with an infill of steel mesh or slats, or they may be made of cast iron sections bolted together, while simple short benches may be supplied as one-piece units. Seats and benches are bolted down or set into the ground. Some types are supplied for cantilever fixing to walls; they may be permanently fixed or removable by means of keys or special tools. Sizes range from individual seats to 2.0 m long units. It is advisable to install a group of seats at slightly different heights from the ground, as it is important that people can rest their feet on the ground while sitting, and certain types of physical handicap are best served by high seats and others by low seats. Varying-height furniture should be used only in exceptional circumstances, as irregular furniture may look accidental rather than intentional. All seats and benches must drain quickly after heavy rain, and they should not hold water, ice or snow, which means that solid seats are not feasible unless slightly sloped.

Tables and picnic benches

These are most often required in inn gardens, country park centres, roadside picnic areas and similar locations. The surface of the tables has to remain flat and stable at all times without water pools, and the type with a closely slatted top will drain better than a solid surface. The tables must be resistant to food chemicals, cigarettes and spirits, and not liable to splinter or chip. A proportion of picnic tables should have an unobstructed cantilever area at the end to allow for wheelchairs.

Litter bins

Bins must stand up to any kind of chemical or burning rubbish, and withstand large items being pushed into them, children climbing on them, and impact from cycles and trolleys. Depending on the situation, ground-based bins may be more stable than wall-mounted types, but they can form an obstruction to cleaning machinery. The inner liners may be plastic sacks in supervised areas, or metal liners that are locked into place. Dog-waste bins are now usually provided in areas where children play, or around the perimeter of public sports fields.

Grit and salt bins

Public roads are usually gritted and salted by special machinery, but superstores, loading yards and private roads may need a supply of grit and salt for hand spreading. The bins should contain enough material to treat the whole area, and a surplus to allow for extra spreading. They must be robust enough to hold the weight of the material, waterproof and lockable, and proof against damage by trolleys or forklift trucks.

Signs

Illuminated signs are covered in V41. Signs can be fixed on two supports rather than one where there is a risk of vandalism. The surfaces should be graffiti proof and easily cleaned unless they are high enough to be safe, and they should be fixed with special tools. Wooden carved signs are attractive in rural areas, but are very vulnerable to theft and damage; metal enamelled signs can be chipped and may rust; plastic signs can be cut and burnt. The sturdiest material is concrete or stone with incised lettering, which is difficult to damage, though the slab or post must be well bedded or it can be removed or overturned. Traditional Victorian cast iron signs are very strong

Fig. 4.69 Victor Stanley S35 90 litre bin. (CU Phosco Street Furniture)

but expensive, and they must be fixed to posts or walls. Raised or sunken lettering can be read even if covered with paint.

Cycle stands

Provision must be made for locking the cycle, preferably at waist height, and also for supporting the machine without straining the frame or wheels. The most suitable type of stand is the plain steel or cast iron post-and-rail to which the cycle can be padlocked; this type looks well when empty or full of cycles, and can be arranged to fit in with most street layouts. Unless cycle rails are provided, the adjoining seats and railings will be used to secure cycles. Where there are a large number of cycles left for a long period, as at universities and other student or shopping facilities, some shelter should be provided; this must keep rain and snow off the machines without obstructing supervision by security staff, and should be well lit.

Flag-poles

Flag-poles come under the control of the Planning Acts, and permission must be sought if they form part of the design. A hard-standing space must be left at the foot of the pole to raise and lower flags, and very tall poles are lowered to the ground for flag changing, so that a space the height of the pole must be left clear.

Planters

This section deals with prefabricated planters only. They may be made of wood, stone, precast concrete, glass fibre, plastic or terracotta, but they have to be robust enough to stand the weight of soil and plants, and to resist accidental or deliberate damage. The planter must be stable enough to prevent overturning by wind action on mature trees and shrubs or by vandalism, and it may be advisable to anchor it to the ground. Drainage is necessary for plants, and irrigation may be needed in big planters for establishing and maintaining large plants, though water bowsers may be used where regular watering is not critical. Holes in the sides and base of small free-standing planters, together with a watering point, are satisfactory, but *in situ* planters should be treated as building construction and serviced accordingly. A water point should be easily accessible for planters, as otherwise they will not be properly maintained.

Bases

Street furniture can be bolted down to sockets cast in the paving, padlocked to ring-bolts, fixed with special ground anchors, cast into a concrete base, or bolted to a wall. The type of fixing selected will depend on the degree of vandalism expected and the need to remove the items at intervals; some street furniture is so heavy that it is self-securing. All street furniture should be placed on a hard surface, which can give a secure hold to fixings and provide a dry clean surface for users. The base should extend into grassed areas far enough round the furniture so that mowing and street cleaning can easily be carried out.

BRITISH STANDARDS

There are no British Standards specifically for street furniture.
4848 Part 2 : 1991   Steel: hot-finished hollow sections

DATA

Typical benches

All dimensions in mm:

Hardwood with concrete supports set in ground:	720 high	× 670 wide	× 1670 long
Hardwood slats on cast iron frame:	490	× 440	× 1840

Typical seats

All dimensions in mm:

Hardwood with concrete supports set in ground:	760 and 450 high	× 655 wide	× 1800 long
Steel slats and frame:	810 and 430	× 620	× 1400
Hardwood slats on cast iron frame:	860 and 450	× 680	× 1840
High back: hardwood slats on powder-coated steel frame:	900 and 460	× 590	× 1800

Typical tables

Hardwood on cast iron: 730 mm high × 640 mm wide × 1840 mm long

Typical picnic benches

Hardwood table and seats: 698 mm high × 1500 mm wide × 1670 mm long
Hardwood on powder-coated steel with fold-up seats: 800 and 470 mm high × 1850 mm wide × 1800 mm long

Typical litter bins

Cast iron or ductile iron with metal liner. Ground fixing:
    800 mm high × 500 mm square
    800 mm high × 400 mm dia.
Polyethylene with covered top and refuse sack. Ground fixing:
    110 litres × 1070 mm high × 500 mm dia. Lockable, fire extinguisher optional.
GRC with covered top and metal liner:
    0.01 m³ × 950 mm high × 480 mm wide × 900 mm long. Fire resistant
    0.01 × 950 mm × 480 mm × 900 with lift-up lid, timber clad
GRC with open top and metal liner. Ground fixing:
    0.13 m³ × 610 mm high × 430 mm wide × 960 mm long
    0.12 m³ × 700 mm high × 530 mm square
Reinforced GRC dog-waste bin, one-way opening, fly-proof. Ground fixing:
    0.11 m³ × 1030 mm high × 480 mm wide × 450 mm long
Powder-coated steel bins. Wall or pole fixing:
    56 litres × 670 mm × 300 mm square. Open top with metal liner.
Precast concrete with exposed aggregate finish, metal liners. Ground fixing:
    45 litres × 760 mm × 430 mm dia.
    300 litres × 850 mm × 600 mm wide × 1200 mm long
All these bins can be supplied with logos or signs on the sides.

Typical salt and grit bins

High-visibility yellow GRC weather-resistant bins.
    0.34 m³ × 760 mm high × 860 mm long × 600 mm deep. Access from base.
    1.01 m³ × 870 mm × 1270 mm × 1017

Typical signs

Signpost: steel shaft, cast aluminium arms, cast lettering, 2300 mm high.
Park sign: powder-coated tubular steel frame, aluminium panel, 1200 mm × 900 mm.

Typical cycle stands

Precast concrete bollard with cycle holders:
660 mm high × 600 mm dia. with 8 holders. Exposed aggregate.
Powder-coated tubular steel post-and-rail unit: (or cast iron)
550 mm high × 1360 mm long set in ground.
750 mm high × 1200 mm long double rail set in ground.
750 mm high × 1000 mm long round corners, bolted to ground.

Typical flag-poles

Glass-fibre, hollow steel, or aluminium poles with galvanized fittings, 6.0–20 m high; high poles have hinged base with lowering system. Set in concrete base.
Reinforced glass fibre: 3.05, 3.81, 4.57, 5.33, 6.10, 7.62, 9.15, 10.67 and 12.20 m high.

Typical prefabricated planters

Polyethylene free-standing:
550 litres × 550 mm high × 1000 mm square. Brown, buff, stone, grey, black.
Precast concrete:
165 litres × 470 mm high × 900 mm dia. Exposed aggregate.
90 litres × 590 mm high × 760 mm dia. White concrete.
135 litres × 515 mm high × 835 mm × 650 mm. Exposed aggregate.

Lifebuoy housing

High-visibility polyethylene frame and cover:
Housing for four buoys: 940 mm × 870 mm × 215 mm deep. Pole mounted.

OUTLINE SPECIFICATION

Seat or bench

a. *In grassed area excavate for base 2250 mm × 1575 mm and lay 100 mm hardcore, 100 mm ST1 concrete, brick pavers in stack bond bedded in 25 mm Group 1 cement:lime:sand mortar.*

b. Supply and fix where shown on drawing 3 no. proprietary seats Type XXX, hardwood slats on black powder-coated steel frame, bolted down with 4 no. 24 mm × 90 mm recessed hex-head stainless steel anchor bolts set into concrete.

Litter bin

Supply and fix to wall where shown on drawing 4 no. wall-mounted proprietary litter bins Type XXX blue powder-coated steel 670 mm × 300 mm square, open top with metal liner, fixed with 2 no. stainless steel recessed hex-head anchor bolts 16 mm × 60 mm. 50 mm high logo 'LITTER' in white on both sides and front.

Cycle stands

Supply and fix 10 no. cycle stands 2.0 m long × 750 mm high 1200 mm apart of 60.3 mm black powder-coated hollow steel sections to BS 4948 Pt 2, one-piece with rounded top corners, set 250 mm into paving.

Prefabricated planter

Supply and locate in positions shown on drawing 6 no. proprietary precast concrete planters Type XXX, 515 mm × 835 mm × 650 mm with white spar exposed aggregate finish. Drainage holes to be positioned away from footway in direction of paving fall.

for planting   *Q31*

R12 Drainage below ground: surface water drainage

GENERAL GUIDANCE

This section deals with surface water drainage and storm water from non-absorbent landscape areas, the gullies, channels, manholes, inspection chambers, and pipework to an outlet. The drainage of tennis courts and other special sports surfaces is not included.

The main drainage problem that the landscape designer has to face is the surface water run-off from non-absorbent surfaces such as paving, *in situ* concrete, and those of little absorbency such as macadam and asphalt roads. All this water has to be drained by effective falls on the surface and efficient channelling to gullies and underground drainage to a specified outlet. In urban areas this is most likely to be the local surface or storm water drainage system. Alternative methods of disposal such as soakaways and outfalls to watercourses are discussed in R13.

Levels

The falls on unit paving and monolithic paving must be sufficient to prevent water from running back against boundary walls and buildings, or ponding on the surface. Even a light film of water that freezes is a danger to the public, particularly to the elderly and disabled; it can also cause spalling, and if it penetrates the joints frost heave may occur. Where paved areas abut buildings, they must, in order to comply with Building Regulations, be at least 150 mm below the damp-proof course. Obviously this is not possible where there are basements and underground car parks; here the architect or civil engineer will have tanked the building to prevent water seepage. Snow provides its own problems where it piles up against buildings through drifting, and this contingency should be allowed for in areas prone to heavy falls of driven snow.

Where, as is often the case in housing work, the road is considerably higher than the entrance and/or garage, a channel to catch the run-off from the drive must be provided, and positioned at least 450 mm from the garage entrance; the channel should have a removable grating so that it can be cleaned out regularly. This channel normally flows to a house gully. Where buildings are level or slightly below the road or paved surface it may be possible to construct a wide flat-bottomed dishing 450 mm away from the building; the slope of the dishing against the wall may be quite steep, but to comply with Building Regulations the top must be at least 150 mm below the DPC.

In order to keep them clean, paving finishes are laid after the building work is completed, and it is important to distinguish between 'finished floor levels' and 'slab levels' when designing paving adjoining the building.

Once the contract is handed over, the landscape designer should provide the owner with a set of 'as built' drawings showing the actual position of all gullies, manholes, inspection chambers, soakaways and so forth in order to ensure that they can be maintained efficiently.

Falls and cambers

Recommended falls given in technical information by advisory organizations and manufacturers are adequate for normal rainfall in carefully constructed and drained small paved areas, but in vast open areas of paving such as tree-lined pedestrian squares more generous falls should be provided, or more frequent outlets should be installed. The regular cleaning and maintenance of the surface, channels and gullies is most important if the paved areas are to be safe and pleasant to use.

The recommended falls are often not sufficient to cope with heavy downpours; paths against houses should fall away from the building, and steps should always have slight falls to shed any surface water likely to freeze. Footways to roads have a slight slope to the kerb/channel, and roads will normally have slight falls from the centre to the kerb/channel on either side. To carry the surface water away effectively, these falls must be accompanied by efficient and well-maintained channels, gullies and drainage outlets.

The gradients for paving given in British Standard 6367 control the spacing of channels, and these recommended gradients allow for quick run-off compatible with safe walking or driving.

Rigid pipes

These include clay, vitrified clay, precast concrete or grey (cast) iron pipes; they do not deform to any significant degree under their design load.

Clay pipes of the traditional thick-walled spigot-and-socket pattern were heavy to handle, and are to a great extent superseded by lighter vitrified clay spigot-and-socket or plain-ended pipes; manufacturers have developed simple sleeved joints and junctions.

Precast concrete pipes are strong and heavy, and most commonly used in the larger sizes, which require mechanical handling.

Grey (cast) iron pipes are heavy; they have comparatively thick walls to prevent them from fracturing, and are rarely used in landscape work. They have been replaced by ductile (flexible) iron pipes.

Flexible pipes

These include PVC-U or high-density polyethylene and ductile iron (now used in preference to grey iron); all will deform under load, and are dependent on the fill around the pipe for rigidity.

Ductile cast iron pipes are not subject to brittle fractures; these pipes have thinner walls and are therefore lighter than grey cast iron pipes.

PVC-U pipes have the advantage of being immune to the aggressive soil conditions (mainly acidic) that can affect metal pipes. They are light, and can be manually handled. They are available in varying grades, each with a large selection of long- and short-radius bends, junctions, couplers, reducing pieces and adaptors for connecting to clay or cast iron pipes.

Channels and gullies

The landscape designer should take care over the position of open channels in paved areas, because if the gullies become blocked, ponding will occur at these points. Open parking areas and yards used for washing vehicles may require trapped yard gullies with removable buckets to catch any sediment likely to block the drain; these have 50 mm water seals. Road gullies or silt pits must be used where mud or sand are often washed into the drainage system; these can be emptied by hand or by mechanical gully-cleaning equipment.

Planting in paved areas

Paving and planted areas of shrubs and trees at the same or even a lower level can be very attractive features, but the landscape designer must decide whether or not the surface water is pure enough to allow surface water from the paving to drain onto the earth. Unlimited run-off should not drain directly onto grassed areas, because it can cause ponding and waterlogging of the root system.

French drains can be laid at the junction between grass and paving to carry run-off to the drainage system; they are described in section R13.

Calculation of paving drainage

The principal guidance for drainage design is contained in BS 6367 *Drainage of roofs and paved areas*. This British Standard gives recommendations for various weather conditions, rates of water flow, and recommended falls for a range of circumstances. It also contains a rainfall map for the British Isles, but it is worth remembering that although there is a higher rainfall in the north of Britain it is more evenly distributed, while the south has less rainfall, which is inclined to fall in heavy bursts. The effect on the drainage system is obvious, and as well as designing for the regional differences, the landscape designer is advised to design for the short sharp burst of rain.

The 'design rates' given in the British Standard for both paved areas and flat roofs represent the maximum rate of run-off in a 2 min period together with the expected frequency; these are given as 50 mm/h, 75 mm/h, 100 mm/h, 150 mm/h and 225 mm/h (the higher 'design rates' are intended for flat roofs). Even 50 mm/h results in temporary ponding after the rain has ceased, but can be used in most situations where this is acceptable; for hospital entrances and so forth the higher rate of 75 mm/h should be used, and where there are basements containing irreplaceable artefacts or instruments the highest rates should be used. The surface water drainage pipes that take the discharge must be adequate for all expected flows of water.

BRITISH STANDARDS

The British Standards in this section cover both the original British and the new European Standards. For this reason some may complicate the situation by contradicting each other, and the landscape designer is advised to contact the manufacturer's technical department before firming up the specification.

EN 124 : 1994 & AMD	Manhole covers, road gully gratings and frames for drainage purposes (replaces BS EN124 : 1986 and BS 497 Pt 1 : 1976)
EN 295-4 : 1996	Vitrified clay pipes and fittings and pipe joints for drains and sewers: special fittings, adaptors and compatible accessories
EN 295-5 : 1994	Vitrified clay pipes and fittings
EN 295-6 : 1996	Vitrified clay pipes and fittings and pipe joints for drains and sewers: vitrified clay manholes
EN 295-7 : 1996	Vitrified clay pipes and fittings and pipe joints for drains and sewers: clay pipes and joints for pipe jacking
EN 598 : 1995	Ductile iron pipes and fittings for sewerage (replaces 4772 : 1988)
65 : 1991 & AMD	Vitrified clay drain and sewer pipes
437 : 1978 & AMD	Cast iron spigot and socket drainpipes and fittings
1247 Pt 1 : 1990	Manhole step irons

4660 : 1989	PVC-U underground drainpipes and fittings 110 and 160 mm
5911 Pt 2 : 1982	Inspection chambers and street gullies
5911 Pt 100 : 1988 & AMD	Unreinforced and reinforced concrete pipes and fittings with flexible joints
5911 Pt 103 : 1994	Precast concrete pipes: prestressed non-pressure pipes and fittings: flexible joints
5911 Pt 110 : 1992	Ogee jointed precast concrete pipes, bends and junctions, reinforced or unreinforced
5911 Pt 114 : 1992	Precast concrete porous pipes
5911 Pt 120 : 1989	Precast concrete pipes; reinforced jacking pipes with flexible joints
5911 Pt 200 : 1994	Precast concrete pipes: unreinforced and reinforced circular manholes and soakaways
5911 Pt 230 : 1994	Precast concrete inspection chambers and gullies
5955 Pt 6 : 1980 & AMD	Code of Practice for installation of unplasticized PVC-U pipework for gravity drains and sewers
6367 : 1983	Code of Practice for drainage of roofs and paved areas
6437 : 1984 & AMD	Polyethylene pipes (type 50) in metric diameters for general purposes
7158 : 1989	Plastics inspection chambers for drains
8000 Pt 14 : 1989	Code of Practice for below ground drainage
8301 : 1985 AMD	Code of Practice for building drainage: including design and construction of groundwater drains

DATA

Calculation of paving drainage

Design rate (mm/h)	Duration in minutes					
	1	2	3	4	5	10
50	0.83	1.67	2.50	3.33	4.17	8.33
75	1.25	2.50	3.75	5.00	6.25	12.50
100	1.67	3.33	5.00	6.67	8.33	16.67
150	2.50	5.00	7.50	10.50	12.50	25.50
225	3.75	7.50	11.25	15.00	18.75	37.50

Rule-of thumb calculation: discharge from paved area receiving 50 mm rainfall per hour.

$$R = A \times I \times 0.0139$$

where R = run-off in litres/seconds (l/s),

        A = area to be drained in m²,

        I is the impermeability factor,

        0.0139 expressed in l/s is the constant conversion factor for rainfall of 50 mm/h.

(The constant conversion factor for 75 mm/h is 0.0208 l/s.)

Permissible gradients limits for paving

Parking or courtyards:	fall 1 in 60 min.
Footpaths or walkways:	fall 1 in 40 min. 1 in 30 max.
Access or estate roads:	longitudinal fall 1 in 15 max. cross-fall 1 in 40 min.

Impermeability factor of paved areas

Solid concrete or asphalt: 0.85–0.95

Paving flags bedded on mortar with mortar joints: 0.75–0.85

Paving units bedded on sand with open joints: 0.5–0.7

Paving blocks with open joints: 0.4–0.5

Macadam (varies by the number of coats and density of material): 0.25–0.6

Gravel or hoggin (varies over thickness and consolidation): 0.15–0.3

Grass (depending on density of soil): 0.05–0.25

Example based on the above:

Sand-bedded plain blocks area 20 m × 20 m	= 400 m² run-off
Impermeability factor	= 0.65
Constant for 50 mm/h	= 0.0139
Therefore 400 × 0.65 × 0.0139	= 3.61 l/s.

Flow of water that can be carried by various sizes of pipe

Clay or concrete pipes

	Gradient of pipeline							
	1:10	1:20	1:30	1:40	1:50	1:60	1:80	1:100
Size of pipe Flow (l/s)								
DN 100	15.0	8.5	6.8	5.8	5.2	4.7	4.0	3.5
DN 150	28.0	19.0	16.0	14.0	12.0	11.0	9.1	8.0
DN 225	140.0	95.0	76.0	66.0	58.0	53.0	46.0	40.0

DN sizes are convenient round figures expressing the internal diameter of the pipe without taking note of manufacturers' variations.

Plastic pipes

| Size of pipe | Gradient of pipeline | | | | | | | |
	1:10	1:20	1:30	1:40	1:50	1:60	1:80	1:100
	Flow (l/s)							
82.4 mm i.d.	12.0	8.5	6.8	5.8	5.2	4.7	4.0	3.5
110 mm i.d.	28.0	19.0	16.0	14.0	12.0	11.0	9.1	8.0
160 mm i.d.	76.0	53.0	43.0	37.0	33.0	29.0	25.0	22.0
200 mm i.d.	140.0	95.0	76.0	66.0	58.0	53.0	46.0	40.0

The figures given are for simple projects only, and where very large or irregular flows are expected, the services of a drainage engineer should be engaged. Small-bore pipes with a low flow cannot be laid at a shallower gradient, as the velocity of the water is insufficient to clean the pipe out. It is undesirable to lay pipes at a shallower gradient than 1:40, as any ground settlement may reduce the gradient to an unacceptable level.

Bedding rigid pipes: clay or concrete

(For vitrified clay pipes the manufacturer should be consulted.)

Class D
Pipe laid on natural ground with cut-outs for joints, soil screened to remove stones over 40 mm and returned over pipe to 150 mm min. depth. Suitable for firm ground with trenches trimmed by hand.

Class N
Pipe laid on 50 mm granular material of graded aggregate to Table 4 of BS 882, or 10 mm aggregate to Table 6 of BS 882, or as-dug light soil (not clay) screened to remove stones over 10 mm. Suitable for machine-dug trenches.

Class B
As Class N, but with granular bedding extending halfway up the pipe diameter.

Class F
Pipe laid on 100 mm granular fill to BS 882 below pipe, minimum 150 mm granular material fill above pipe: single-size material. Suitable for machine-dug trenches.

Class A
Concrete 100 mm thick under the pipe extending halfway up the pipe, backfilled with the appropriate class of fill. This is used where there is only a very shallow fall to the drain, and the alignment of the pipes is therefore important, because class A bedding allows the pipes to be laid to an exact gradient.

Concrete surround

25 mm sand blinding to bottom of trench, pipe supported on chocks, 100 mm concrete under the pipe, 150 mm concrete over the pipe. It is preferable to bed pipes under slabs or wall in granular material.

Bedding flexible pipes: PVC-U or ductile iron

Type 1 = 100 mm fill below pipe, 300 mm above pipe: single-size material
Type 2 = 100 mm fill below pipe, 300 mm above pipe: single-size or graded material
Type 3 = 100 mm fill below pipe, 75 mm above pipe with concrete protective slab over
Type 4 = 100 mm fill below pipe, fill laid level with top of pipe
Type 5 = 200 mm fill below pipe, fill laid level with top of pipe
Concrete = 25 mm sand blinding to bottom of trench, pipe supported on chocks, 100 mm concrete under the pipe, 150 mm concrete over the pipe.

This method should be used where pipes pass close to buildings or walls. If the side of the pipe trench is less than 1 m from the foundation of a wall, Building Regulations require that the pipe be bedded in concrete to the level of the underside of the wall foundation.

Sizes of granular fill material

10 mm granular material for 100 mm pipes
10 mm or 14 mm granular material for 150 mm pipes
10 mm, 14 mm, 20 mm granular material for 200–300 mm pipes
14 mm or 20 mm granular material for 375–500 mm pipes

Protecting pipes

Rigid pipes
Less than 1000 mm below road foundation   must be protected
900 mm below fields and gardens   must be protected

Ductile iron pipes
Less than 1000 mm below road foundation   no protection
300 mm below road foundation   concrete protection
1000–300 mm below road foundation   granular surround

Flexible pipes
Less than 600 mm below soft landscape, footways   must be protected
900 mm below road foundation   reinforced concrete
Pipe protection under major highways must be specified by the highway engineer.

Manhole covers, road gully gratings and frames for drainage purposes

BS EN 124 : 1994

Manhole and gully tops:	Materials can be cast iron, cast steel or rolled steel.
Gratings:	Materials can be cast iron or cast steel.
Cover fittings:	Can be of concrete or to suit the area in which used.
Manhole opening:	Minimum size for safe access 600 mm.
Hinged covers:	To open at least 100° to the horizontal.
Slotted manholes:	Require dirt pans, and the drainage and ventilation must work when the pan is full.

Solid cast iron or cast steel manhole tops to have a raised pattern:

class A15, B125, C250 2–6 mm high over a maximum of 70% of the top's area

class D400, E600, F900 3–8 mm high over a maximum of 70% of the top's area.

Gullies: may have dirt buckets; slot size for road gullies:

class A15 8–18 mm wide; no limit on length

class B125 18–25 mm wide; length 170 mm max.

Note: Manufacturers may make the slots 5 mm wide in pedestrian areas.

Suitable uses by class

A15   Pedestrian and pedal cyclists.

B125   Footways, pedestrian areas, car parks.

C250   Gully tops in kerbside channels of roads (when installed max. 500 mm into carriageway and 200 mm into the footway; both dimensions measured from the kerb).

D400   Carriageways, pedestrian streets, hard shoulders, parking areas for all types of vehicle.

E600   High wheel loads, docks and aircraft pavements.

F900   Very high wheel loads, aircraft pavements.

Note: The figure relates to the test load in kN.

Identification marks

EN 124, class or classes, manufacturer's name/symbol.

Precast concrete pipes: prestressed non-pressure pipes and fittings: flexible joints to BS 5911 Pt 103 : 1994

Rationalized metric nominal sizes: 450 mm, 500 mm

Length:	500–1000 mm by 100 mm increments
	1000–2200 mm by 200 mm increments
	2200–2800 mm by 300 mm increments
Angles: length:	450–600 mm angles 45°, 22.5°,11.25°
	600 mm or more angles 22.5°, 11.25°

Identification marks:

BS 5911 Pt 103 : 1994 crushing strength: L for light, M for medium and H for heavy; type of cement etc.: S for sulphate-resisting Portland cement, B for GGBS (ground granulated blast-furnace slag), P for PFA (pulverized fuel ash) and A for admixture added.

Precast concrete pipes: unreinforced and circular manholes and soakaways to BS 5911 Pt 200 : 1994

Nominal sizes:
    shafts: 675, 900 mm
    chambers: 900, 1050, 1200, 1350, 1500, 1800, 2100, 2400, 2700, 3000 mm.
Large chambers: to have either tapered reducing rings or a flat reducing slab in order to accept the standard cover.
Climbing irons or ladders are not included in the standard.
For deep manholes the flat reducing slab can be used as a platform, provided that the access hole is a minimum of 900 mm dia.
Ring depths:
1. 300–1200 mm by 300 mm increments except for bottom slab and rings below cover slab; these are by 150 mm increments.
2. 250–1000 mm by 250 mm increments except for bottom slab and ring below cover slab; these are by 125 mm increments.

Access hole: 750 mm × 750 mm for DN 1050 chamber
                1200 mm × 675 mm for DN 1350 chamber

Identification marks

BS 5911 Pt 200 : 1994, R for reinforced, S or a blue mark for sulphate-resistant Portland cement, B for GGBS (ground granular blast-furnace slag), P for PFA (pulverized fuel ash), A for admixture used.
    Note: Estate road cover slabs must be so marked, and must have access holes for:

    DN 1050 chamber 750 mm × 750 mm
    DN 1350 chamber 1200 mm × 675 mm

Precast concrete inspection chambers and gullies to BS 5911 Pt 230 : 1994

Nominal sizes: 375 mm diameter, 750, 900 mm deep
                450 mm diameter, 750, 900, 1050, 1200 mm deep
Depths: from the top for trapped or untrapped units:
    centre of outlet 300 mm
    invert (bottom) of the outlet pipe 400 mm
Depth of water seal for trapped gullies: 85 mm, rodding eye i.d. 100 mm
Cover slab: 65 mm min.

Identification marks

BS 5911 Pt 230 : 1994, S for sulphate-resistant Portland cement, B for GGBS (ground granular blast-furnace slag), P for PFA (pulverized fuel ash), A for admixture used.

Traffic loads

Approximate figures are given.

Area	Impact factor	Wheel load (kN)	Wheels
Field or garden	2.0	30 (static)	2
Secondary roads	1.5	70 (static)	2
Main roads	included*	112.5	8

*Included in BS 5400: Pt 2 : 1978 Steel, concrete and composite bridges

Ductile iron pipes to BS EN 598 : 1995

Type K9 with flexible joints should be used for surface water drainage.
5500 mm or 8000 mm long

80 mm bore	400 mm bore	1000 mm bore
100	450	1100
150	500	1200
200	600	1400
250	700	1600
300	800	
350	900	

These pipes may be cement mortar or bitumen lined for special applications such as natural gas supplies.

Identification marks

BS no., class no., manufacturer, date of manufacture and the size.

Concrete mixes to BS 5328 Pt 1 & AMD

Application	Standard mix	Designated mix	Slump (mm)
Mass concrete fill and blinding	ST 2 (C10P)	GEN 1	75
Mass concrete foundations in non-aggressive soils	ST 2 (C10P)	GEN 1	75
Trench fill foundations in non-aggressive soils	ST 2 (C10P)	GEN 1	125
Foundations			
in Class 2 sulphate soils	–	FND 2	75
in Class 3 sulphate soils	–	FND 3	75
in Class 4A sulphate soils	–	FND 4A	75
in Class 4B sulphate soils	–	FND 4B	75
Drain pipe support in non-aggressive soils	ST 2 (C10P)	GEN 1	10
Manholes etc. in non-aggressive soils	ST 2 (C10P)	GEN 1	50

Traditional concrete mixes

Application	Name of mix	Cement:aggregate:sand (parts by volume)
Setting posts, filling weak spots in ground, backfilling pipe trenches:	one, two, three or one to five all-in	1:2:3 or 1:5
Paving bases, bedding manholes, setting kerbs, edgings, precast channels:	one, two, four or one to six all in	1:2:4 or 1:6

'All-in' is mixed aggregate as dug from the pit.

OUTLINE SPECIFICATION

Vitrified clay surface water drain

a. *In trench specified in section D20, lay 110 mm vitrified clay drain with flexible joints, bedding Class B.*

b. *Backfill with excavated material screened to remove stones over 40 mm, backfill to be laid in layers not exceeding 150 mm.*

for excavation   D20

Vitrified clay intercepting trap

Supply and fix vitrified clay interceptor trap to manhole bedded in GEN3 10 mm aggregate concrete, complete with brass stopper and chain. Connect to drainage system with flexible joints.

Vitrified clay gully

Excavate hole, supply and set in GEN3 concrete vitrified clay trapped mud (dirt) gully with rodding eye to BS 65, complete with galvanized bucket and cast iron hinged locking grate and frame, flexible joint to pipe. Connect to drainage system with flexible joints.

Concrete inspection chamber

a. *Excavate pit for inspection chamber 1500 mm deep, including earthwork support and disposal of spoil to approved dump on site not exceeding 100 m.*

b. *Lay GEN1 concrete base 1500 mm dia. × 200 mm thick, 110 mm vitrified clay channels, benching in GEN4 concrete allowing one outlet and two inlets for 110 mm dia. pipe.*

c. *Construct inspection chamber with 1200 mm dia. precast concrete rings to BS 5911 Pt 12, bedded and jointed in sulphate-resisting 1:1/4:3 cement:lime:sand mortar, backfill with excavated material, complete with 2 no. cast iron step irons to BS 1247 Pt 1.*

d. *Supply and fix light-duty precast concrete cover slab, access cover and frame to EN 124 : 1994 Class B125, bedded in 1:3 cement:sand mortar.*

for excavation   D20

Concrete road gully

a. *Excavate hole and lay 100 mm GEN4 concrete base 150 mm × 150 mm to suit given invert level of drain.*

b. *Supply and connect trapped precast concrete road gully 375 mm dia. × 750 mm deep with 160 mm outlet to BS 5911, set in GEN4 concrete surround, connect to vitrified clay drainage system with flexible joints.*

c. *Supply and fix straight bar dished-top cast iron grating and frame to BS EN 124 : 1994 Class D400, bedded in 1:3 cement:sand mortar, connect to drainage system.*

for excavation   D20

PVC-U surface water drain

a. *In trench specified in section D20, lay 110 mm PVC-U drainpipe to BS 4660, Type 2 bedding.*

b. *Backfill with excavated material screened to remove stones over 40 mm to 900 mm below road foundation, lay precast concrete slab 75 mm thick 150 mm wider than pipe trench, backfill to road foundation level.*

for excavation   *D20*

PVC-U inspection chamber

a. *Excavate pit for inspection chamber, supply and install prefabricated PVC-U inspection chamber unit 650 mm deep with straight-through 110 mm channel and 2 no. 110 mm entries to BS 7158, complete with cast iron cover and frame to BS EN 124, bedded in DoT Type 1 granular material. Connect to vitrified clay drainage system with flexible joints.*

b. *Backfill excavation with DoT Type 1 granular material, lay 150 mm GEN4 concrete slab 150 mm wider than cover round top of chamber, allowing for paving to be flush with cover.*

for excavation   *D20*

Polypropylene gully

a. *Excavate hole and lay 100 mm GEN4 concrete base 150 mm × 150 mm to suit given invert level of drain.*

b. *Supply and connect trapped polypropylene gully complete with sealed cast iron plate and frame, connect to drainage system, backfill with DoT Type 1 granular fill.*

Ductile iron surface water drain

a. *In trench specified in section D20, lay 100 mm ductile iron pipes to BS EN 598 Type K9 with flexible joints, Type 4 bedding.*

b. *Backfill with excavated material screened to remove stones over 40 mm.*

for excavation   *D20*

R13 Land drainage

GENERAL GUIDANCE

This section covers groundwater drainage of parks, campuses, recreation grassed areas, banks and terraces. It includes soakaways and outfalls for groundwater.

Before embarking on any new drainage, the existing site must be surveyed and levelled to check the position, size and depth of existing ditches, culverts, ponds and so forth. The annual rainfall must also be checked, together with any appreciable run-off from the surrounding land, either from subsoil drainage or from springs. Where possible, main drainage lines should follow the natural slope of the land. The landscape designer is advised to hand over any large or complex drainage system to a drainage engineer.

Once a decision is made to install land drains, the required depth of the water-table must be fixed, but remember that groundwater levels vary according to the season; this is particularly important in areas near tidal rivers and at or near the coast. The amount and depth of land drainage will depend on the purpose of the proposed landscape scheme and the need for any building work.

Playing fields that form part of parkland (as opposed to those sports grounds constructed to stringent specifications or constructed by specialists) require main drains that should be DN 75 min. with branches DN 50 min. Elsewhere main drains should be DN 100 min. with branches DN 75 min., with the depth varying from 800 mm in heavy soils to 1500 mm in light soils. The gradients of the drains are to a great extent dictated by the natural fall of the land, and may not always be of sufficient velocity to be self-cleansing.

Once the drainage system is completed, the landscape designer should make sure that the client is given a set of 'as built' drawings showing the position of all drains, inspection chambers and access points for rodding any drains.

Soakaways

In urban areas where the local authority drains lack the capacity to take all the surface water, the occasional soakaway is built on site. Although Approved Document H3 to the Building Regulations does not give constructive guidance on this subject, it does say that, if a sewer operated as a combined system does not have sufficient capacity, the rainwater should be run in a separate system with its own outfall. The landscape designer should seek advice from the local building control office on the required distance that the soakaway should be from any building. Trial pits should be dug equal to the diameter of the soakaway and filled with water, and the amount of seepage should be noted. The water remaining in the pit will give a good estimate of the water-table. The soakaway should then be designed so that the base of the soakaway is above the water-table. In rural locations where large areas are to be drained to soakaways it is advisable to divide the area into sections and provide a soakaway for each one; if the conditions are suitable, they may be linked together. To be effective, they do need to be built in reasonably permeable soil. An alternative is the trench soakaway, which is a long, narrow trench following the contour. A rough soakaway can be made by digging a pit and simply filling it with rubble, but the sides must be supported during excavation.

Land drains

The most commonly specified drainage systems have pipes that can be of clayware, precast concrete or plastics material. The main runs are normally 100 mm and the laterals 60 mm, and the layout is designed on a herring-bone, grid or fan system. All pipes are either solid laid with open joints, perforated, or semi-perforated. Semi-perforated pipes have holes in one half of the pipe. When the collected water must be taken to an outfall, the pipe is laid with the perforated side up. Conversely, where water is to be dispersed throughout the subsoil (as from the outfall of a septic tank), the pipes are laid perforated side down. Where the pipes are near trees they should be enclosed in solid pipes to prevent the tree roots from penetrating the open pipework and blocking them. In normal circumstances, pipes should have 350 mm cover to ensure that the pipes will not be disturbed by standard cultivation, though this may not be acceptable on archaeological sites. Where the slope of the land allows, the landscape designer should allow for sufficient falls to prevent silting up inside the pipe. This will depend on the material of the pipe and the smoothness of the inside; the normal velocity required for self-cleaning is 0.75 m/s. The manufacturer will advise on the self-cleaning velocity for the particular pipe specified, and on the need for filter-wrapping of pipes in areas where the soil contains fine particles. In all cases, silt traps should be specified at regular intervals; these should be clearly marked on the 'as built' drawings and on site so that they can be easily identified and cleaned out as part of the regular maintenance.

Open ditches

Once the main way of draining land, these are not now so commonly used, probably because they are not considered safe for public amenity areas. They are now mainly recommended for woodland and forestry work. Ditches can also have nature conservation benefits.

Outfalls

Where the water is to drain into a ditch, lake or pond, the outfall must be protected from erosion and the entry of vermin. The end of the land drain should project slightly through the bank; it is fitted with a pipe-flap valve, which allows water to pass outwards under pressure but closes when the pipe is empty. The bank round the outfall and the bottom of the watercourse should be protected from erosion by concrete or brick aprons, and the position of the outfall should be marked to prevent damage by machinery. The design and construction of outfalls and headwalls should be as sympathetic and unobtrusive as possible.

Rubble drains

These are the simplest type, and are constructed by excavating a trench and backfilling with 20–75 mm selected rubble or stone, which allows water to percolate through it. The drain may be topped with a geotextile material and finished with gravel or grass. As these drains may become clogged after a time, they may have to be dug out and the filter material replaced. As they do not contain any visible depth of water they are suitable for parks and other public areas. Although not the most efficient type of land drain, they may be suitable in some circumstances.

French drains or catchwater drains

These are mainly used to control seepage in motorways, but they do have a place in ordinary landscape work, where they can drain water from behind embankments and walls. They consist of a trench taken down below the level of the water to be drained, which is filled with medium aggregate up to finished ground level, where they may be capped with gravel or grass. Catchwater drains may be plain trenches, or they may have a land drainpipe laid at the bottom to carry off the seepage. Where the soil to be drained contains fine particles, the landscape designer should specify that a geotextile material of the correct gauge to keep out the fine particles is wrapped round the pipe before the drain is backfilled.

Fin drains

These are a version of the French drain that have a proprietary water-conductive material of sandwich construction set vertically in the trench and then wrapped round the perforated pipe. The outer layers of the material prevent the silt from entering the drain while the core allows the water to flow freely into the pipe.

Mole drainage

This system has a maximum lifespan of ten years; after this it has to be reworked, and its use is limited to draining clay soils. A mole plough has an adjustable blade with a bullet-shaped tool at the end, which forms the drains to the required depth, and the excess water drains into them. Where possible, they should be cut in March or April, but other times may be acceptable as long as the clay soil is moist and pliable.

Sand/gravel slitting

This is a suitable method for draining existing grass areas. Slits 50 mm wide and 300 mm deep are cut with a spade or a special tool from 1 to 3 m apart, depending on the severity of saturation. These narrow slits connect to a wider lateral trench, and are then drained to the specified outlet. The slits are filled with coarse sharp sand.

Table 4.6   Relationship between soil permeability and subsoil drainage (BS 4428 : 1989)

Soil types	Soil permeability	Recommended lateral drainage	
		Distance apart of laterals (mm)	Depth of laterals (m)
Clay	Nearly impervious extremely slow	Sand slitting for topsoil drainage or mole drains may be appropriate 10–15	0.9–1.0
Silty clay/loam	Very slow	Mole drains may be appropriate 10–15	0.9–1.0
Sandy clay/clay loam	Adequate permeability	12–20	0.9–1.2
Medium loam		18–30	1.0–1.2
Fine sandy loam	Moderate	30–40	1.0–1.2
Medium sandy loam	Rapid	30–60	1.35–1.5
Chalk or large mineral particles	Rapid	Subsoil drains not needed	

Note (1) This table is not applicable to playing fields.
Note (2) For schemes that are to be supplemented by a slit drainage scheme drain spacing can be increased.
Note (3) Drains should be omitted only where both subsoil and topsoil are free-draining and where the water table dos not come close to the surface.
Note (4) Deeper drains may be needed if the water-table comes close to the surface.
Note (5) Drain spacings are for guidance only, and much will depend on the individual site, annual rainfall and level of use.

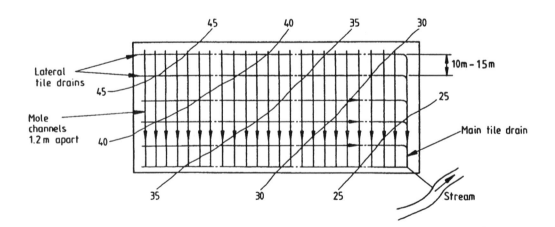

Fig. 4.70 Mole drains: plan. (BS 4428 : 1989)

BRITISH STANDARDS

EN 295-5 : 1994   Vitrified clay pipes and fittings
1196 : 1989   Clayware field drainpipes and junctions
4428 : 1989 & AMD   Code of practice for general landscape operations
4962 : 1989 & AMD   Plastics pipes and fittings for use as subsoil field drains
5911 Pt 200 : 1989   Precast concrete manholes, soakaways of circular
               cross-section

DATA

Soakaways: calculation of required size

The correct size can be calculated by a very complicated method described in BRE (Building Research Establishment) Digest No. 365, or by assuming a depth of 18 mm of water over the area to be drained and selecting a soakaway that contains this amount. The figure of 18 mm is in excess of the British Standard recommendations, but gives a allowable tolerance for unusually heavy rainfall.

Example

Area drained: $= 20 \text{ m} \times 20 \text{ m} = 400 \text{ m}^2$
Depth of water over area: $= 20 \text{ m} \times 20 \text{ m} \times 0.018 \text{ m} = 7.2 \text{ m}^3$
Soakaway must have 7.2 m^3 min. capacity below the inlet invert.

Alternative solutions

a. Precast concrete ring or structural brick soakaway not filled with rubble:
    Assume soakaway diameter 1.2 m $= 0.6 \text{ m} \times 0.6 \text{ m} \times 3.142 \text{ (pi)}$
                   $= 1.13 \text{ m}^2$
Then effective depth $= 7.2 \text{ m}^3 / 1.13 = 6.36 \text{ m}$
Select 11 no. 600 mm rings $= 6.6 \text{ m}$
Total capacity 1.13 × 6.6 $= 7.458 \text{ m}^3$ to inlet invert, so this
                   soakaway will be satisfactory
b. Non-structural brick built soakaway rubble filled (effective capacity 60%)
Assume soakaway diameter 1.5 m $= 0.75 \text{ m} \times 0.75 \text{ m} \times 3.142 \text{ (pi)}$
                   $= 1.77 \text{ m}^2$
Required capacity 7.2 m^3
Actual capacity to inlet invert $= 7.2 \text{ m}^3 \times 100 / 60 = 12.0 \text{ m}^3$
Then effective depth $= 12.0 \text{ m}^3 / 1.77 \text{ m}^2 = 6.78 \text{ m}$
Effective capacity 60% $= 7.65 \text{ m}^3$ actual capacity to inlet
                   invert

Spacing of branches to field drains in different soils

Soil type	Distance between branches related to main drain invert	
	800–1000 mm	1000–1500 mm
Clay	2–6 m	–
Sandy clay	6–12 m	–
Clay loam	12–16 m	15–20 m
Loam	16–20 m	20–30 m
Sandy loam	–	30–45 m
Sand	–	45–90 m

Vitrified (perforated) clay pipes and fittings to BS EN 295-5 1994

Length not specified. All dimensions in mm.

75 bore	250 bore	600 bore
100	300	700
125	350	800
150	400	1000
200	450	1200
225	500	

Type A has minimum of 3 mm² x mm bore per metre length area of perforations.

Type B has minimum of 10 000 mm² per metre length area of perforations.

Preferred bends: 15°, 22.5°, 30° and 45°.

Preferred branch angle of junctions: 45° and 90°.

Perforations: holes   Type A 10 mm diameter max.

Type B 13 mm diameter max.

slots 8 mm wide max.

Note: At least 90° of the pipe's circumference to be unperforated; last 100 mm need not be perforated.

Marking: EN 295-5, DN size, crushing strength in kN/m, type.

Plastics pipes and fittings for subsoil field drains to BS 4962

Coilable pipes

These, together with their joints, are intended to be laid by machine. They have corrugated walls, and must withstand tensile forces without extending unduly during the laying process. Coilable pipes must state the installation practice to be followed.

Non-coilable pipes
These pipes are not intended to bend appreciably, or to withstand tensile forces. They normally have smooth surfaces internally and externally.

Sizes
Any size up to 400 mm external diameter.

Perforations
Not less than 0.8 mm nor greater than 2.0 mm, with at least one perforation in every 50 mm along the length of pipe, and not less than one perforation per 120° segment in every 150 mm length of pipe, except for unperforated zones. (In practice this means that coiled pipes have slotted holes 4 mm long by 0.9 mm wide, which go round the pipe; rigid pipes have holes of about 8 mm diameter at 75 mm centres.)

Laying procedure
Minimum depth of soil cover is 600 mm from top of pipe to top of the open ground, and 900 mm below road foundation under roads. Neither pipes or fittings should be installed when the temperature is below 0 °C, and only with care when it is below 5 °C.

Marking
Either by shallow indentation in the pipe, printing on the pipe, or on a label to include BS 4962 : 1989, manufacturer's name, type reference and date manufactured. They must be marked 'unperforated' or, if perforated, the perforation length must be given. Coilable pipes must state the min. bending radius together with the installation practice to be followed.

Concrete mixes to BS 5328 Pt 1 & AMD

Application	Standard mix	Designated mix	Slump (mm)
Mass concrete fill and blinding	ST 2 (C10P)	GEN 1	75
Mass concrete foundations in non-aggressive soils	ST 2 (C10P)	GEN 1	75
Trench fill foundations in non-aggressive soils	ST 2 (C10P)	GEN 1	125
Foundations			
in Class 2 sulphate soils	–	FND 2	75
in Class 3 sulphate soils	–	FND 3	75
in Class 4A sulphate soils	–	FND 4A	75
in Class 4B sulphate soils	–	FND 4B	75
Drain pipe support in non-aggressive soils	ST 2 (C10P)	GEN 1	10
Manholes etc. in non-aggressive soils	ST 2 (C10P)	GEN 1	50

Traditional concrete mixes

Application	Name of mix	Cement:aggregate:sand (parts by volume)
Setting posts, filling weak spots in ground, backfilling pipe trenches:	one, two, three or one to five all-in	1:2:3 or 1:5
Paving bases, bedding manholes, setting kerbs, edgings, precast channels:	one, two, four or one to six all in	1:2:4 or 1:6

'All-in' is mixed aggregate as dug from the pit.

Mortar mixes

Mortar group	Cement:lime:sand	Masonry cement:sand	Cement:sand with plasticizer
1	1:0–0.25:3		
2	1:0.5:4–4.5	1:2.5–3.5	1:3–4
3	1:1:5–6	1:4–5	1:5–6
4	1:2:8–9	1:5.5–6.5	1:7–8
5	1:3:10–12	1:6.5–7	1:8

Group 1: strong inflexible mortar. Group 5: weak but flexible. All mixes within a group are of approximately similar strength. Frost resistance increases with the use of plasticizers. Cement:lime:sand mixes give the strongest bond and greatest resistance to rain penetration.

Bedding rigid pipes: Clay or concrete

(For vitrified clay pipes the manufacturer should be consulted.)

Class D
Pipe laid on natural ground with cut-outs for joints, soil screened to remove stones over 40 mm and returned over pipe to 150 mm min depth. Suitable for firm ground with trenches trimmed by hand.

Class N
Pipe laid on 50 mm granular material of graded aggregate to Table 4 of BS 882, or 10 mm aggregate to Table 6 of BS 882, or as-dug light soil (not clay) screened to remove stones over 10mm. Suitable for machine-dug trenches.

Class B

As Class N, but with granular bedding extending halfway up the pipe diameter.

Class F

Pipe laid on 100 mm granular fill to BS 882 below pipe, minimum 150 mm granular material fill above pipe: single-size material. For machine-dug trenches.

Class A

Concrete 100 mm thick under the pipe extending halfway up the pipe, back-filled with the appropriate class of fill. This is used where there is only a very shallow fall to the drain, and the alignment of the pipes is important, because class A bedding allows the pipes to be laid to an exact gradient.

Concrete surround

25 mm sand blinding to bottom of trench, pipe supported on chocks, 100 mm concrete under the pipe, 150 mm concrete over the pipe. It is preferable to bed pipes under slabs or wall in granular material.

Bedding flexible pipes: PVC-U or ductile iron

Type 1 =   100 mm fill below pipe, 300 mm above pipe: single-size material.
Type 2 =   100 mm fill below pipe, 300 mm above pipe: single-size or graded material.
Type 3 =   100 mm fill below pipe, 75 mm above pipe with concrete protective slab over.
Type 4 =   100 mm fill below pipe, fill laid level with top of pipe.
Type 5 =   200 mm fill below pipe, fill laid level with top of pipe.
Concrete = 25 mm sand blinding to bottom of trench, pipe supported on chocks, 100 mm concrete under the pipe, 150 mm concrete over the pipe.

This method should be used where pipes pass close to buildings or walls. If the side of the pipe trench is less than 1 m from the foundation of a wall, BuildingRegulations require that the pipe be bedded in concrete to the level of the underside of the wall foundation.

Sizes of granular fill material

10 mm granular material for 100 mm pipes.
10 mm or 14 mm granular material for 150 mm pipes.
10 mm, 14 mm or 20 mm granular material for 200–300 mm pipes.
14 mm or 20 mm granular material for 375–500 mm pipes.

Protecting pipes

Rigid pipes

Less than	1000 mm below road foundation	must be protected
	900 mm below fields and gardens	must be protected

Ductile iron pipes

Less than	1000 mm below road foundation	no protection
	300 mm below road foundation	concrete protection
	1000–300 mm below road foundation	granular surround

Flexible pipes

Less than	600 mm below soft landscape, footways	must be protected
	900 mm below road foundation	reinforced concrete

Pipe protection under major highways must be specified by the highway engineer.

OUTLINE SPECIFICATION

Agricultural rubble drain

a. *Excavate trench 225 mm wide average depth 800 mm, spoil stored alongside trench, topsoil and subsoil to be kept separate.*

b. *Lay 100 mm bore perforated clay pipes to BS EN 295-5 Type A to given falls and inverts, drains to be laid herring-bone pattern 12.0 m between laterals, lay 225 mm bore main drain and form junctions.*

c. *Bed pipes on 50 mm granular material 10 mm maximum, backfill trench with excavated subsoil screened to remove all stones over 40 mm, remove surplus subsoil to approved dump on site, complete back-fill to finished ground level with excavated topsoil.*

for excavation   *D20*

French or catchwater drain

a. *Excavate trench 300 mm wide average depth 900 mm, remove spoil to approved dump on site, including earthwork support.*

b. *Lay 110 mm half-perforated PVC-U pipe to BS 4962 with perforations uppermost to given falls and inverts, filter-wrap pipe with high-permeability geotextile material, bedding Type 2, backfill with excavated material screened to remove all stones over 40 mm to within 150 mm of finished ground level.*

c. *Lay geotextile material as above across top of drain, backfill with approved topsoil screened to remove stones over 25 mm, supply and lay turf to BS 3969 type general-purpose utility turf without perennial rye-grass, beat down and water lightly.*

for excavation   *D20*
for turfing   *Q30*

Fin drain to rear of retaining wall

a. *Clean out trench 300 mm wide to rear of wall specified in section F10, excavate to foundation level if necessary.*

b. *Supply and install proprietary fin drain system comprising vertical geo-textile/drain/geotextile compound material fixed to wall, wrap material round 160 mm perforated PVC-U pipe to BS 4962 laid to given falls, bedding Type 1.*

c. *Fill gap between wall and bank with 200 mm aggregate 40 mm, backfill with excavated material as the work proceeds.*

   for excavation   D20
   for brickwork   F10

Mole drainage

Drain area shown on drawing with mole plough 50 mm dia. 450 mm deep at 1500 mm centres in parallel lines. Ends of plough runs to be connected to watercourse by hand.

Site-built non-structural brick soakaway

a. *Excavate pit for soakaway including earthwork support, remove spoil to specified dump on site.*

b. *On 250 mm hardcore base 1650 mm dia. construct FND2 concrete circular foundation strip 300 mm wide × 150 mm deep, construct circular half-brick-thick wall 2.0 m high × 1200 mm internal dia. of Class FL bricks laid in SRPC Group 1 cement:lime:sand mortar with perpends left open below 600 mm beneath cover slab.*

c. *Connect inlet pipe and rodding eye to soakaway with jointing appropriate to pipe material, fix 150 mm perforated plastic inspection pipe 2.8 m long below inspection cover.*

d. *Fill soakaway with clean hard broken brick or stone aggregate 40 mm minimum size to invert level of inlet pipe.*

e. *Supply and lay reinforced precast concrete cover slab of C25P concrete 100 mm thick with cast iron inspection cover and frame 610 × 457 opening Class A15 to BS EN 124.*

f. *Supply and fit vertical precast concrete marker 600 × 150 × 50 mm to edge of soakaway.*

   for excavation   D20
   for reinforcement   E30
   for brickwork   F10

Precast concrete soakaway

a. *Excavate pit for soakaway including earthwork support, remove spoil to specified dump on site.*

b. *On 150 mm hardcore base lay FND2 concrete circular foundation strip 300 mm wide × 150 mm deep, supply and install 8 no. precast perforated concrete rings 600 mm × 1200 mm dia. to BS 5911 Pt 200 jointed in Group 1 SRPC cement:sand 1:2 mortar.*

c. *Supply and lay high-permeability geotextile filter fabric round outside of soakaway as the work proceeds, backfill with excavated spoil screened to remove all stones over 50 mm.*

d. *Supply and fix flat precast concrete cover slab with access hole 750 mm × 750 mm and cast iron cover and frame to BS EN 124 Class A15.*

e. *Supply and fit vertical precast concrete marker 600 mm × 150 mm × 50 mm to edge of soakaway.*

   for excavation   *D20*

Outfall

a. *Supply and fix pipe flap valve to end of main land drainpipe, jointing to suit material of drainpipe, lay in situ GEN3 aggregate concrete apron 100 mm thick × 900 mm wide × 1500 m deep round outfall, pipe to be 250 mm below top of slab on centre line of apron.*

b. *Supply and fit vertical precast concrete marker 600 mm × 150 mm × 50 mm to top of outfall.*

S10 Piped supply systems

GENERAL GUIDANCE

This section includes the water supply to water features in landscape work.

The landscape designer is often asked to provide a water feature within the landscape design, and should be aware of the more simple supply systems required to feed them. He or she may also be required to provide a simple irrigation system to a conservatory or greenhouse; any complex irrigation system for lawns, shrubberies and so forth will be designed and set out by specialist contractors, and at most the landscape designer will be expected to provide the pipework from the mains to the start of the system.

As the landscape designer will normally be connecting any water supply to the mains, he or she should be aware of the local company by-laws, which are based on the Department of the Environment Model Byelaws in so far as they affect the supply to landscape work. Part III deals with the prevention of water contamination by back syphonage. This requires the use of a mechanical device; the most commonly used device for this purpose is known as a check valve.

On large schemes, water may be supplied from another source that is described in the Model Byelaws as being 'not fit for human consumption or supplied from any other source other than the undertaker's main'. This could be water from a borehole or well, or simply surface water run-off fed to a pond or lake. Any new borehole made in England and Wales will require a drilling licence. Water abstraction will require permission from the National Rivers Authority, depending on the form of irrigation. Scotland does not require the owner to obtain permission to drill a borehole, and riparian owners have a limited right to abstract river water.

Laying pipes

Pipes should be laid at least 750 mm below ground to avoid freezing, but in severe conditions they should be laid deeper, with a maximum depth of 1350 mm. Local water by-laws lay down the method of bedding mains water service pipes, and for landscape work it is good practice to bed the pipes in a granular material, as the cost of repairing leaks can be considerable. Standpipes or branches above ground must be insulated against frost.

Mains water pipes, and preferably all other pipes, should be laid with no joints in the running lengths, so as to avoid weak points where leakage might occur. It is common practice to lay marker tape at a safe distance above the pipe to prevent diggers and other plant from damaging the pipe; the colours match those of the pipes.

Controls

Stop valves must be provided to cut off all external water services if necessary for repairs or for winter drain-down of pipes, and non-return check valves must be fitted. If a number of water features are to be supplied, it is advisable to provide a stop valve to each one. These can be hand operated for convenience or key operated for security. Stop valves and all other controls for external water services are best fitted inside a building where they are not liable to frost damage, but if they must be fitted below ground, they should be placed in a pit able to withstand ground movement, tree roots and so forth. Traditional pits are made of half-brick walls and concrete base, but a small PVC-U inspection chamber about 250 mm diameter can be used with the pipework sealed into it with a flexible seal, as this is large enough for hand operation of the stop valve; it should have a sealed plate cover.

BRITISH STANDARDS

BE EN 545 : 1995	Ductile iron pipes and fittings for water pipelines (replaces BS 4772 : 1988)
417 Pt 2 : 1987	Galvanized low-carbon steel cisterns
1427 : 1993	Guide to field tests and on-site test methods for the analysis of water

3284 : 1967 & AMD	Polyethylene pipe type 50 for cold water services
3505 : 1986	PVC-U pressure pipes for cold potable water
5433 : 1976	Underground stop valves for water services
1010 Pt 2 : 1973	Draw-off taps and above-ground stop valves
6572 : 1985	Blue polyethylene pipes to size 63 for below ground use for potable water
6730 : 1986	Black polyethylene pipes to size 63 for above ground use for potable cold water
6700 : 1987	Design, installation, testing and maintenance of services supplying water for domestic use
7491 Pt 1 : 1991	Glass reinforced plastic cisterns for cold water storage: one-piece costerns of capacity up to 500 litres

DATA

Identification of pipes

When excavating in or near buildings in order to provide a water supply, the following table will be found useful.

Identification of service tubes from utility to dwellings

Utility	Colour	Size (mm o.d.)	Depth (mm)
British Telecom	grey	54	450
Electricity	black	38	450
Gas	yellow	42 rigid	450
		60 convoluted	
Water	may be blue	(normally untubed)	750

Notes:

a. This table is based on National Joint Utilities Group: *Service Entries to New Dwellings.* The recommendations are not always adhered to.

b. If the water supply is laid in the same trench as other services, it should be 450 mm wide and 750 mm deep. Water dictates the route into the house.

c. Where possible, service trenches should be made centrally in the footway. This prevents disturbance to boundary walls and kerbs, and protects the trench from vehicle overrun.

d. Pipes should not be laid under private drives or paved areas.

Guide to the sizes of cold water cisterns

Nominal capacity (litres)	(gals)	Height to top of cistern (mm)	Nominal capacity (litres)	(gals)	Height to top of cistern (mm)
18	4	310	227	50	660
38	8	380	273	60	660
68	15	430	318	70	660
91	20	510	455	100	760
114	25	560	500	115	
182	40	610			

For specific dimensions, refer to manufacturers' catalogues.

Blue polyethylene pipes to BS 6572 : 1985

Pipes up to nominal size 63 for below-ground use for potable water.
Nominal DN sizes: 20 25 32 50 63
Limits pressure up to 12 bar (this equals 10^5 N/m^2 or 0.1 MPa) at 20 °C when used below ground or ducted to protect it from sunlight.

Identification markings
Include BS 6572 : 1985, nominal size, manufacturer's name or logo, and the word 'WATER' three times per metre run; all information to be printed at intervals along the pipe.

Black polyethylene pipes to BS 6730 : 1986

Pipes up to nominal size 63 for above-ground use for cold potable water.
Nominal DN sizes: 20 25 32 50 63 and specified as straight or coiled pipe
Pressure limited to 12 bar at 20 °C

Identification markings
Include BS 6730 : 1986, nominal size, manufacturer's name or logo, and the word 'WATER' three times per metre run; all information to be printed at intervals along the pipe.

Cisterns to BS 7491 Pt 1 : 1991

Glass-fibre-reinforced plastics cisterns for cold water storage.
   These tanks are designed to have covers, and can store water at a maximum of 39 °C. The volume stated on the tank is to the top of the cistern; the water line is 100 m below this.

Identification marking
BS 7491 Pt 1 : 1991/GRP/A350 ('A' refers to Class A potable water, the figure to capacity in litres).

BS 7491 Pt 1 : 1991/GRP/B50 ('B' refers to Class B non-potable water, the figure to the capacity in litres).

As the British Standard gives no guidance on the size of the cisterns, the landscape designer must refer to manufacturers' technical literature for this information.

Cisterns to BS 417 Pt 2 : 1987

Galvanized low carbon steel cisterns

The British Standard gives capacities from 18 to 3364 l minimum capacity; sample sizes up to 841 l are given below.

Minimum capacity (litres)	(gals approx)	Length × Width (mm)	Minimum capacity (litres)	(gals approx)	Length × Width (mm)
18	4	482 × 330	191	42	787 × 610
38	8	635 × 330	227	50	940 × 635
54	12	635 × 432	264	58	940 × 686
68	15	635 × 457	423	93	991 × 787
86	19	635 × 482	491	108	1118 × 889
114	25	711 × 533	709	156	1194 × 914
159	35	762 × 584	841	185	1549 × 940

Insulation of pipework above ground

Insulating material values				
W/(m/K)	0.035	0.04	0.055	0.07
Pipes up to 22 mm	27 mm	38 mm	63 mm	100 mm
22–42	27	38	63	89
42–54	19	32	50	75
54–76.1	16	25	44	63
over 76.1	16	25	32	50

Bedding flexible pipes: PVC-U or ductile iron

Type 1 =   100 mm fill below pipe, 300 mm above pipe: single-size material.
Type 2 =   100 mm fill below pipe, 300 mm above pipe: single-size or graded material.
Type 3 =   100 mm fill below pipe, 75 mm above pipe with concrete protective slab over.
Type 4 =   100 mm fill below pipe, fill laid level with top of pipe.
Type 5 =   200 mm fill below pipe, fill laid level with top of pipe.
Concrete = 25 mm sand blinding to bottom of trench, pipe supported on chocks, 100 mm concrete under the pipe, 150 mm concrete over the pipe.

This method should be used where pipes pass close to buildings or walls. If the side of the pipe trench is less than 1 m from the foundation of a wall, Building Regulations require that the pipe be bedded in concrete to the level of the underside of the wall foundation.

OUTLINE SPECIFICATION

Laying cold water supply pipe

a. *Excavate trench 800 mm deep as specified in section P30, supply and lay 50 mm blue polyethylene pipe to BS 6572 : 1985, bedding Type 3, backfill with excavated material screened to remove all stones over 40 mm.*

b. *Excavate hole, supply and install PVC-U shallow inspection chamber 250 mm dia. × 900 mm deep with cast iron sealed cover, supply and fit screw-down stop valve to BS 5433.*

for trenches   P30

S15 Water features

GENERAL GUIDANCE

This section includes reservoirs, lakes, ponds, pools, waterfalls, and fountains, excluding swimming pools.

Where any water features, particularly lakes, ponds or pools, are frequented by young children, steep slopes into the water should be avoided and, if necessary, fencing should be provided to segregate the children from the water. All water features should have a shallow sloping section, which will provide an access and escape route for animals and amphibians.

Natural run-off water, rainwater, springs and streams can be used to supply ponds and lakes, but the source of the water should be under the client's control so that contamination does not occur. Rain may be acidic, and run-off from concrete may be alkaline; the pH of the supply should be checked, and if necessary the water should be passed through filters and chemically corrected. Run-off from roads, car parks, farmsteads and stables can be dangerously polluted or too rich in nutrients, and should not be used.

Definitions

Lakes

Large informally shaped areas of water made from a natural basin or formed by excavation or dredging. Unless they are excavated below the constant level of the water-table they will need to have a flexible liner or clay puddle base and be fed by a stream, spring or from the run-off from buildings and hard landscape. If deep enough a lake may be used for coarse fishing, but the landscape designer should get a clear brief from the client, as the type of liner and depth of water may be critical. An overflow drainage system may be required to prevent flooding, and provision must be made for keeping the water level at a safe height in drought conditions. Lakes and even large ponds are a hazard for children, and they should not have unguarded steep sides adjoining deep water where it is impossible to scramble out; deep areas for fish should be placed in the centre of the water and preferably guarded from paddlers and boaters. Lifebuoys should always be provided at lakesides, and instructions on reviving apparently drowned persons are usually included.

Natural ponds

Natural or naturally shaped water features with gently sloping sides and one or more areas of the water at least 1000 mm deep. They are intended to promote wild life of all types, including amphibians that need to be able to enter and leave the pond from gently sloping sides or shallow steps. The deep portion of the pond is required to keep fish below the ice level in the most severe winter. A eutrophic zone will encourage even more flora and fauna, and may be requested by a client who is a nature enthusiast. Ponds may be fed from surface water run-off, but they will probably require a water supply to top them up and oxygenate the water. An overflow drainage system will be required to prevent flooding during heavy rainfall.

Formal pools

Water features of a formal nature, with preformed or concrete linings and surrounded by hard landscaping. They may be required to hold specimen fish and aquatic plants, and will require a water supply and probably an overflow drainage system to prevent heavy rain from flooding the surrounding area. Such pools are usually cleaned regularly, and provision should be made for draining the water away to an approved disposal point such as a watercourse or soakaway.

Fountains

Large fountains, such as those in civic situations with complex mechanisms that alter the height of the water and coordinate the movement through a colour cycle and/or music, are the province of specialists, who will normally design the system and carry out the work as a subcontract. These are not included here. Garden fountains of modest performance can be installed by a contractor to the landscape designer's drawings and specification; it is these that are considered here. All require a water supply, electrical supply (low voltage) and a pump to recycle the water, and a drain for overflow disposal. It is difficult to make a waterproof entry to a pool for electric cables, and unless the fountain housing is designed to take cables from beneath the pool it is preferable to lead them over the top edge, concealed by planting or stone features.

Cascades and waterfalls

These can be used when there is a distinct change in level. Water falls over an inclined masonry wall, natural rocks or steps. The depth of water over the top edge must be carefully worked out to give a natural effect. Specific aeration requirements would be the province of a specialist.

Materials

The materials used in pools are subject to freeze–thaw cycles, ultraviolet rays, thermal expansion and contraction, and abrasion from wave action, animals and people, chemical attack from water chemicals used in controlling weeds, and natural ageing.

Butyl or EPDM synthetic rubber liners are reputed to have the longest life, although various types of reinforced polyethylene are available for cheaper work. One-piece liners can be installed by site labour up to about 750 m², but large lake areas are usually supplied and laid by specialist subcontractors, as the heat welding required to join sheets and the method of laying large sheets is a highly skilled job. Slightly cheaper than butyl are flexible polypropylene alloy (FPA) liners, which must be installed by specialist subcontractors. Butyl, EPDM and FPA have a 20-year guarantee.

All liners should be laid on a smooth firm base gently graded to contour without sudden changes of level and direction, which may strain the fabric. The ground should be brought to grade, organic material and all stones and debris over 50 mm removed, the surface rolled or compacted to a smooth finish, and 150 mm of coarse sand spread evenly over the surface and compacted. A 300 g/m² geotextile membrane laid on top of the liner will help to retain the soil, though a plastic geogrid may be necessary on steep slopes. The liner is laid loose, and a layer of low-fertility topsoil or gravel 150–300 mm deep is spread over the bottom of the basin; the water basin is then filled gently with water to prevent disturbance of the soil covering, and the top edges are laid in a perimeter trench and held in position with soil. Paving may be laid on top of the edging. Design options for 'safe sandwiching' must suit the circumstances.

Liners need protection from puncturing by:

- underlying stones or debris;
- any hard objects in the soil covering;
- exposure or piercing through the soil covering.

Concrete pools may be constructed with ordinary dense sulphate-resisting concrete, and the addition of a waterproofer is not necessary if the workmanship is of good quality. Waterproofers are not totally effective for water-containing structures, and unless a small degree of leakage can be tolerated it is better to tank the pool. The concrete must be laid on a firm foundation of hardcore or granular fill, and the sides must be thick enough to withstand earth pressure, as they are in effect retaining walls; reinforcement may be needed if the substrate is not stable. Movement joints will have to be provided in any but the tiniest pools, and these should be formed with special flexible seals cast into the concrete. However, cracking will always be a possibility unless the pool is constructed to swimming pool standards, and it is preferable to construct a base and walls in concrete, tank them with mastic asphalt, liquid bitumen waterproofing, or a loose flexible membrane, and to construct a protective and decorative brick, stone or concrete skin on the inside. Small concrete pools can 'float' if the groundwater pressure is greater than the weight of the pool, with consequent cracking and leakage; the outside of the pool may then have to be drained. Although flexible membranes may float in the same way, they do not crack. However, floating may lead to other problems with the integrity and protection of the liner, such as undetected variations in the level of the water. If floating is likely to occur, specialized engineering advice should be sought.

Small brick or concrete pools may be waterproofed with acrylic coating, cement-based paint, chlorinated rubber paint or epoxide coating, but these coatings need regular maintenance and re-painting.

Prefabricated glass-fibre forms may be used for small ornamental pools up to 2–3 m²; the excavation must be carefully shaped to fit the pool, and the glass fibre should be bedded on a 100 mm sand layer.

Clay puddle, composite liners, or bentonite granule matting lined ponds are very difficult to construct satisfactorily, and the advice of the British Trust for Conservation Volunteers should be sought if such a pond is required. The type of subsoil and the system of natural drainage are important factors in constructing these ponds, and only certain sites will be suitable. This type of construction is most often used for the restoration of traditional ponds such as dew ponds.

BRITISH STANDARDS

882 : 1992   Aggregates from natural sources for concretes
3882 : 1994   Specification for topsoil (bought and sold)
8007 : 1987   Code of practice for design of concrete structures for retaining
              aqueous liquids

There is no British Standard for butyl rubber sheeting or EPDM.

DATA

Types of flexible membrane

Butyl
- Vulcanized synthetic rubber.
- Elongation at breaking point 300%.
- Elastic from –35 °C to 120 °C.
- Resistant to acidic and alkaline water, brine, lime, concrete and cement.
- Not resistant to petrol, oil, timber impregnation agents or turpentine.

Reinforced bitumen

- 3 mm thick in rolls 5 m wide and 100 m long.
- 5 mm thick in rolls 5 m wide and 50 m long.
- Can be site joined or factory welded.

EPDM (ethylene propylene diene monomer)

- 0.75 mm, 1.0 mm, 1.15–2.3 mm in rolls 2.3–15.25 m wide, length 30.5–60.1 m.
- Laid by specialist subcontractor.
- Can be used on flat or shallow slope, laid loose and ballasted.

FPA (flexible polypropylene alloy)

Must be installed by specialist subcontractors.

HDPE (high-density polyethylene)

Often combined with LDPE (low-density polyethylene). Proprietary fabrics may vary in their composition.

Soils: angle of repose

Earth,	loose and dry	35–40° from horizontal
	loose and naturally moist	45°
	loose and saturated	27–30°
	consolidated and dry	42°
	consolidated and naturally moist	37°
Loam,	loose and dry	40–45°
	loose and saturated	20–25°
Gravel,	medium coarse and dry	30–45°
	medium coarse and wet	25–30°
Sandy gravel		35–45°
Sand,	compact	35–50°
	loose and dry	30–35°
	saturated	25°
Shale (blaes) filling		30–35°
Broken brick		35–45°
Broken rock		35–45°
Clay, loose and wet		20–25°
	solid naturally moist	70°

Topsoil quality to BS 3882: 1994

Topsoil grade	Properties
Premium	Natural topsoil, high fertility, loamy texture, good soil structure, suitable for intensive cultivation.

Topsoil quality to BS 3882: 1994 *continued*

Topsoil grade	Properties
General Purpose	Natural or manufactured topsoil of lesser quality than Premium, suitable for agriculture or amenity landscaping, may need fertilizer or soil structure improvement.
Economy	Selected subsoil, natural mineral deposit such as river silt or greensand. The grade comprises two subgrades: 'low clay' and 'high clay', which is more liable to compaction in handling. This grade is suitable for low-production agricultural land and amenity woodland or conservation planting areas.

Concrete mixes to BS 5328 Pt 1 & AMD

Application	Standard mix	Designated mix	Slump (mm)
Mass concrete fill and blinding	ST 2 (C10P)	GEN 1	75
Mass concrete foundations in non-aggressive soils	ST 2 (C10P)	GEN 1	75
Trench fill foundations in non-aggressive soils	ST 2 (C10P)	GEN 1	125
Foundations			
in Class 2 sulphate soils	–	FND 2	75
in Class 3 sulphate soils	–	FND 3	75
in Class 4A sulphate soils	–	FND 4A	75
in Class 4B sulphate soils	–	FND 4B	75

Traditional concrete mixes

Application	Name of mix	Cement:aggregate:sand (parts by volume)
In-situ concrete paving:	one, one & half, two & half	1:1.5:2.5

'All-in' is mixed aggregate as dug from the pit.

Fountains with lights

Floating lights require a minimum water depth of 1000 mm to accommodate the equipment below water level. Lights are grouped in threes to keep their balance; they can be clear, or coloured yellow, red, blue and green. The lighting system can be used with any pump and fountain head. All installation work must be carried out by a qualified electrician.

Pumps come in many sizes from 5 W with an output of 5 l/min up to 1250 W with an output of 750 l/min. The landscape designer is advised to contact the manufacturer for detailed information on all aspects of fountain technology.

Typical fountain nozzles

		Height (m)	Width (m)	Pump (kW)
Sky cascades	from	2.8	1.4	0.3
	up to	4.7	1.2	1.5
Heavy water sky cascades		1.5	0.8	0.3
	up to	4.0	1.2	1.2
Geyser	from	1.5	2.0	0.3
	up to	4.5	2.5	1.5
Arch	from	2.1	4.0	0.3
	up to	5.2	7.0	1.5
Tall sky	from	10.0	–	3.7
cascades	up to	21.0	–	11.1
Fog jet	from	5.5	3.5	3.7
	up to	11.0	4.5	11.1

OUTLINE SPECIFICATION

Informal pond or lake

a. *Excavate area shown on drawings to given depths and contours, spread spoil on site where shown, bring ground to grade, dig out and backfill weak spots with consolidated excavated material, remove all organic matter and stones and debris over 40 mm. If any groundwater appears in the excavation, refer to the landscape designer for instructions.*

b. *Excavate retaining trench at perimeter 600 mm deep for edge of liner and deposit spoil alongside.*

c. *Lay 150 mm sand to BS 882 Grade C spread evenly and compacted, supply and lay EPDM liner 1.15 mm thick, factory-welded sheets, lay edge of liner loosely in perimeter trench. Lay geotextile membrane 300 g/m² over liner, lay 200 mm topsoil to BS 3882 : 1994 Economy grade spread evenly over surface of geotextile.*

d. *Fill pond gently with water, and backfill perimeter trench with excavated material.*

for excavation	*D20*
for aquatic planting	*Q31*

Formal pool

a. *Excavate area for pool to levels and falls shown on drawing and remove spoil to dump on site not exceeding 100 m, dig out and backfill weak spots with consolidated excavated material, remove all organic matter and stones and debris over 40 mm. Trim sides and bottom of excavation to true surfaces.*

b. *Construct formwork and pour 200 mm FND4 concrete base and sides to pool, supply and cast in drainage outlet at position shown on drawing, cover pool during curing of concrete.*

c. *Face base and sides with mastic asphalt 30 mm thick in three-coat work, blind with sand, lay 50 mm cement:lime:sand screed 1:½:4 on base, construct half-brick protective walls in engineering brick Class B laid in Group 1 cement:lime:sand mortar, edging to pool in bull-nose engineering brick as above.*

d. *Supply and install arch pattern fountain 2.1 m high in position shown on drawing complete with pump and electrical connections. Fill and drain pool twice with water after concrete is cured.*

for excavation	*D20*
for concrete	*E10*
for formwork	*E20*
for brickwork	*F10*
for tanking	*J*
for electrical work	*V41*

V41 Electrical supply/power/lighting systems

GENERAL GUIDANCE

This section deals with power and lighting circuits to outside appliances, types of lighting appliance, wall lights, floodlights, bollards and low-level lighting, street lighting and light signs. All work for the supply of electricity and the cables to fittings has to be installed by a qualified electrician to the current requirements of the Institution of Electrical Engineers (IEE) Regulations for Electrical Installations: therefore only very general data is given on the installation of appliances.

Most manufacturers of lighting fittings and columns will provide an advisory service, as the calculation of lighting levels for specific purposes (especially floodlighting) and the selection of the appropriate fittings are fairly complicated.

Safety precautions for electrical installations

The first priority is to protect people and animals from electric shocks, and the second is to protect the installation from overloading its capacity or short-circuiting. The three requirements for this are:

Fig. 4.71 Shelter and lighting fixtures, Rhyl seafront. (Woodhouse Co.)

- All live parts must be insulated.
- Circuit breakers must cut off the whole or part of supply when overloaded.
- All metalwork must be earthed, to prevent shocks to people or animals.

Wall lights

These comprise: recessed fittings, which can be set flush with the surface of the wall, mainly used for step lights and locations where a projecting fitting would be unsafe; bulkhead fittings, which fit closely to the wall; and bracket fittings of various ornamental designs, which project to a greater or lesser distance.

Floodlights

Floodlights are mounted at high level to illuminate a wide area of ground such as a sports field or car park. They are usually mounted on high masts, whose construction is designed to withstand high winds and severe weather conditions, and the selection of the appropriate luminaires and lamps is a specialized job. They are covered by BS 4533:102.5 : 1990.

Fig. 4.72 Custom-built river wall lighting fixtures, Tower Bridge. (Woodhouse Co.)

Illuminated bollards

Most illuminated bollards are used in private locations where vandalism is not a problem, but some heavy vandal-proof concrete bollards are manufactured for public areas, though the light they give is restricted because of the small openings in the concrete. Bollards are best suited for indicating pathways and courtyards traversed by regular users, as they do not illuminate anything but the immediate surroundings. They can be made in almost any material – steel, concrete, timber, glass fibre or cast iron – and the range of designs is very wide, though the basic specification is given in BS 873 Pt 3 : 1980.

Low-level lighting

There is no definition of low-level lighting, but it is usually taken to cover strong waterproof fittings set into the ground, and dwarf lamps that are used in much the same circumstances as illuminated bollards.

Spotlights

Many buildings and specimen trees are illuminated at night by carefully placed spotlights, which can either pick out a particular feature or provide a

Fig. 4.73 Gateway lighting feature, Windsor. (Woodhouse Co.)

general illumination to the whole subject. They must be sited so that they are not obtrusive in the daytime, and as they are often concealed in vegetation they must be of sturdy construction to withstand accidental damage.

Street lighting

The design and layout of public street lighting for local areas is covered by BS 5489 : Pts 1, 3 and 9. Other parts deal with highway lighting, which is a

Fig. 4.74 Trojan lighting bollards. (Kingfisher Lighting)

specialized subject. The height and spacing of the columns, the type of luminaire and the lamps are all specified.

Lighting columns

These are covered in BS 5649 Pts 1–9 : 1989, which deals with plain columns for highway contracts. Columns can be made of steel, aluminium or concrete, though as street luminaires tend to become higher concrete columns are seldom specified. Many manufacturers supply ornamental columns, which are intended for conservation areas and prestige projects; these include reproduction Victorian columns, some of which are designed to be gas lit.

BRITISH STANDARDS

873 Pt 3 : 1980	Internally illuminated bollards
873 Pt 7 : 1984 & AMD	Lighting posts and fittings
4343 : 1968 & AMD	Industrial plugs, socket outlets, couplers for AC and DC: outdoors and agriculture
4533:102.5 : 1990 & AMD	Floodlights
4533:s102.7 : 1990	Portable luminaires for garden use
5489 Pt 1 : 1992 & AMD	Road lighting. General principles
5489 Pt 3 : 1992 & AMD	Code of practice for lighting for subsidiary roads and associated pedestrian areas
5489 Pt 9 : 1992	Code of practice for lighting for urban centres and public amenity areas
5649 Pts 1–9 : 1989	Street lighting columns: steel, aluminium and concrete
7671 : 1992	Requirements for electrical installations. The IEE Wiring Regulations 16th edition
BS EN 60529 : 1992	Classification for degrees of protection provided by enclosures

DATA

Regulations and publications

Institution of Electrical Engineers (IEE) Regulations for Electrical Installations
The current regulations at the time of writing are the 16th edition dated 1991. It is essential that the latest edition is referred to in any specification.

The On-Site Guide to the 16th Edition Wiring Regulations dated 1992
This is a useful guide, which covers small-scale work.

Use and colour temperature of lamps

Type of lamp		Control gear	Colour	Colour rendering	Exterior lighting: suitable uses
Tungsten filament (GLS)		No	Warm white	Good	Amenity lighting Security lighting
Compact fluorescent	SL	No	Warm white	Fair	Amenity lighting and security lighting
	PLCE	No	Warm white	Fair	
	PLC	Yes	Warm white	Fair	
	PLL	Yes	Intermediate	Fair	

Use and colour temperature of lamps *continued*

Type of lamp	Control gear	Colour	Colour rendering	Exterior lighting: suitable uses
High-pressure discharge mercury – – fluorescent	Yes	Intermediate white	Moderaate to good	Amenity lighting Public lighting General floods
High-pressure mercury discharge – metal halide	Yes	Cool white	Good to excellent	Amenity lighting Sports lighting Building floods Area floods
High-pressure sodium discharge	Yes	Golden white	Moderate	Amenity lighting Sports lighting Building floods Area floods Security lighting Public lighting High mast lighting
Low-pressure sodium discharge	Yes	Yellow	Bad	Security lighting Public lighting

Notes:

a. Tungsten filament and compact fluorescent lamps give instant full lighting as soon as they are switched on.

b. Low-pressure sodium discharge lamps are not recommended, as the light they give is of poor quality.

c. High-pressure discharge lamps have a warming-up time when switched on, and once turned off should be given a cooling-down period before being switched on again.

Electrical insulation class EN 60.598 BS 4533

Class 1 luminaires comply with class 1 earthed electrical requirements.
Class 2 luminaires comply with class 2 (II) double insulated electrical requirements.
Class 3 luminaires comply with class 3 (III) electrical requirements.

Protection to light fittings

BS EN 60529 : 1992 *Classification for degrees of protection provided by enclosures* (IP Code – International or ingress Protection)

IP code: indicates the degree of protection provided by an enclosure against:

First characteristic: ingress of solid foreign objects

The figure   2 indicates that fingers cannot enter

         3 that a 2.5 mm diameter probe cannot enter

         4 that a 1.0 mm diameter probe cannot enter

         5 the fitting is dust proof (no dust around live parts)

         6 the fitting is dust tight (no dust entry)

Second characteristic: ingress of water with harmful effects

The figure   0 indicates unprotected

         1 vertically dripping water cannot enter

         2 water dripping 15° (tilt) cannot enter

         3 spraying water cannot enter

         4 splashing water cannot enter

         5 jetting water cannot enter

         6 powerful jetting water cannot enter

         7 proof against temporary immersion

         8 proof against continuous immersion

Optional additional codes

A--D   Protects against access to hazardous parts

   H   High-voltage apparatus

   M   Fitting was in motion during water test

   S   Fitting was static during water test

   W   Protects against weather

Marking code arrangement

Example: IPX5S = IP (International or ingress Protection); X (denotes omission of first characteristic); 5 = jetting; S = static during water test.

Spacing of lighting columns: based on BS 5489 Pt 3

Pedestrian precincts, footways and bicycle paths

Minimum horizontal illuminance 10 lux.

Lighting on one side only with 70 W high-pressure sodium lamp:

Height of lamp (m)	Width of path (m)	Spacing of lighting columns (m apart)	Average lux
3.0	3.0	15	42
		20	32
3.5	3.0	20	29
		25	23
4.0	3.0	25	21
		30	18

Local roads in residential areas

Minimum horizontal illuminance 3.5 lux.

Lighting on one side only with 100 W high-pressure sodium lamp:

Height of lamp (m)	Width of road (m)	Spacing of lighting columns (m apart)	Average lux
4.0	4.0	15	55
		20	44
		25	37
6.0	4.0	25	28
		30	24
		35	22

Lighting on both sides staggered with 100 W high-pressure sodium lamp:

Height of lamp (m)	Width of road (m)	Spacing of lighting columns (m apart)	Average lux
4.0	5.0	30	28
		40	22
		50	19

Car parks, squares, shopping precincts and parks

Minimum horizontal illuminance 20 lux for pleasant lighting.

Lighting all round with 125 W high-pressure mercury lamp:

3.5 m	area 14 × 14 m²	24 lux
	area 20 × 20 m²	14 lux

Note: Shields should be used to prevent lights from annoying residents in nearby houses.

Floodlights

Precautions based on BS 4533 102.5: 1990 & AMD *Floodlights*. Floodlights for outdoor use require protection against ingress of moisture to IP X3.
Cables to high floodlights must withstand winds up to 150 km/h.
Glass to high floodlights must be either:

- film-coated glass;
- safety glass;
- caged in fine wire mesh.

Typical luminaires and columns

Wall luminaires

Step light recessed cast aluminium with louvres	236 mm × 100 mm deep 9 W PL
Bulkhead circular cast aluminium	245 mm dia. × 100 mm deep 75 W GLS
Rectangular cast aluminium	250 mm × 160 mm × 95 mm deep 60 W GLS

Bollards

Concrete bollards vandal resistant 870 mm above ground 420 mm dia. 70 W SON

	480	420	70 W SON
Black cast aluminium	780	122	60 W GLS
Tubular steel and polycarbonate	1217	195	70 W SON

Low-level 120 W

Spotlight ground fixed	300 mm high 350 mm wide	100 W
Path lighter cast aluminium body ground fixed	75 mm 250 mm dia.	100 W
Path lighter hooded spun copper	450 mm 160 mm dia.	60 W
Uplighter ground fixed	200 mm 250 mm wide	60 W

Street luminaires

Clear acrylic globe		460 mm dia. 120 W SON
As above but with cut-off		
Drum-shaped opal high-density polyethylene	500 mm high	390 mm dia. 55 W SOX
Traditional lantern black aluminium HD polyethylene opal	1085 mm high	475 mm dia. 150 W SON

Columns

Aluminium reproduction style with lantern top	5.0 m high above ground	76 mm dia.
Tapered aluminium self-colour	3–6.0 m	
Tapered tubular steel galvanized	4–12.0 m,	114–168 mm dia. at base
Prestressed concrete	4.8 m	

OUTLINE SPECIFICATION

Precinct lighting

Supply and install 30 no. general lighting units at locations shown on drawings: 3.0 m tapered aluminium columns to BS 5649 Ref. XXX, set 300 mm into paving, 500 mm dia. clear acrylic globe luminaires IP54 Ref. XXX complete with control gear and electrical supply, installed all in accordance with IEE Regulations.

Car park

Supply and install 10 no. floodlighting units at locations shown on drawings: 5.0 m galvanized painted tubular steel columns to BS 5649 Ref. XXX, set 450 mm into paving, low light pollution aluminium luminaires to BS 4533 : 102.5 IP65 Ref. XXX complete with control gear and electrical supply, installed all in accordance with IEE Regulations.

Courtyard

a. *Supply and install 30 no. illuminated bollards at locations shown on drawings: 900 mm high blue powder-coated steel bollard to BS 873 Pt 3 set 450 mm into ground, louvred polycarbonate luminaire IP54, Ref. XXX, complete with control gear and electrical supply, installed all in accordance with IEE Regulations.*

b. *Supply and install 11 no. 120 V groundlighters at locations shown on drawings, cast bronze casing 250 mm dia. IP67W set into paving, polycarbonate diffuser, Ref. XXX, complete with control gear and electrical supply, installed all in accordance with IEE Regulations.*

c. *Supply and install 4 no. 120 V ground-mounted spotlights at locations shown on drawings: black powder-coated diecast aluminium casing 220 mm dia. IP55 bolted to paving, polycarbonate diffuser, Ref. XXX, complete with control gear and electrical supply, installed all in accordance with IEE Regulations.*

Suggested further reading

History

Banks, E. (1991) *Creating Period Gardens*, Phaidon Press, Oxford.

Brookes, J. (1987) *Gardens of Paradise: The History and Design of the Great Islamic Gardens*, Weidenfeld & Nicolson, London.

Brown, J. (1986) *The English Garden in Our Time, from Gertrude Jekyll to Geoffrey Jellicoe*, Antique Collectors' Club, Woodbridge.

Elliott, B. (1986) *Victorian Gardens*, Batsford, London.

Jacques, D. (1983) *Georgian Gardens: The Reign of Nature*, Batsford, London.

Jellicoe, G. (1994–) *Collected Works*, 4 vols, Garden Art Press, Woodbridge.

Jellicoe, G. and Jellicoe, S. (1995) *The Landscape of Man, Shaping the Environment from Prehistory to the Present Day*, 3rd edn, Thames & Hudson, London.

Lambert, D. *et al.* (1995) *Researching a Garden's History: A Guide to Documentary and Published Sources*, Landscape Design Trust, Reigate with Institute of Advanced Architectural Studies, University of York.

Landscape Institute (1994) *A Visitor's Guide to 20th Century British Landscape Design*, Landscape Institute, London.

Mosser, M. and Teyssot, G. (1991) *The History of Garden Design: The Western Tradition from the Renaissance to the Present Day*, Thames & Hudson, London.

Spens, M. (1994) *The Complete Landscape Designs and Gardens of Geoffrey Jellicoe*, Thames & Hudson, London.

Turner, T. (1986) *English Garden Design, History and Styles Since 1650*, Antique Collectors' Club, Woodbridge.

Landscape design theory

Appleton, J. (1975, repr. 1986) *The Experience of Landscape*, Wiley, Chichester.

Francis, M. and Hester, R.T. (1990) *The Meaning of Gardens, Idea, Place and Action*, M IT, Cambridge, MA.

Hough, M. (1990) *Out of Place: Restoring Identity to the Regional Landscape*, Yale University Press, New Haven.

McHarg, I.L. (1992) *Design With Nature*, rev. edn. John Wiley, New York.

Nasar, J.L. (1988) *Environmental Aesthetics: Theory, Research, and Applications*, Cambridge University Press, Cambridge.

Spirn, A.W. (1984) *The Granite Garden: Urban Nature and Human Design*, Basic Books, New York.

Turner, T. (1995) *City as Landscape: A Post-Postmodern View of Design and Planning*, E. & F.N. Spon, London.

Landscape design: general

Baljon, L. (1992) *Designing parks: An Examination of Contemporary Approaches to Design in Landscape Architecture (based on Parc de la Villette)*, Architectura & Natura Press, Amsterdam.

Bell, S. (1993) *Elements of Visual Design in the Landscape*, E. & F.N. Spon, London.

Crowe, S. (1994) *Garden Design*, Packard, Chichester.

Fieldhouse, K. and Harvey, S. (1992) *Landscape Design: An International Survey*, Laurence King, London.

Filor, S.W. (1991) *The Process of Landscape Design*, Batsford, London.

Hill, W.F. (1995) *Landscape Handbook for the Tropics*, Garden Art Press, Woodbridge.

Jellicoe, G. *et al.* (eds) (1986) *The Oxford Companion to Gardens*, Oxford University Press, Oxford.

Lancaster, M. (1994) *The New European Landscape*, Butterworth Architecture, London.

Lyall, S. (1991) *Designing the New Landscape*, Thames & Hudson, London.

Marcus, C.C. and Francis, C. (1990) *People Places: Design Guidelines for Urban Open Space*, Van Nostrand Reinhold, New York.

Miyagi, S. and Kokohari, M. (1990) *Contemporary Landscapes in the World*, Process Architecture Co. Ltd, Tokyo.

Phillips, A. (1993) *The Best in Science, Office and Business Park Design*, Batsford, London.

Preece, R.A. (1991) *Designs on the Landscape: Everyday Landscapes, Values and Practice*, John Wiley, New York.

Reis, G.W. (1993) *From Concept to Form in Landscape Design*. Van Nostrand Reinhold, New York.

Turner, T. (1987) *Landscape Planning*. Hutchinson, London.

Landscape practice (including CAD, drawing and management)

Batty, M. (1987) *Microcomputer Graphics: Art, Design and Creative Modelling*, Chapman & Hall, London.

Clamp, H. (1988) *Landscape Professional Practice*, Gower, Aldershot.

Clamp, H. (1995) *Spon's Landscape Contract Handbook*, E. & F.N. Spon, London.

Clamp, H. and Cox, S. (1989) *Which Contract? Choosing the Appropriate Building Contract* (+ supplement June 1990), RIBA Publications, London.

Cobham, R. (1990) *Amenity Landscape Management: A Resources Handbook*, E. & F.N. Spon, London.

Environment Council (1995) *Who's Who in the Environment*, Environment Council, London (also available on CD-ROM).

External works 7 (1996) Stirling: Landscape Promotions (Guide to products, services and suppliers for the external environment).

Harte, J.D.C. (1985) *Landscape, Land Use and the Law: An Introduction to the Law Relating to the Landscape and Its Use*, E. & F.N. Spon, London.

Hitchmough, J.D (1994) *Urban Landscape Management*, Inkata Press, Sydney.

Hughes, D.(1992) *Environmental Law*, 2nd edn, Butterworths, London.

Lin, M.W. (1993) *Drawing and Designing with Confidence: A Step-By-Step Guide*, Van Nostrand Reinhold, New York.

Lovejoy, D. & Partners and Davis Langdon & Everest (1995) *Spon's Landscape and External Works Price Book 1996*, E. & F.N. Spon, London.

Macdougall, E.B. (1983) *Microcomputers in Landscape Architecture*, Elsevier, New York.

Murdoch, J. and Hughes, W. (1992) *Construction Contracts, Law and Management*, E. & F.N. Spon, London.

Nurcombe, V.J. (ed.) (1996) *Information Sources in Architecture and Construction*, 2nd edn, Bowker Saur, London.

Parker, J. and Bryan, P. (1989) *Landscape Management and Maintenance: A Guide to its Costing and Organization*, Gower Technical Press, Aldershot.

Ritter, J. (1989) *Quality Assurance: Quality Systems in Building Design Firms*, RIBA Indemnity Research Ltd, London.

Walker, T.D. (1989) *Perspective Sketches*, Van Nostrand Reinhold, New York.

Walker, T.D. and Davis, D.A. (1990) *Plan Graphics*, Van Nostrand Reinhold, New York.

Wester, L.M. (1990) *Design Communication for Landscape Architects*, Van Nostrand Reinhold, New York.

Hard landscape (including engineering)

Aurand, C.D. (1987) *Fountains and Pools: Construction Guidelines and Specifications*, E. & F.N. Spon, London.

Cheetham, C.comp. (1994) *Dealing with Vandalism: A Guide to the Control of Vandalism*, CIRIA and Thomas Telford, London

Coppin, N.J. and Richards, I.G. (1990) *Use of Vegetation in Civil Engineering*, CIRIA and Butterworths, London.

Department of the Environment (1992) *Handbook of Estate Improvement*, Vol. 1: *Appraising Options*, Vol. 2: *External Areas*, Vol. 3: *Dwellings*, HMSO, London.

Gibbons, J. and Oberholzer, B. (1991) *Urban Streetscapes: A Workbook for Designers*, BSP Professional Books, Oxford.

Hall, M.J. *et al.* (1993) *Design of Flood Storage Reservoirs*, CIRIA and Butterworth-Heinemann, London.

Hemphill, R. and Bramley, M. (1989) *Protection of River and Canal Banks: A Guide to Selection and Design*, CIRIA and Butterworths, London.

Lisney, A. and Fieldhouse, K. (1990) *Landscape Design Guide*, Vol.2: *Hard Landscape*, Gower, Aldershot.

Littlewood, M. (1993) *Landscape Detailing*. Vol. 1: *Enclosures*, Vol. 2: *Surfaces*, Butterworth-Heinemann, London.

McCluskey, J. (1987) *Parking: A Handbook of Environmental Design*, E. & F.N. Spon, London.

McCluskey, J. (1992) *Road Form and Townscape*, 2nd edn, Butterworth Architecture, London.

National Rivers Authority (1992) *Thames Environment Design Handbook*, National Rivers Authority, Reading.

Natural Stone Directory, Ealing Publications

Pinder, A. and Pinder, A. (1990) *Beazley's Design and Detail of the Space between Buildings*, 2nd edn, E. & F.N. Spon, London.

Strom, S. and Nathan, K. (1993) *Site Engineering for Landscape Architects*, **2nd edn**, Van Nostrand Reinhold, New York.

Soft landscape (including planting design)

Bradshaw, A. *et al.* (1995) *Trees in the Urban Landscape: Principles and Practice*, E. & F.N. Spon, London.

Clouston, B. (ed.) (1990) *Landscape Design with Plants*, Butterworth-**Heinemann,** London.

Department of the Environment (1944) *Tree Preservation Orders: a guide to law and good practice*

Erhardt, **A.** and Erhardt, W. (1996) *PPP Index: The European Plant Finder Book* (including CD-ROM), Eugen Ulmer Verlag, Stuttgart.

Johnston, J. and Newton, J. (1993) *Building Green: A Guide to Using Plants on Roofs, Walls and Pavements*, London Ecology Unit, London.

Lisney, A. and Fieldhouse, K. (1990) *Landscape Design Guide*, Vol. 1: *Soft Landscape*, Gower Technical Press, Aldershot.

Littlewood, M. (1988) *Tree Detailing*, Butterworth Architecture, London.

Philip, C. (ed.) (1996) *The RHS Plant Finder 1996/7*, Royal Horticultural Society, London.

Walker, T.D. (1991) *Planting Design*, 2nd edn, **Van** Nostrand Reinhold, New York.

Special applications

Adie, D.W. (1984) *Marinas: A Working Guide to Their Development and Design*, 3rd edn, Architectural Press, London.

Baines, C. and Smart, J. (1991) *A Guide to Habitat Creation*, Packard, Chichester (for London Ecology Unit).

Balogh, J.C. and Walker, W.J. (1992) *Golf Course Management and Construction: Environmental Issues*, Lewis, Boca Raton.

Buckley, G.P. (1992) *Ecology and Management of Coppice Woodlands*, Chapman & Hall, London.

Burman, P. and Stapleton, H. (1988) *The Churchyards Handbook: Advice on the History and Significance of Churchyards, Their Care, Improvement and Maintenance*, 3rd edn, Church House Publications, London.

Department of the Environment (1992) *Amenity Reclamation of Mineral Workings: Report by Land Use Consultants*, HMSO, London.

Department of the Environment (1992) *The Use of Land for Amenity Purposes: A Summary of Requirements*, HMSO, London.

Dobson, M.C. and Moffat, A.J. (1993) *Potential for Woodland Establishment on Landfill Sites*, Department of the Environment, HMSO, London.

Emery, M. (1986) *Promoting Nature in Cities and Towns: A Practical Guide*, Croom Helm, London (for the Ecological Parks Trust).

Ferris-Kaan, R. (ed.) (1995) *The Ecology of Woodland Creation*, Wiley, Chichester.

Hampshire County Planning (1993) *The Highways Environment: Design Guidelines for Special Areas*, Hampshire County Council, Winchester.

Hawtree, F.W. (1983) *The Golf Course: Planning, Design, Construction and Maintenance*, E. & F.N. Spon, London.

Heseltine, P. and Holborn, J. (1987) *Playgrounds: The Planning, Design and Construction of Play Environments*, Mitchell, London.

James, N.D.G. (1989) *The Forester's Companion*, 4th edn, Blackwell, Oxford.

John, G. and Campbell, K. (1993) *Handbook of Sports and Recreational Building Design*, Vol. 1: *Outdoor Sports*, 2nd edn, Butterworth Architecture, London (for the Sports Council).

Johnston, J. (1990) *Nature Areas for City People: A Guide to the Successful Establishment of Community Wildlife Sites*, London Ecology Unit, London.

Lucas, O.W.R. (1991) *The Design of Forest Landscapes*, Oxford University Press, Oxford.

Marcus, C.C. and Sarkissian, W. (1986) *Housing as if People Mattered*, University of California Press, Berkeley.

Moore, R.C. (1986) *Childhood's Domain: Play and Place in Child Development*, Croom Helm, London.

Moyer, J.L. (1992) *The Landscape Lighting Book*, John Wiley, New York.

Moyer, J.L. (1992) *The Landscape Lighting Book*, John Wiley, New York.

Purseglove, J. (1988) *Taming the Flood: A History and Natural History of Rivers and Wetlands*, Oxford University Press, Oxford (in association with Channel 4 Television).

Stoneham, J. and Thoday, P. (1994) *Landscape Design for Elderly and Disabled People*, Packard, Chichester.

Wylson, A. (1986) *Aquatecture: Architecture and Water*, Architectural Press, London.

Environmental assessment (including EIA)

Countryside Commission (1993) *Landscape Assessment Guidance*, CCP 423.

Department of the Environment and Welsh Office (1989) *Environmental Assessment: A Guide to the Procedures*, HMSO, London.

Essex Planning Officers' Association (1992) *The Essex Guide to Environmental Assessment*, Essex CC, Chelmsford.

Fortlage, C.A. (1990) *Environmental Assessment: A practical guide*, Gower Publishing, Aldershot.

Glasson, J. *et al.* (1994) *Introduction to Environmental Impact Assessment: Principles and Procedures, Process, Practice and Prospects*, UCL Press, London.

Institute of Environmental Assessment (1995) *Guidelines for Baseline Ecological Assessment*, E. & F.N. Spon, London.

Landscape Institute and Institute of Environmental Assessment (1995) *Guidelines for Landscape and Visual Impact Assessment*, E. & F.N. Spon, London.

Morris, P. and Therivel, R. (1995) *Methods of Environmental Impact Assessment*, UCL Press, London.

Smardon, R.C. *et al.* (1986) *Foundations for Visual Project Analysis*, John Wiley, New York.

Other sources

Landscape Design, the monthly journal of the Landscape Institute, publishes special theme issues periodically: e.g. EA updates in May 1995 and September 1996; a CAD update in July/August 1994; and technical sections covering subjects such as Sports (Sept.1994), Water (Sept.1995) and Paving (June 1995). Enquiries to *Landscape Design* at 13a West Street, Reigate, Surrey RH2 9BL. Tel. 01737 223294.

The Landscape Institute publications list covers mainly practice-related documents, and can be obtained from Landscape Services Ltd, 6–8 Barnard Mews, London SW11 1QU. Tel. 0171-738 9166.

See also information about obtaining Countryside Commission, Forestry Commission, MAFF publications, etc.

Index

Page numbers shown in **bold** refer to illustrations